D1661431

Notes on Numerical Fluid Mechanics and Multidisciplinary Design

Volume 129

About this Series

Notes on Numerical Fluid Mechanics and Multidisciplinary Design publishes state-of-art methods (including high performance methods) for numerical fluid mechanics, numerical simulation and multidisciplinary design optimization. The series includes proceedings of specialized conferences and workshops, as well as relevant project reports and monographs.

More information about this series at http://www.springer.com/series/4629

Michael Klaas · Stefan Pischinger
Wolfgang Schröder
Editors

Fuels From Biomass: An Interdisciplinary Approach

A collection of papers presented at the Winter School 2011 of the North Rhine Westphalia Research School "Fuel production based on renewable resources" associated with the Cluster of Excellence "Tailor-Made Fuels from Biomass", Aachen, Germany, 2011

Springer

Editors
Michael Klaas
Chair of Fluid Mechanics and Institute of Aerodynamics
RWTH Aachen University
Aachen
Germany

Wolfgang Schröder
Chair of Fluid Mechanics and Institute of Aerodynamics
RWTH Aachen University
Aachen
Germany

Stefan Pischinger
Institute for Combustion Engines
RWTH Aachen University
Aachen
Germany

ISSN 1612-2909 ISSN 1860-0824 (electronic)
Notes on Numerical Fluid Mechanics and Multidisciplinary Design
ISBN 978-3-662-45424-4 ISBN 978-3-662-45425-1 (eBook)
DOI 10.1007/978-3-662-45425-1

Library of Congress Control Number: 2015932426

Springer Heidelberg New York Dordrecht London

Printed on acid-free paper

Springer-Verlag GmbH Berlin Heidelberg is part of Springer Science+Business Media
(www.springer.com)

Preface

In 2008, the Ministry of Innovation, Science, and Research of the German state North Rhine-Westphalia launched the "NRW Research Schools" which offer the possibility of studying for a PhD/doctorate within a structured doctoral program for PhD candidates with superior academic track records. Currently, 17 Research Schools are funded by this intiative.

The international NRW Research School "Energy Generation Based on Renewable Energy Resources" (BrenaRo) is a joint initiative of more than 15 institutes from RWTH Aachen University. Based on an interdisciplinary PhD program, the main long-term objective of this research school is to train highly qualified experts in the field of energy generation with a special focus on renewable energy resources.

This interdisciplinary approach combines the research fields of Biology, Chemistry, and Engineering. The close cooperation of the Research School with the cluster of excellence "Tailor-Made Fuels from Biomass" and the DFG Collaborative Research Centre "Model based control of homogenized low-temperature combustion" provides the opportunity to work in an interdisciplinary scientific environment.

This volume contains the papers presented at the "BrenaRo Winterschool 2011" held at the "Erholungs-Gesellschaft Aachen 1837" in Aachen, Germany on November 21–22, 2011. The symposium was organized by the Institute Cluster IMA/ZLW & IfU and the Institute of Aerodynamics of RWTH Aachen University, Germany.

In the name of all scientists involved in the NRW Research School BrenaRo, the speaker of the program, Wolfgang Schröder (RWTH Aachen University, Aachen Germany), would like to express gratitude to all the participating scientists for their contributions.

The present monograph is a snapshot of the state of the art of the joint initiative of biology, chemistry, and engineering to develop new fuels. The volume gives a broad overview of the ongoing work in this field in Germany. The order of the papers in this book corresponds closely to that of the sessions of the Symposium.

The editors are grateful to Prof. Dr. W. Schröder as the General Editor of the "Notes on Numerical Fluid Mechanics and Multidisciplinary Design" and to the Springer-Verlag for the opportunity to publish the results of the Symposium.

Aachen, December 2011

Michael Klaas
Stefan Pischinger
Wolfgang Schröder

Contents

Spray Phenomena of Surrogate Fuels and Oxygenated Blends in a High Pressure Chamber

M.M. Aye, J. Beeckmann, N. Peters and H. Pitsch

Abstract In this study, we investigate oxygenated blends and Diesel surrogate fuels under engine-like conditions in a high-pressure chamber. The investigated surrogate fuels are composed of n-decane and alpha-methylnaphthalene with different compositions according to the reference cetane numbers (CN) 53, 45, 38 and 23. In addition to the two-component surrogate fuel mixtures, we examine a three-component mixtures composed of n-decane, alpha-methylnaphthalene, and di-n-butyl ether with a reference cetane number of 53 to highlight the influence of adding di-n-butyl ether to the surrogate fuel at constant cetane number. Further, four blends with DNBE contents of 0, 10, 20 and 100 % in EN590 Diesel and corresponding cetane numbers of 53, 57.7, 62.4, and 100 were studied. We examine fuel spray characteristics in the liquid and vapor phases and the relationship between ignition quality and lift-off length. Vapor pressure is observed to significantly affect spray characteristics in the liquid phase. Vapor penetration lengths of the different fuels with the same injection pressure are found to be similar, because the differences of fuel density and viscosity in the vapor phase are too small to considerably affect the momentum flux. However, changing the injection pressures affects the vapor penetration lengths. Results show that CN is a good indicator for ignition delay. Furthermore, we discuss the fuel overlap number (OL) to indicate the separation between the liquid spray core and the reaction zone in engine-like conditions. It is found for the surrogate mixtures that OL generally increases with decreasing CN, while for the DNBE/Diesel mixtures, the opposite trend is observed. The OL number is found to be caused by a combination of cetane number and vapor pressure effects, where CN has the stronger effect for the surrogate mixtures, while the vapor pressure effect is dominant for the DNBE/Diesel blends. In the latter case, the high vapor pressure leads to short liquid penetration length and thereby larger OL number.

M.M. Aye (✉) · J. Beeckmann · N. Peters · H. Pitsch
Institute for Combustion Technology, RWTH Aachen University, Aachen, Germany
e-mail: mmaye@itv.rwth-aachen.de

M. Klaas et al. (eds.), *Fuels From Biomass: An Interdisciplinary Approach*,
Notes on Numerical Fluid Mechanics and Multidisciplinary Design 129,
DOI 10.1007/978-3-662-45425-1_1

1 Introduction

Compression Ignition (CI) engines are known to have particularly high thermal efficiencies. Further, it is possible to use various non-petroleum-based fuels in CI engines. Nowadays, Diesel engines occupy a vital role in the automobile market due to their many advantages such as high performance, long life, and high reliability among others. However, the Diesel process can lead to high emissions of soot and NO_x, caused by locally fuel rich incomplete combustion of hydrocarbons and large regions of high temperature, respectively.

In order to solve this problem, it is necessary to understand the fundamental issues such as fuel spray formation and the effect of different fuels on the combustion behavior. The spray characteristics affect spray formation and air-fuel mixing processes inside the engine, which have a major influence on ignition, the combustion process, and the formation of emissions.

Ignition delay of fuel in air is one of the most important characteristics for combustion in compression ignition (CI) engine. The cetane number is a rating to indicate how quickly a fuel will ignite in a CI engine [1]. A fuel with a longer ignition delay allows longer mixing time for fuel and air before combustion [2]. Fuel mixtures with different ignition quality can be obtained with a combination of good and poor ignition quality fuels in different proportions.

In terms of combustion and spray formation, it has been widely discussed in the past, which surrogate fuel is best suited to replace Diesel fuel in fundamental theoretical and computational studies [3]. The reference IDEA fuel, composed of 70 % n-decane and 30 % alpha-methylnaphthalene (AMNL) was found to be a good surrogate for diesel in conventional Diesel combustion mode [4]. In the present study, there will be a focus on oxygenated fuels. The oxygen content in the fuel strongly influences particulate emissions. A higher percentage of oxygen in fuel blends contributes to greater overall reduction in particulate emissions [5]. Liota et al. [6] found the oxygen content in ethers to be more effective than the oxygen content in alcohols and more details on the effect of different oxygen groups in the fuel were given by Pepiot-Desjardins et al. [7].

The fuel spray characteristics have been examined in a high-pressure chamber for different variables such as injector-type, nozzle-type, injection-pressure, energizing-time, ambient pressure and temperature in the high pressure chamber, for instance by Kim et al. [8].

To study the influence of the fuel properties on the spray characteristics, different single component fuels and fuel mixtures are investigated here for given injection system and ambient conditions. AMNL is used as a cetane reducer to control the cetane number in the present study. Di-n-butyl ether (DNBE) has been found to be a cetane enhancer for Diesel fuels [9].

2 Experimental Apparatus

2.1 Pressurized Spray Chamber

In order to study fuel spray characteristics and combustion details, a constant volume, constant flow-rate, high-pressure chamber was used that was built at the Institute for Combustion Technology (ITV) at RWTH Aachen University.

Experiments were performed using Mie scattering imaging technique for the liquid phase and shadowgraphy imaging technique for the vapor-phase to detect spray penetration and spray cone angle. Furthermore, OH* chemiluminescence was used for characterization of lift-off length (LOL) and ignition probabilities of the injected fuels.

A schematic picture of the spray chamber, its dimensions, and the cross sectional views of the measurement area are shown in Fig. 1. The principle of the set-up has been described in detail by Spieckermann et al. [10]. Essentially, the chamber has a constant high pressure and high temperature air through-flow, where pressure and temperature can be adjusted independently up to 50 bars and 800 K. The design of the chamber enables three-sided optical access through quartz-glass windows, which are orientated perpendicularly (90°) to each other.

Compressed air is sent through micro filters to the chamber by means of an external compressor. The air then enters the heater, which has a maximum power of 24 kW, where it is heated to the desired temperature. Following the heating, the hot air passes through the bottom flange into the measuring area of the chamber. For these conditions, at 17.42 kg/m^3 and 800 K, the hot air velocity in the chamber is 0.12 m/s. This is very small in comparison to the spray tip velocity during fuel injection.

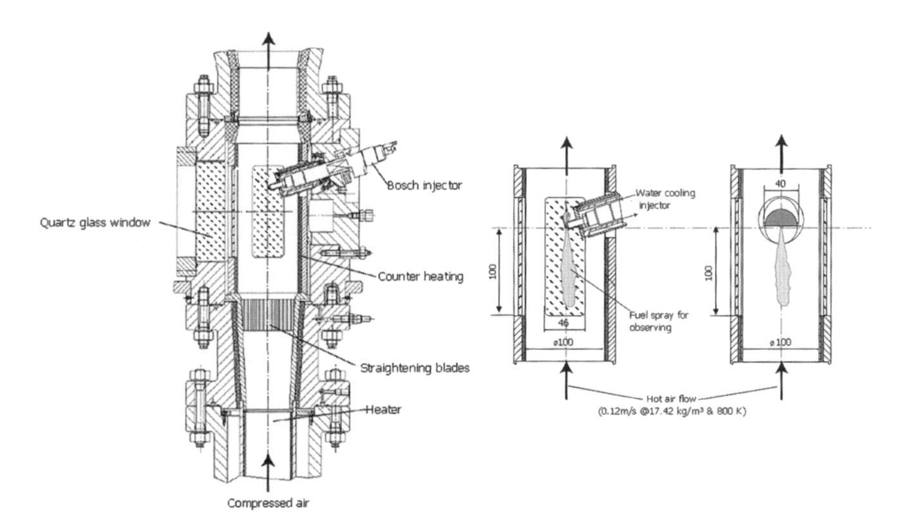

Fig. 1 Constant flow spray chamber and *cross sectional view* of observing section

Hot pressurized air passing through the chamber in the time between the end of combustion and the next injection process helps to remove the combustion residue and unburned hydrocarbons. This is then cooled down in a water-cooled external heat exchanger and throttled to ambient pressure using a Degussa air-flow controller with a constant flow rate of 60 kg/h.

2.2 Injection System

The principle of the injection system has been described in detail by Aye et al. [2]. A Bosch solenoid injector with a three hole nozzle specifically designed for research is used. This nozzle is similar to a sample injector nozzle 0 433 171 838 (Bosch), but it is has a cone angle of 148° and 3 spray holes, and the spray hole diameter is 0.141 mm. Measurements have been conducted with an energizing time of 700 μs and rail pressure of 600 bar for surrogates and an energizing time of 1,000 μs and rail pressures of 700, 1,000, and 1,300 bar for oxygenated blends.

2.3 Mie and Shadow Imaging Technique

The optical set-up for the simultaneous measurements using Mie, shadowgraphy, and OH* imaging techniques is shown in Fig. 2. An optically simultaneous Mie and shadowgraphy imaging technique is used for the investigation of liquid and vapor penetration of the investigated sprays. The optical set-up and all specifications have been described by Aye et al. [2]. For the image-data acquisition, a Davis LaVision image acquisition system is used.

2.4 OH-Measurement Technique

The isolation of the OH signal is achieved by applying a filter combination with a transmission range of 290–325 nm and a maximum at around 313 nm. The principle of the optical set-up for measuring the OH* is described in detail by Pauls et al. [9]. The OH* spectral range is from 275 to 350 nm [11]. Soot luminosity also emits broadband emission with significant emission at around 310 nm, but OH* dominates in the LOL region, where soot has not been formed yet [12]. The OH chemiluminescence image in Fig. 3 shows that probable soot luminosity (red color) occurs only in the downstream part of the flame. The measuring procedure in this paper therefore is consistent with the LOL measurement from [13]. The LOL of the flame and the probability of ignition delay are extracted from time series images of OH*.

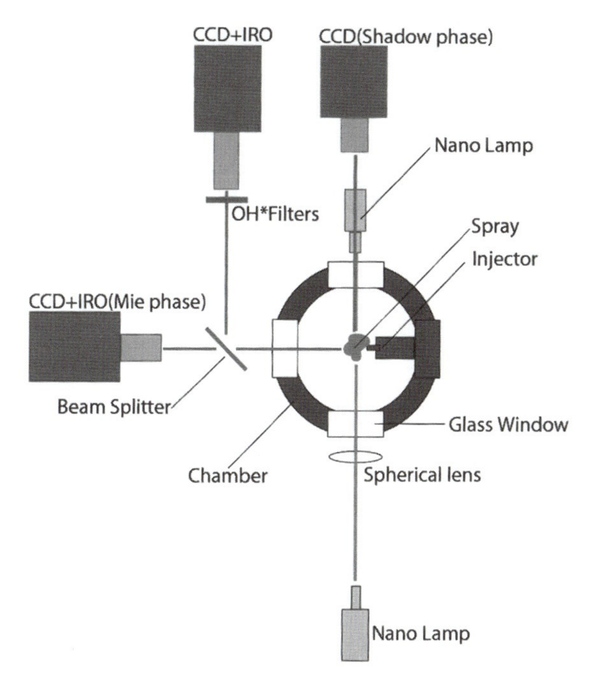

Fig. 2 Sketch of imaging set-up

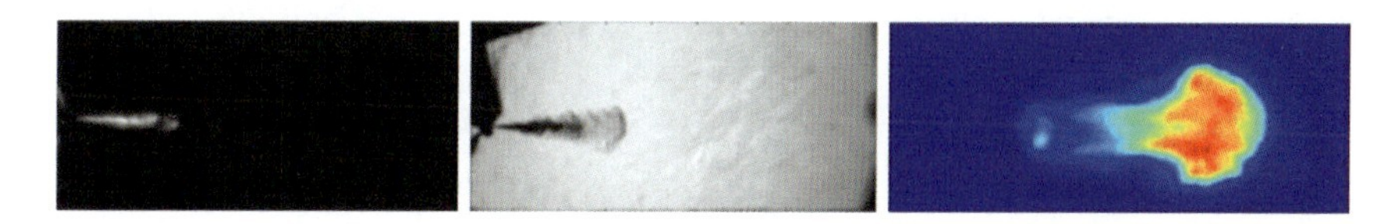

Fig. 3 Measured liquid, liquid + vapor, and OH* signals (from *left* to *right*)

3 Definition of Spray Characteristic

For the image processing, a Matlab code was developed. The procedures of the code have been described in previous work [2]. The spray penetration length S for liquid and vapor is defined as the distance between the nozzle tip and the final pixel of spray along the axial line of the nozzle for Mie and Shadowgraphy images, respectively. The spray cone angle ß is calculated from the upstream half of the spray area assuming a triangle for Mie and shadow images. LOL is defined as the distance between the nozzle tip and the first pixel in a vertical profile of OH*. Spray penetration length, spray cone angle, and LOL are shown in Fig. 4. The formulation for spray cone angle is as follow:

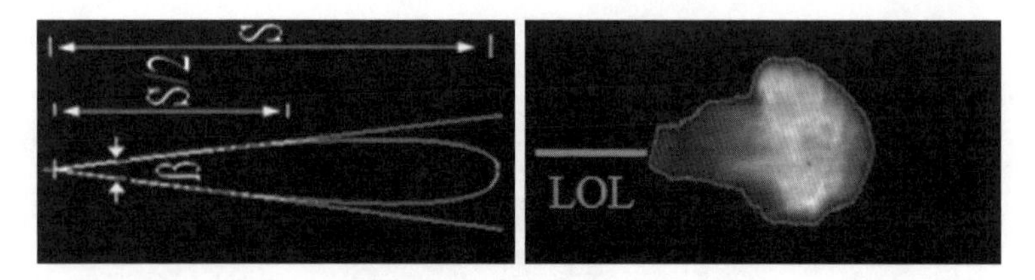

Fig. 4 Definition of spray characteristics (S = Penetration Length, LOL = flame lift-off, ß = spray angle)

$$\tan\left({}^{ß}\!/_{2}\right) = \frac{Area_{,s/2}}{\left({}^{S}\!/_{2}\right)^{2}}$$

and can be evaluated for both the liquid and vapor phase.

4 Fuel Properties

Thermodynamic and fluid properties of the tested fuels are shown in Fig. 5 and Table 1. These data help to understand the characteristics of the formation of spray, its dynamics, and the behavior of the vapor phase. The order of vapor pressure of the considered fuels is DNBE > n-Decane > AMNL. Figure 5 shows that densities and viscosities at 1 atm. differ among the fuels in the liquid phase, but are similar in the gas phase. Density and viscosity of n-decane, DNBE, and AMNL show a discontinuity caused by the phase change at 450, 410, and 515 K (respective boiling temperatures at 1 atm.).

5 Fuel Blends Preparation

The investigated fuels and fuel blends along with the respective cetane numbers are shown in Table 2. The formulation to determine the cetane numbers of the mixtures used here is described in Ref. [2]. European standard EN 590, which has low sulphur content and is oxygen-free, was used as reference Diesel fuel (CN = 53). The fuel was obtained from Haltermann Products and complies with the CEC reference fuel standard, RF-06-99 [14]. To obtain the required cetane numbers for the blends, AMNL and n-decane were mixed in different proportions.

Additionally, the oxygenated fuel DNBE is added in one of the mixtures to highlight differences of CN = 53 surrogate fuels. For further oxygenated blends, DNBE was mixed in different proportions with Diesel.

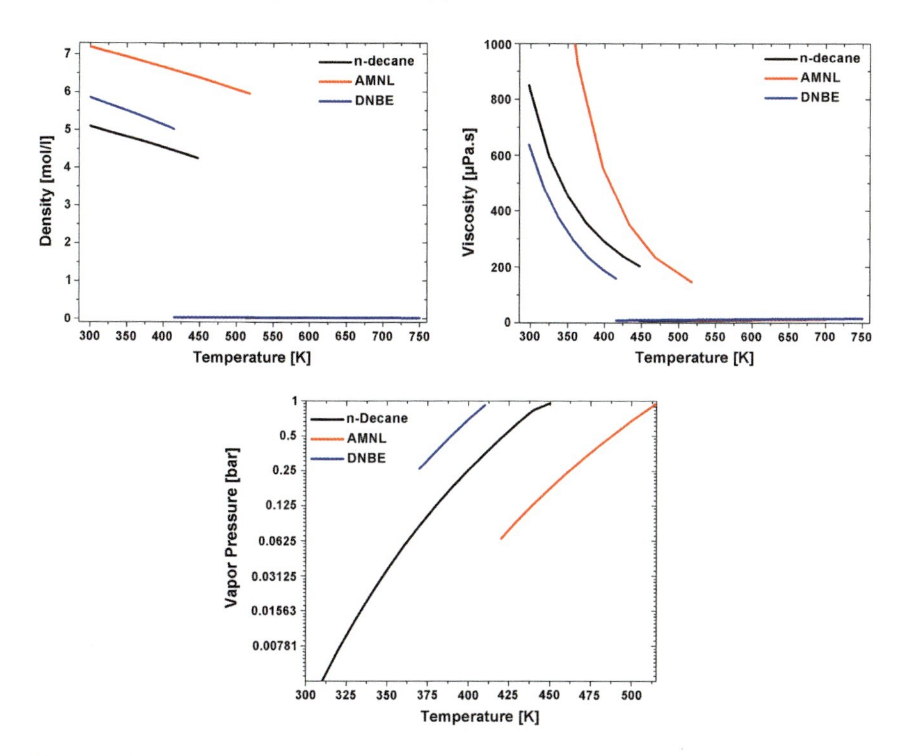

Fig. 5 Fuel properties

Table 1 Fuels properties [14, 20]

	Diesel	n-Decane	AMNL	DNBE
Density @ 15 °C kg/m^3	833.8	734 @ STP	1001 @ STP	771.3
Viscosity	2.892	0.969	3.436 @ 25 °C	0.7198
Vapor pressure @ 40 °C	–	0.255	0.03	0.692
FBP °C	359.3	174	241	142
Dist. 50 % BP °C	275.5	–	–	–
Dist. 95 % BP °C	346.1	–	–	–
Cetane number	53	76	0	100

6 Results and Discussion for Fuel Spray Process

This section is divided into three parts. In the first part, the mixture formation of Diesel, DNBE, and the oxygenated fuel blends are discussed. Mixture formation of surrogate fuels is then examined in the second part. The parameters that influence mixture formation of the oxygenated fuel blends and surrogate fuels are also analyzed. In the final part, we will discuss the flame lift-off and ignition of all tested fuels. Further, we will examine fuel overlap to estimate emissions for all tested fuels.

Table 2 Fuel compositions

Fuels	CN	Remark
Diesel	53	EU-fuel EN 590
30 % AMNL + 70 % n-decane	53	Surrogate
34 % AMNL + 46 % n-decane + 20 % DNBE	53	Surrogate
40 % AMNL + 60 % n-decane	46	Surrogate
50 % AMNL + 50 % n-deacne	38	Surrogate
70 % AMNL + 30 % n-decane	23	Surrogate
DNBE	100	Ether group (O_2 12.29 %wt)
20 % DNBE + 80 % diesel	62.4	Oxygenated blend (O_2 2.31 %wt)
10 % DNBE + 90 % diesel	57.7	Oxygenated blend (O_2 1.15 %wt)

6.1 Spray Formation of DNBE and Oxygenated Blends

We investigated DNBE in mixtures of 0, 10, 20, and 100 % (by liquid volume) with EN590 Diesel for different injection pressures to illuminate the influence of the fuel properties and injection pressure on fuel spray.

The average spray penetration lengths of DNBE, Diesel, and their blends of both liquid and vapor phases are shown in Fig. 6, which clearly indicates that the fuels with higher DNBE content have smaller penetration lengths in the liquid phase. The figure indicates for instance a difference between liquid penetration lengths of Diesel and DNBE by more than a factor of two.

Here, it may be noted that DNBE has a substantially higher value of vapor pressure. Figure 6 shows that vapor penetration lengths of all tested fuels are similar even though they are of different fuel compositions. It is clear from Fig. 5, that most of the fluid properties of the vapor phases for given conditions are not very different. This means that, for a given injection pressure, the overall spray lengths for all test fuels are similar.

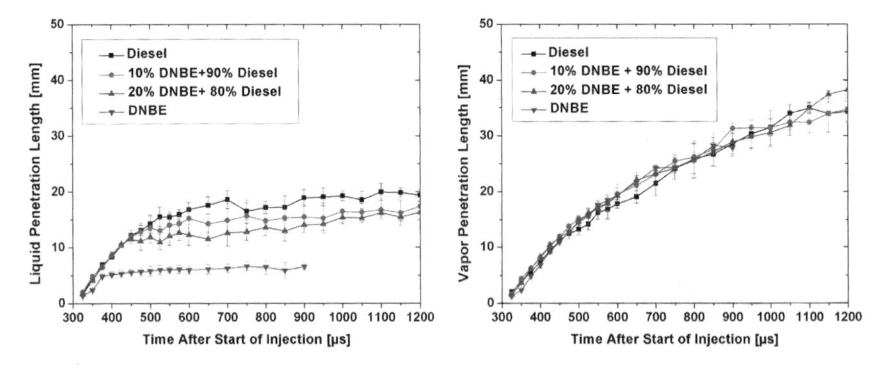

Fig. 6 Penetration length of oxygenated blends ($T_{cham:}$ = 800 K, $P_{cham:}$ = 50 bar, $P_{inj:}$ = 700 bar)

However, differences can be observed in the vapor phase volume (or area in the two-dimensional images). Figures 7, 8 and 9 show exemplarily instantaneous images of the liquid phase (dark color) and vapor phase (grey color) for Diesel, DNBE, and 20 % DNBE in Diesel. It is observed that a fuel with a higher DNBE content provides for better mixture formation as indicated by the larger vapor phase area compared to a fuel with lower DNBE content. This can be observed, for instance, at 575 μs after start of injection (ASOI). Note that in Fig. 8 there is no DNBE image at 1,000 μs ASOI, since by this time the spray has already ignited.

Instantaneous images of 20 % DNBE and averaged penetration lengths for both liquid and vapor phases of DNBE are shown for varying injection pressures in Figs. 10 and 11, respectively. The results show that liquid spray penetration lengths of DNBE with different injection pressures are almost the same. This is interesting, since the higher pressure injection leads to higher injection velocities, which in turn would suggest higher liquid penetration length. But the high vapor pressure of DNBE seems to lead to fast vaporization soon after droplets are formed. Therefore,

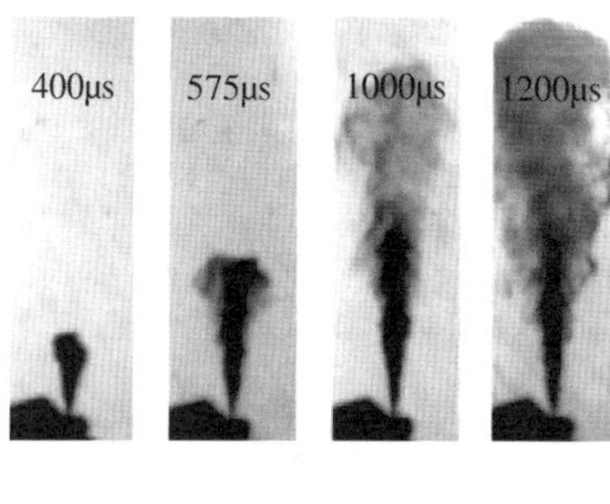

Fig. 7 Diesel spray formation

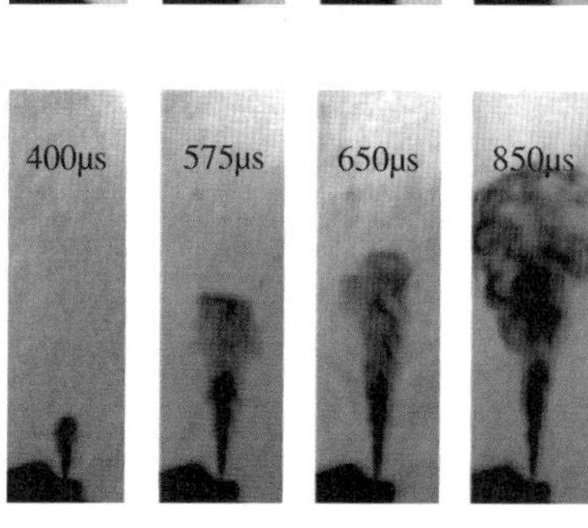

Fig. 8 DNBE spray formation

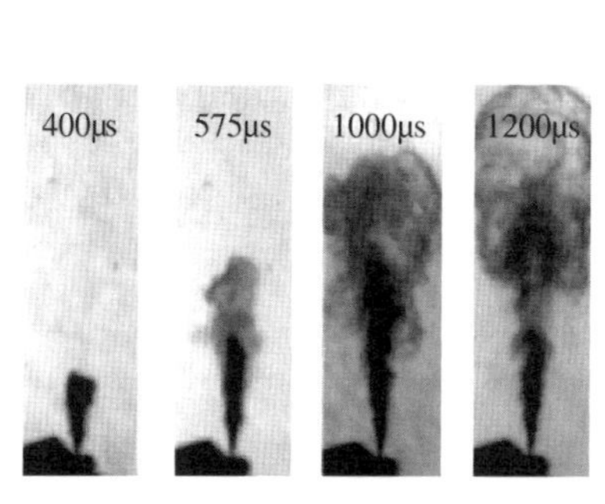

Fig. 9 20 % DNBE spray formation

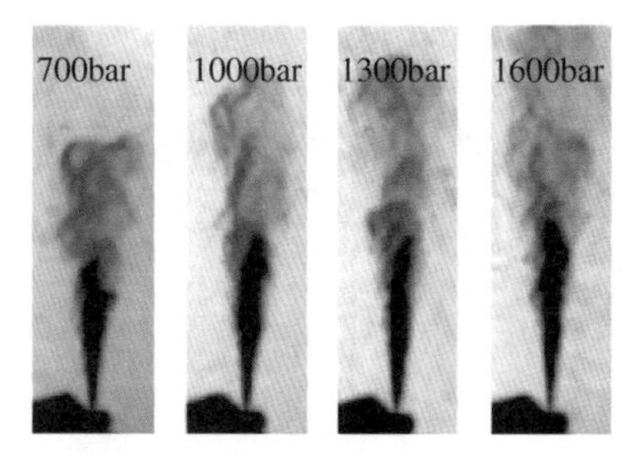

Fig. 10 20 % DNBE spray for varying injection pressures @ ASOI 800 μs

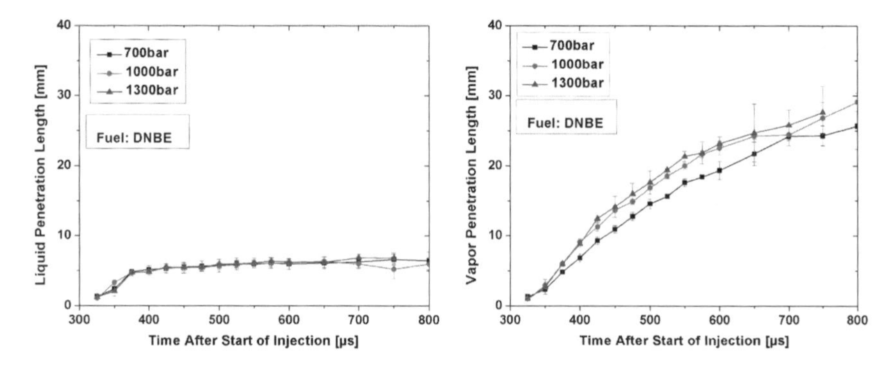

Fig. 11 Spray penetration lengths of DNBE ($T_{cham:}$ = 800 K, $P_{cham:}$ = 50 bar)

in this case, it is noted that vapor pressure is more influential for mixture formation than the injection pressure. However, it is clear that due to momentum conservation, the total injection pressure affects the overall fuel penetration length, which is shown in Fig. 11 (right).

The following figures illustrate the influence of the fuel's vapor pressure and the injection pressure on the atomization characteristics more clearly. Figures 12 and 13 show the penetration lengths of 20 and 10 % DNBE in the fuel, respectively, for various injection pressures. Both liquid and vapor penetration gradually increase with increasing injection pressure for these fuels. This shows that reducing the percentage of DNBE content (reducing vapor pressure) in the fuel leads to an increasing influence of injection pressure on the liquid penetration.

6.2 Spray Formation of Surrogate Fuels

The surrogate fuels are investigated under the same chamber conditions. The comparison of liquid penetration lengths in Fig. 14 clearly shows that the tested surrogate fuels have markedly different spray phenomena. For the tested surrogate

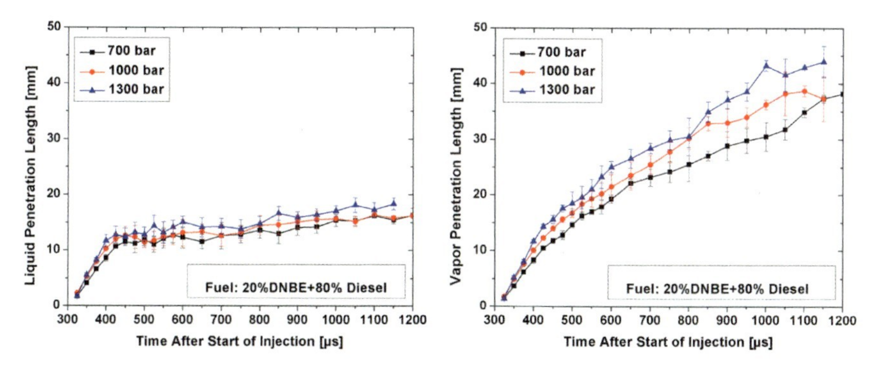

Fig. 12 Penetration length of 20 % DNBE blend ($T_{cham:}$ = 800 K, $P_{cham:}$ = 50 bar)

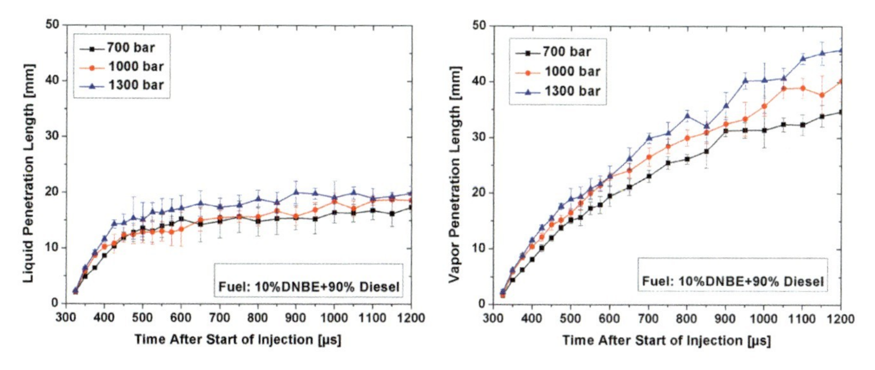

Fig. 13 Penetration length of 10 % DNBE blend ($T_{cham:}$ = 800 K, $P_{cham:}$ = 50 bar)

fuels, higher AMNL (reduced vapor pressure) content in the fuel gradually increases the penetration length in the liquid phase (Fig. 14, left). This shows that vapor pressure under Diesel conditions is one of the most important fuel properties for spray development and subsequent mixing for both oxygenated or non-oxygenated fuels. The right panel of Fig. 14 showing liquid penetration for different fuel compositions at the same cetane number demonstrates that this effect does not necessarily correlate with cetane number, because liquid penetration lengths of the same CN fuels are substantially different.

Figure 15 (left) shows that all tested surrogate fuels with varying AMNL contents have similar vapor penetration lengths before ignition takes place. The start of ignition is visible in Fig. 15 (left) by the plateauing vapor penetration, and is at around 975, 1125, and 1275 μs for the 30 % AMNL in the fuel, 40 % AMNL in the fuel, and 50 % AMNL in the fuel mixtures, respectively. Ignition was not seen in the observation area for the 70 % AMNL in the fuel mixture. After start of ignition, the exact value of vapor penetration length cannot be measured accurately anymore and should therefore not be further discussed.

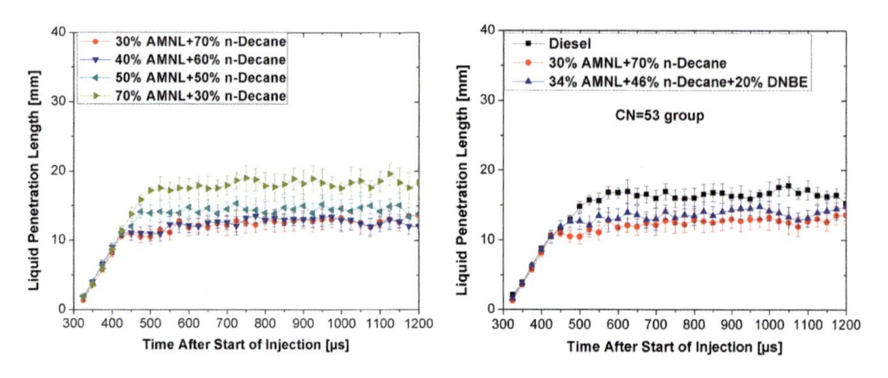

Fig. 14 Liquid penetration lengths of surrogates ($T_{cham:}$ = 800 K, $P_{cham:}$ = 40 bar, $P_{inj:}$ = 600 bar)

The right panel of Fig. 15 shows the overall spray cone angle, which is an indication for the vapor phase area. This quantity has a local peak for all tested fuels at 450 μs after start of injection, which is at the instance of full injector opening, at which the droplet cloud is struck with a higher velocity jet of following droplets [15]. As for the oxygenated blends, fuels with higher AMNL content, and therefore lower vapor pressure, have a smaller vapor phase area, while the vapor penetration length for all surrogates is nearly equal.

In conclusion, a better mixture preparation, indicated here by a larger vapor phase area can be realized by increasing the DNBE content in the oxygenated blends and decreasing the AMNL content in the surrogate fuels at the same injection pressure.

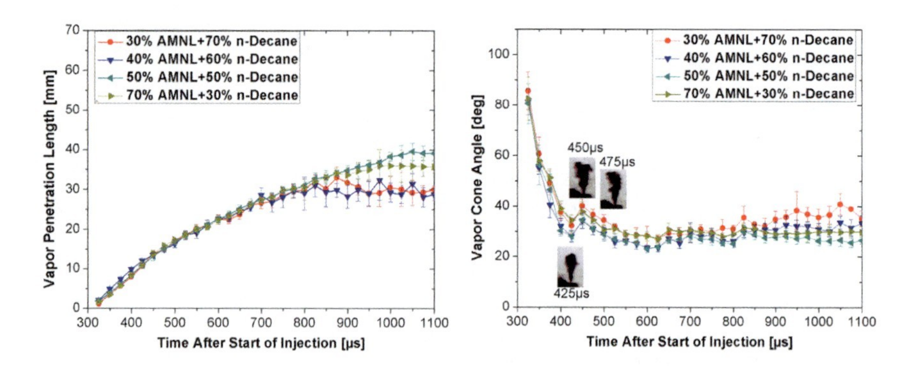

Fig. 15 Vapor characteristics of surrogates ($T_{cham:}$ = 800 K, $P_{cham:}$ = 40 bar, $P_{inj:}$ = 600 bar)

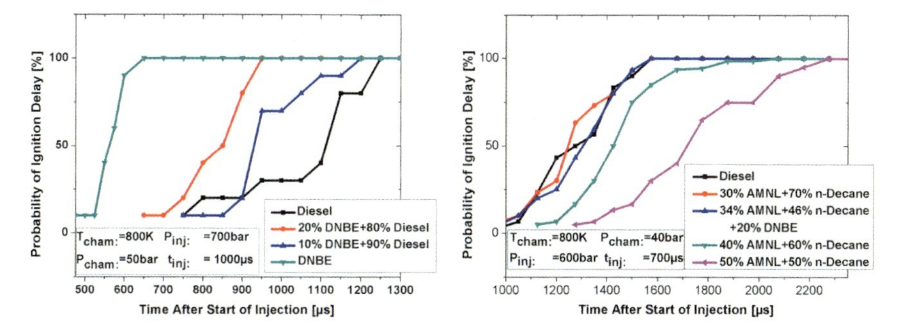

Fig. 16 Ignition qualities of oxygenated fuels and surrogate fuels

6.3 Spray Combustion

6.3.1 Probability of Ignition Delay

The definition of ignition probability has been described in [2]. The ignition quality could not be compared between the surrogates and oxygenated fuels due to the different injection and chamber pressure conditions. However, in the results for both fuel groups, the ignition delay increases with decreasing CN. In Fig. 16 (left), the transition time of 0–100 % ignition probability is 150 μs for DNBE and is increased up to 500 μs for Diesel. The oxygenated blends show a significant reduction in the transition time for increasing oxygen ratio in the blend. The interesting observation in Fig. 16 (right) is that the fuels with the same cetane number of CN = 53 have a similar trend for the ignition probability. The fuel containing 40 and 50 % AMNL have increased ignition delay by factor of 1.1 and 1.4 compared with the fuels with CN = 53 (30 % AMNL, 20 % DNBE, and Diesel), respectively. It follows that CN is a good indicator for ignition delay for both oxygenated blends and surrogates.

6.3.2 Flame Lift-Off Length

The flame lift-off length (LOL) is an important characteristic in spray combustion. A high LOL can lead to reduced soot emissions, since a high lift-off leads to increased time for mixing before combustion. The flame LOLs of oxygenated blends and surrogate fuels are shown in Fig. 17.

The left panel of Fig. 17 shows the influence of the injection pressure on LOL. It is observed that the fuels with higher oxygen content have a smaller LOL dependence on pressure than fuels with lower oxygen content. There is also generally a reduction in LOL with increasing oxygen content in the fuel. It is also well known that under Diesel engine combustion conditions, particulate emissions are typically significantly reduced using oxygenated fuels [16]. In Fig. 17 (right), LOL increases with increasing AMNL content in the mixture. Higher CN fuels produce a shorter

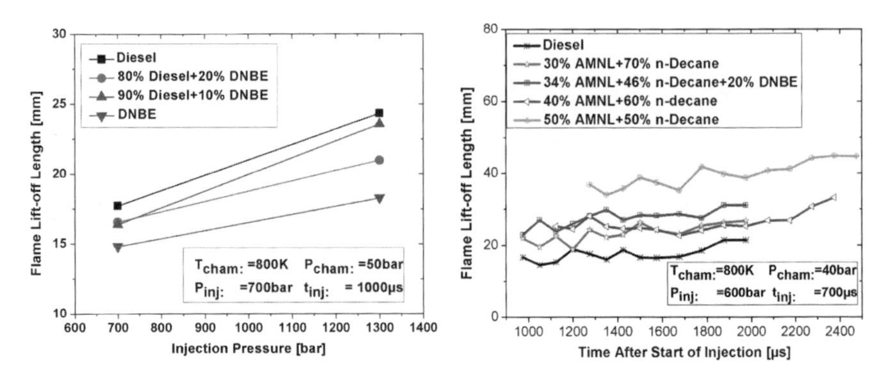

Fig. 17 Flame lift-off lengths

LOL [17]. But Fig. 17 (right) shows that even different fuels with the same CN have different LOL depending on their composition. Therefore, ignition quality expressed by CN and LOL, does not necessarily correlate. It is interesting to see that the fuels, which are rich in oxygen have shorter LOL and those rich in aromatic compounds have longer LOL.

6.3.3 Overlap Number

The OL number is defined in Ref. [18] as the difference between LOL and liquid penetration length (LPL) divided by LOL

$$OL = \frac{(LOL - LPL)}{LOL}.$$

A higher OL number can therefore be caused by the attenuation of the reactivity of the fuel leading to a larger LOL, and it leads due to the increased separation of liquid spray and reaction zone to reduced burning under very rich conditions. As shown in [18], the OL number correlates with the tendency of the fuel to form soot, such that a higher OL number was found to result in less soot. The definition of the OL number is similar in its meaning to the percentage of stoichiometric air (ζ_{st}) introduced by Siebers and Higgins [19] that expresses the trend of soot formation as function of the LOL and a characteristic length of the spray that represents the liquid penetration length. This quantity ζ_{st} is the average percentage amount of air at the lift-off location to stoichiometrically burn the evaporated fuel. Therefore, the higher the entrainment of air mixing upstream of the lift-off location, the higher is its value. Interestingly, the authors give an analytical expression, which has only the LOL as an unknown. The authors point out that the oxygen in the entrained air reacts with the fuel in a rich reaction zone in the central region of the spray just downstream of the LOL. The products of this rich reaction zone are ideal for forming soot, suggesting a strong link between soot formation and the amount of air

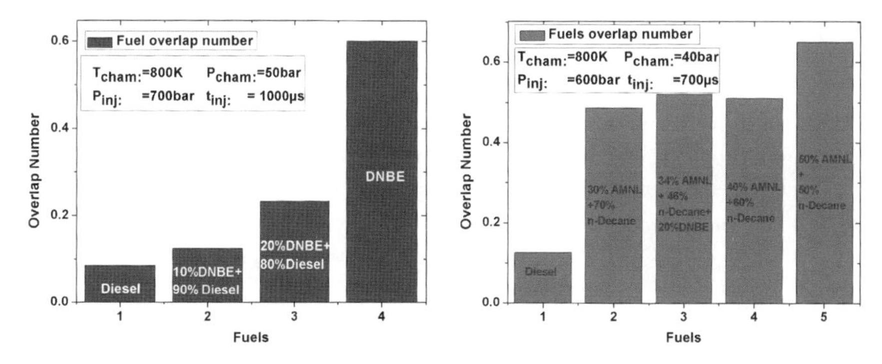

Fig. 18 Fuel overlap number

entrainment that occurs upstream of location of LOL. Figure 18 provides the OL numbers for the different fuels considered here. It should be noted here that the OL numbers cannot be compared among the two fuel groups examined here, because the experiments for both were performed with different ambient conditions and injection pressures.

Figure 18 shows that both fuel groups with higher oxygen content and higher aromatic content show an increase in OL number. Although LOL of DNBE is short, the separation between liquid core and reaction zone is enough to result in a higher overlap number due to its very short liquid penetration length. It is also found that an increase in AMNL, and therefore in aromatic content, leads to an increase in OL number which indicates a reduction in soot formation. This is an interesting trend, since higher aromatic content, especially of naphthalene, is known to increase soot in basic kinetic experiments. Here, however, higher AMNL content also reduces the reactivity of the fuel. This is evident in Fig. 17 (right), where an increase in LOL is observed with increasing AMNL in the fuel. For increased LOL, combustion takes place in a region where the liquid phase is mostly or completely evaporated and fuel and air are already well mixed.

7 Conclusions

In this paper we investigated the spray phenomena of oxygenated fuel mixtures, surrogate fuels, and Diesel. For the liquid penetration length it is found that lower vapor pressure of the fuel leads to higher maximum values in all fuel groups. The vapor penetration lengths of either blends with oxygenates or aromatic compounds are not influenced by fuel properties.

The vapor penetration lengths of both fuel groups increase with increasing injection pressures. By investigating the pure DNBE, 10 % DNBE in the fuel and 20 % DNBE in the fuel with different injection pressures, it is also found that vapor pressure is more influential for mixture preparation than injection pressure. The results demonstrate that vapor pressure is a very important fuel property irrespective of the fuels containing oxygenated or aromatic compounds.

The results show that CN is a good indicator of ignition delay for both oxygenated blends and surrogates because fuels with the same CN have similar trends in ignition probability and fuels with different CN show that their ignition trends correlate with the value of CN. In the oxygenated blends study, LOL is increased by either decreasing oxygen content or increasing injection pressure.

The trends in ignition quality expressed by CN do not necessarily correlate with LOL, because LOLs can be substantially different for fuels with the same CN depending on their composition. Interestingly, it is found that fuels rich in DNBE have shorter LOL and those rich in aromatic compounds have longer LOL.

Oxygenated compound fuels are well known to reduce particulate emissions, and AMNL addition is expected to produce more soot due to its aromatic nature. The OL number indicates the separation between the liquid spray core and reaction zone, which provides an indication for the tendencies of the tested fuels to form soot. Higher OL number indicates less overlap between liquid spray core and reaction zone, which would lead to diminished soot formation. The experimental results show that both the fuel groups with higher oxygen content and with higher aromatic content have an increase in OL number, which indicates less soot. The reason for this trend is that higher AMNL content fuel reduces the reactivity and increases LOL, which leads to stronger separation of liquid spray core and the flame, and therefore to enhanced mixing of fuel and air before combustion.

In view of this study, vapor pressure and CN should be considered as important parameters in selecting the fuel for CI combustion. We can conclude that better spray atomization could be realized not only with modified injection systems such as nozzle, injector, pressure, etc. but also with fuel composition and fuel molecular structure.

Acknowledgments The authors would like to thank "BrenaRo" for the scholarship grant. This work was performed as part of Cluster of Excellence "Tailor-Made Fuels from Biomass", which is funded by the Excellence Initiative of German federal state governments to promote science and research at German University.

References

1. Parkash, S.: Petroleum Manufacturing Hand Book. The McGrew-Hill Companies, Inc, USA (2010). ISBN 978-0-07-163240-9
2. Aye, M.M., Beckmann, J., Vanegas, A., Peters, N., Pitsch, H.: Experimental investigation of diesel and surrogate fuels: spray and ignition behavior. SAE 2011-01-1921, JSAE 20119071 (2011)
3. Farrell, J.T., Cernansky, N.P., Dryer, F.L., Friend, D.G., Hergart, C.A., Law, C.K., McDavid, R.M., Müller, C.J., Patel, A.K., Pitsch, H.: Development of an experimental database and kinetic models for surrogate diesel fuels. SAE Paper, SAE 2007-01-0201 (2007)
4. Antoni, C.: Untersuchung des Verbrennungsvor im direkteinspritzenden Dieselmotor mit zyklusaufgelöster emissionsspektroskopie. Ph.D. thesis, RWTH Aachen (1998)

5. Choi, C.Y., Reitz, R.D.: An experimental study on the effects of oxygenated fuel blends and multiple injection strategies on di diesel engine emissions. Elsevier, Fuel **78**(1999), 1303–1317 (1999)
6. Liotta, Frank J., Montalvo, Daniel M.: The effect of oxygenated fuels on emissions from a modern heavy-duty diesel engine. SAE Paper **932734**, 18–21 (1993)
7. Pepiot-Desjardins, P., Pitsch, H., Malhotra, R., Kirby, S.R., Boehman, A.L.: Structural group analysis for soot reduction tendency of oxygenated fuels, comb. Flame **154**(1–2), 191–205 (2008)
8. Kim, H.J., Park, S.H., Chon, M.S., Lee, C.S.: A comparison of effects of ambient pressure on the atomization performance of soybean oil ester and dimethyl ether sprays. Oil Gas Sci. Technol. Rev. IFP Energies nouvelles, **65**, 883–892 (2010)
9. Pauls, C., Grünefeld, G., Vogel, S., Peters, N.: Combined simulations and oh-chemiluminescence measurements of the combustion process using different fuels under diesel-engine like conditions. SAE Paper, 2007-01-0020 (2007)
10. Spieckermann, P., Jerzembeck, S., Felsch, C., Vogel, S., Gauding, M., Peters, N.: Experimental data and numeriacl simulation of common-rail ethanol sprays at diesel engine-like conditions. Atomization Sprays **19**(4), 357–386 (2009)
11. Gessenhardt, C., Reichle, R., Pruss, C., Osten, W., Schalz, C.: In-cylinder imaging diagnostics with highly efficient UV-transparent endoscopes. In: Seventh International Symposium Towards Clean Diesel Engine, TCDE (2009)
12. Higgins, B., Siebers, D.: Diesel-spray ignition and premixed burn behavior. SAE Paper, 2000-01-0940 (2000)
13. Siebers, D., Higgins, B.: Flame lift-off on direct-injection diesel sprays under quiescent conditions. SAE Paper, 2001-01-0530 (2001)
14. Haltermann Products, Product information, Certificate of Analysis (2007)
15. Baumgarten, C.: Mixture Formation in Internal Combustion Engines. Springer, New York (2006). ISSN-1860-4846
16. Bowman, C.T., Golden, D.M., Hanson, R.K., Pitsch, H.: Optimization of systhetic oxygenated fuels, energy research at Stanford (2005–2006)
17. Beckmann, J., Aye, M.M., Gehmlich, R., Peters, N.: Experimental investigation of the spray characteristics of di-n-butyl ether (DNBE) as an oxygenated compound in diesel fuel. SAE Paper, 2010-01-1502 (2010)
18. Janssen, A., Kremer, F., Pischinger, S., Reddemann, M., Kneer, R., Hottenbach, P., Grünefeld, G.: Potential of tailor-made fuels for diesel combustion engine. In: 3rd TMFB International Workshop (2010)
19. Siebers, D., Higgins, B.: Flame lift-off on direct-Injection diesel sprays under quiescent conditions. 2001-01-0530 (2001)
20. Chemical and physical property data of fuels. http://webbook.nist.gov/chemistry/

GIS-Based Model to Predict the Development of Biodiversity in Agrarian Habitats as a Planning Base for Different Land-Use Scenarios

M. Ernst, J. Oellers, A. Toschki, H. Hollert and M. Roß-Nickoll

Abstract Biodiversity is defined in the political sphere as an important subject of protection and the extinction of species to be listed here should not only be held in acknowledged protected areas, but rather also in areas used agriculturally. The situation in agriculture is currently strongly changing through influencing factors, such as climate change and restructuring in favour of renewable resources. Since this development on the one hand holds the risk of a loss of biodiversity, and on the other redevelopments also offer opportunities, this is indispensable for a well-founded evaluation to operably forecast the effects of land use change on organisms. Suitable methods for this are currently lacking. These loopholes should be closed by the model worked out here. Based on species collections from earlier studies, autecological organism data from the bibliography, as well as suitable map material, a Geographical Information System will be created which makes it possible to forecast biodiversity in a regional context.

1 Historic Development and Current Situation of Biodiversity in the Agricultural Landscape

Except for a few, island-like confined locations, the landscape existing in central Europe today can by no means be described as "natural", i.e. not influenced by people. Moreover, it has already been greatly changed through anthropogenic use in pre-industrial times. Without this influence, beech forests and mixed beech forests would above all dominate in Germany on the basis of the given climatic and

M. Ernst (✉) · H. Hollert · M. Roß-Nickoll
Institute for Environment Research (Biology V), RWTH Aachen University, Worringerweg 1, 52074 Aachen, Germany
e-mail: marit_ernst@bio5.rwth-aachen.de

J. Oellers · A. Toschki
Research Institute for Ecosystem Analysis and Evaluation E.V. (Gaiac), Kackertstaße 10, 52072 Aachen, Germany

M. Klaas et al. (eds.), *Fuels From Biomass: An Interdisciplinary Approach*, Notes on Numerical Fluid Mechanics and Multidisciplinary Design 129, DOI 10.1007/978-3-662-45425-1_2 edaphic conditions. Deforestation has however produced a mosaic of forest and non-forested land, over-use has led to a degradation of many regions, in which open brush-land and moorland have come into being. It is to be assumed that this increase in biotope diversity is also accompanied by a significant increase in biodiversity. This development was reversed however with industrialisation in the middle of the 19th century, which in agriculture led to mechanisation, growing farm sizes and an increase in fertilisation and use of pesticides; the biological diversity of the artificial landscape again decreased. This trend intensified in the middle of the 20th century through another drive of technology and innovation, strong growths in productivity accompanied further devastation of habitats and decreasing diversity in the agricultural landscape. It was not until the 1970s that increased efforts to protect nature and biotopes were made [6].

In spite of this, loss of biodiversity has also continued. The main cause of this is mentioned by Möckel [35] as being the change from an extensive, varied agriculture and forestry to the intensive farming of today, which levels out the landscape through land consolidation and melioration, extensive mono-cultures, ploughing up, intensive use of fertilisers, pesticides, high grazing density, drainage of wet locations, locations with marginal yields becoming fallow etc. Accordingly, the loss of biodiversity in agricultural lands is significantly larger than in areas close to nature [6, 21, 47, 49, 53].

In an international context and also in Germany, pressure on agricultural ecosystems is currently intensifying through changed framework conditions in agriculture. Alongside changes through direct and indirect effects of global climate change (Federal Office of Environmental Protection [7]), the growing competition through globalisation, high demand for and promotion of biomass from the energy sector, obligatory laying fallow of land ceasing to apply, increasing land competition and higher competitiveness of field use, increased intensification of field and grassland use, stronger utilisation of lands set aside, as well as ploughing grass lands for field use are to be mentioned in this regard [Cf. 35, 37]. The risk of loss of biodiversity through the extension of renewable resources, which are used as energy (production of heat/cold, power, fuel) and industrial plants (fibres, chemical and pharmaceutical products) in particular is referred to many times (e.g. [13, 16, 20, (Panel of Experts for Environmental Issues) 48]). It is therefore the declared goal of the Federal Government of Germany to increase the proportion of renewable energies in gross final energy consumption to 18 % by 2020, 30 % by 2030, 45 % by 2040 and 60 % by 2050 (German Federal Ministry for Economics and Technology and German Federal Ministry for the Environment, Environmental Protection and Reactor Safety [11]). Renewable energy sources should replace the only limited fossil and nuclear energy sources and therefore make a large contribution to conservation of resources and to controlling the greenhouse effect (Federal Environmental Agency [55]). Furthermore, the addition of bio-components to fuels should be increased [11]. For this purpose, large areas will be needed for the cultivation of renewable resources. If the proportion of agricultural areas used in such a way still amounted to 11.8 % in 2005 (German Federal Ministry for Nutrition, Agriculture and User Protection [8]), forecasts assume that the proportion of the areas available for this will increase by more than 50 % by the year 2050 (Institute for Technical Thermodynamics (German Aerospace Centre) [26]).

This current and still to be expected process of restructuring the agricultural landscape does not only bring risks, but can also bring opportunities for biodiversity, since reform and land use change also always offer scope for design [53]. Agriculture is at a crossroads in this regard, and it now depends on which direction of development will be taken, and how far insights from research on the support of ecological objectives are integrated in systematic agricultural use (e.g. increase in structure variety).

At the same time biodiversity is defined as a subject of protection in a political framework. Protection and sustainable use of biodiversity are therefore objectives of the Convention on Biological Diversity, CBD, which was negotiated by the United Nations in Rio de Janeiro in 1992, and includes the European Union as one of its signatories. The term biodiversity here includes the levels of ecosystems, species and genes. The Convention enjoins the participating states to develop their own national biodiversity strategies [56]. This was done in Germany in 2007 with the National Strategy for Biodiversity (NBS). Here the goal was stated as stopping the extinction of species by 2010 and reversing the process, which applies to the entire landscape, including intensively used areas. Up to 2020 biodiversity in agricultural ecosystems should be significantly increased [9]. In 2010 the goal of maintaining biodiversity was however unsuccessful, and since the gap between the actual and intended values was still large, the goal for 2010 can only be reached under very great efforts [10]. Alongside the agreements mentioned, biodiversity as a subject of protection or conservation of a habitat function is also anchored in numerous other laws on national and international levels, such as the Cartagena Protocol [12], Directive 91/414/EEC (EU 1991, Pest Control), Directive 2001/18/EC (EU 2001, Use of GMOs), GenTG (German law on genetic engineering [22], Use of GMOs), BBodSchG (German federal law on soil protection [2], § 1, Soil Protection), PflSchG (German law on plant protection [41], § 6, Pest Control), Natura 2000 [50], BNatSchG (German law on nature protection [4], § 1 and § 5, Nature Protection), UVPG (German law on environmental impact assessment [58], Impact Regulation), USchadG (German law on environmental damages [57], Avoidance and Remediation of Environmental Damages).

Decisions on land use are however often made on the basis of short-term, mostly economic considerations, and higher sustainability concerns often remain disregarded.

2 State of Current Research

There are currently no suitable methods to operably forecast biodiversity in agricultural landscapes, and, according to the Commission for Soil Protection at the "Umweltbundesamt" (Federal Environmental Agency [29]), there is currently no knowledge about evaluating the landscape in the case of a change in the systems of use; it is therefore recommended to investigate the development of biodiversity when there are conversions in the relevant model regions. Evaluation of risk for biodiversity outside of authorising materials is generally based on the protection, conservation and/or development goals of nature protection law rules. Corresponding protection targets are predominantly defined for all species or respectively habitat types to be protected, but not for all species and habitat types not considered worthy of protection. The landscape in Germany is not however made up entirely of nature reserves, but rather primarily from anthropogenically used areas (for agriculture, settlements and traffic) which are not under any strict nature protection [47, 54].

Models which deal with damages to species and changes in habitats mostly only focus on climate change. In order to assess sensitivities, dynamic approaches such as bioclimatic modelling e.g. [3, 38] or vulnerability analyses e.g. [25] are used. The dynamic models currently available in science are to date still restricted to global issues, which cannot be transferred to concrete local situations [27]. In the area of practical nature protection, primarily indicator-based sensitivity analyses are carried out, which in their structure correspond to the approach of a vulnerability analysis. So, for example, the risk assessment of possible changes to FFH areas in Germany as a result of climate change follows the methodology of indicator-based sensitivity analyses [40]. Similar studies have already been carried out for individual federal states, or are currently being carried out in North Rhine-Westphalia [31]. Within the framework of these studies, the plant species of individual habitats are grouped and defined in sensitivity levels according to the proportions of hygrophytes and plants growing in cold conditions:

<10 % cold or respectively humidity indicators 1 (low sensitivity), 10–20 % cold or respectively humidity indicators 2 (medium sensitivity) and >20 % cold or respectively humidity indicators 3 (high sensitivity). Climatic changes are not however the only influence on species or habitats. Other environmental changes related to humans such as the deposition of pollutants or changes in land use are significant factors involved which are not considered within the framework of the studies described above.

In addition, the studies are mostly limited to the analysis of legally protected types of habitat, for which there are clear requirements and conservation goals. The typical habitats, predominantly less worthy of protection, of an agricultural landscape are not considered in such studies. A conceptual approach to the evaluation of biodiversity in the agricultural landscape is lacking.

Moreover, no positive development can be represented through the approaches currently available, but rather only the risk of negative consequences is forecast.

Within the framework of the national strategy for biodiversity, 19 indicators were developed which in their monitoring also cover endangered species (red list) as well as protected species within the FFH areas alongside sustainability indicators. Moreover, the states of preservation of areas with special protection status form the focus of the monitoring. As an indicator for recording sustainability, a total of 59 types of bird from six different types of habitat were brought in Achtziger et al. [1], which were determined in the course of the national sustainability strategy. For the part habitat of agricultural land, ten types of bird are representative for biodiversity in this habitat. The fact that only such a small number of species are incorporated into the indicator is due to the fact that the information should be evaluated with an overview. This inevitably produces an area of tension between scientific exactitude and political usability [59]. The principle of representative species is used also in the area of risk analysis of pesticides [46]. It poses the question, however, as to whether this concept is sufficient for a well-founded evaluation.

Collecting monitoring data, above all of animals, is very complex and cost-intensive. There are therefore only few such data, and those which are collected often remain unused, since they are not centrally held and evaluated. An operationalization of biodiversity is therefore required urgently in order to make the data comparable [28]. The ecological area referred to must in doing so always be observed however, i.e. the local, regional natural states and potentials have to be sufficiently taken into consideration. If this does not happen and general nature protection areas are transferred to all locations unconsidered, overall this does not amount to promoting, but rather to levelling out natural diversity [42].

3 The Principle of the Model to Forecast the Development of Biodiversity in Agricultural Habitats

A model such as that created within the framework of this work represents biodiversity in a spatial relationship and makes it possible to operationalise. It is therefore possible to integrate and therefore make usable monitoring data already collected in management concepts for land use.

On the basis of regional prevalence data on plants and animals, biodiversity should be forecast in various agricultural habitats. If a land use change is taken into consideration for a particular area, this allows the model to assess the effects of this change on the composition of the local agricultural biocenoses. The model should initially be developed using the two organism groups of vascular plants and ground beetles (as representatives of ground-proximate arthropods). The model systematic can later be extended to other groups such as birds, spiders, butterflies or grasshoppers. It will initially be developed on the example of the Rhineland. It can however be transferred to other regions later.

4 Selected Organism Groups

Animal and plant organisms are adapted to different degrees to biotic and abiotic location factors, such as temperature, rainfall, the type, quality and wetness of soil, as well as food supply, and form a characteristic community of a habitat through inter- and intraspecific interdependencies, such as, inter alia, competitive behaviour, symbioses or predator-prey relationships [30]. Both the different animal groups and also the vegetation therefore also give indications of the environmental influences at the location through the species composition of their biocenoses and thereby serve as biological indicators for these ecological conditions [15, 30]. Requirements for the suitability of particular animal groups, as well as vegetation as indicator organisms are a broad ecological distribution, a characteristic reaction of the biocenoses to the different biotic and abiotic parameters, as well as comprehensive knowledge about the autecology and the incidence of species [32, 44].

A frequently studied animal group is the ground beetle family (coleoptera, carabidae). It is distributed globally, the autecology of native species is widely known and the influences of different local factors on the carabidae community has been described many times [19, 33, 36, 51]. In particular there is comprehensive knowledge about the incidence, the living and adaptation strategies of the carabidae in agricultural landscapes [23, 24, 34, 52]. Carabidae are therefore well suited for answering ecological questions [44].

Plants also form characteristic biocenoses under the influence of local conditions. If these species communities can be reproduced under defined environmental conditions, they are described as plant communities [43]. The incidence of individual plant communities is also marked by local influences so that particular indicator properties can be attached to the incidence of these species in a habitat [17]. As a result both the plant community encountered in a biotope and the evidence of characteristic indicator species indicate the environmental parameters in a defined habitat [5, 14]. On the basis of its ecological demands, vascular plants at a particular location again possess indicator properties for particular preferred environmental conditions. In order to be able to quantify these properties, the plant communities have been organised according to their location demands for a special local parameter (e.g. shade or pH value) with an integer value (Ellenberg number) between 1 and 9 [15, 17]. Although the Ellenberg numbers are scaled ordinally, it is common in practice to form the arithmetic mean from the Ellenberg numbers of the plants of a defined location in order to be able to derive statements about the environmental conditions at the location [18]. Ellenberg numbers which are important for the model are the temperature number and the moisture number, which are influenced by the climatic conditions. The nitrogen number is also relevant, which allows for statements to be made about the intensity of land use. A high moisture number means that a plant species predominantly prefers moist soils, a high temperature number indicates a preference for higher temperatures and high nitrogen numbers indicate fertiliser indicators [15, 17].

5 Development of the Model

The model is made up of organism databases on plants and ground beetles on the one hand, and cartographical material on different subjects (types of biotope, soil parameters, climate data) on the other. These are brought together in a Geographical Information System (GIS), from which also the forecasts about the expected species are ultimately made.

In order to create the organism database, for one part real distribution data of plant and ground beetle species shall be used, which has already been collected in earlier work in the Aachen area [45]. These data are particularly suitable since, with 93 vegetation records and almost 50,000 carabidae individuals from 85 species, they are very comprehensive. Besides that they were collected at 37 locations in several different types of biotope (forest, hedge, meadow, as well as park-like areas), which are also partly relevant for the agricultural landscape. Since when collecting data for all recordings the exact coordinates of the place where things were found were recorded, it is now possible to combine the species found with the factors present at the place they were found (type of biotope, soil parameter, climate data). In the database produced from this, the preferences of each species regarding their abiotic and biotic surroundings are included. Should it emerge that, based on the Roß-Nickoll data collected, no sufficiently precise model can yet be created to forecast the incidence of organisms subject to the environmental factors mentioned, there are further species collections for different projects and bachelors theses which can be integrated into the model and contribute to its refinement and sharpening of the prognosis. Furthermore, autecological data on the preference of the named environmental parameters in the individual taxa are compiled from the literature.

The cartographical material on the subjects of the types of biotope, soil parameters and climate data are blended with a Geographical Information System. In this way all spatial units can be assigned property combinations, that is to say, that for each point of the area worked on it is known which type of biotope there is and how the soil and climate are constituted.

The organism databases are likewise read into the Geographical Information System. For the connection between organism data and environmental factors it is indispensable that the particular categories match. Now a first sample forecast on the species composition for selected locations can be made to verify the model. For each species, one obtains a probability of its incidence. It is checked by experiment whether the forecast species are also actually to be encountered in the areas. If this is not the case to a sufficiently precise extent, the model must be further optimised by either integrating even more distribution data of the organisms already collected within the framework of other projects, or by parameterising the autecological data in such a way that they fit the distribution data. This is an iterative process, i.e. settings are changed step by step until one reaches realistic results. For this verification and optimisation stage, data should be used which has been collected in 2011 within the framework of a university thesis [39]. The areas of study are located in two different broad expanse areas, one in the Lower Rhine Bay (Jülicher Börde) at Niederzier-Selhausen (Düren administrative district, ordnance survey map 5104 Düren), the other in Nordeifel at Simmerath-Rollesbroich (Aachen administrative district, ordnance survey map 5303 Roetgen). In each of the study areas the biotope types grassland, field and hedge were tested. Beside 30 vegetation records, more than 5,500 carabidae individuals from 54 species were recorded. It could be shown in the work that there are very specific regional patterns in the distribution of ground beetles, which on the one hand depend heavily on use, and on the other hand probably also react sensitively to expected climate changes.

References

1. Achtziger, R., Stickroth, H., Zieschank, R.: Nachhaltigkeitsindikatoren für die Artenvielfalt—ein Indikator für den Zustand von Natur und Landschaft in Deutschland. In: Bundesamt für Naturschutz (ed.) Angewandte Landschaftsökologie, vol 63 (2004)
2. BBodSchG: Gesetz zum Schutz vor schädlichen Bodenveränderungen und zur Sanierung von Altlasten, Bundes- Bodenschutzgesetz vom 17. März 1998 (BGBl. I S. 502), zuletzt geändert durch Artikel 3 des Gesetzes vom 9. Dezember 2004 (BGBl. I S. 3214) (1998)
3. Berry, P.M., Dawsen, T.P., Harrison, P.A., Pearson, R., Butt, N.: The sensitivity and vulnerability of terrestrial habitats and species in Britain and Ireland to climate change. J. Nat. Conserv. **11**, 15–23 (2003)
4. BNatSchG: Gesetz über Naturschutz und Landschaftspflege. Bundesnaturschutzgesetz vom 25. März 2002 (BGBl. I S. 1193), zuletzt geändert durch Artikel 3 des Gesetzes vom 22. Dezember 2008 (BGBl. I S. 2986) (2002)
5. Braun-Blanquet, J.: Pflanzensoziologie, 3rd edn. Springer, Berlin (1964)
6. Bredemeier, M.: Landnutzungswandel als Treiber von Biotopwandel und Veränderungen des landwirtschaftlichen Stoffhaushaltes. In: Herrmann, B. (ed.) Beiträge zum Göttinger Umwelthistorischen Kolloquium 2004–2006. Universitätsverlag Göttingen, Göttingen (2007)
7. Bundesamt für Naturschutz: Biodiversität und Klima - Vernetzung der Akteure in Deutschland. BfN Skripten 131 (2005)
8. Bundesministerium für Ernährung, Landwirtschaft und Verbraucherschutz: Die EU-Agrarreform – Umsetzung in Deutschland (2006)
9. Bundesministerium für Umwelt, Naturschutz und Reaktorsicherheit: Nationale Strategie zur biologischen Vielfalt (2007)
10. Bundesministerium für Umwelt, Naturschutz und Reaktorsicherheit: Indikatorenbericht 2010 zur Nationalen Strategie zur biologischen Vielfalt (2010)
11. Bundesministerium für Wirtschaft und Technologie, Bundesministerium für Umwelt, Naturschutz und Reaktorsicherheit: Energiekonzept für eine umweltschonende, zuverlässige und bezahlbare Energieversorgung (2010)
12. Cartagena, P.: Cartagena Protocol on Biosafety to the Convention on Biological Diversity: Text and Annexes. Secretariat of the Convention on Biological Diversity, Montreal (2000)
13. Dale, V.: Interactions between biofuel choices and landscape dynamics and land use. In: America ESo (ed.) Conference on the Ecological Dimensions of Biofuels, Washington, D.C (2008)
14. Dierschke, H.: Synopsis der Pflanzengesellschaften Deutschlands 3: Molinio-Arrhenateretea. Kulturgrasland und verwandte Vegetationstypen. Teil 1: Arrhenatheretalia - Wiesen und Weiden frischer Standorte. Göttingen (1997)
15. Dierschke, H., Briemle, G.: Kulturgrasland. Wiesen, Weiden und verwandte Staudenfluren. Ulmer, Stuttgart (2002)
16. Ecological Society of America: Policy Statements: Biofuel Sustainability. http://www.esa.org/pao/policyStatements/Statements/biofuel.php (2009)
17. Ellenberg, H.: Vegetation Mitteleuropas mit den Alpen in ökologischer, dynamischer und historischer Sicht, 5th edn. Ulmer, Stuttgart (1996)
18. Ewald, J.: Ansprache von Waldstandorten mit Zeigerarten- Ökogrammen - eine graphische Lösung für Lehre und Praxis. Allgemeine Forst- und Jagdzeitung **10**(11), Ewald, J.: Ansprache von Waldstandorten mit Zeigerarten- Ökogrammen - eine graphische Lösung für Lehre und Praxis. Allgemeine Forst- und Jagdzeitung **10**(11) (2003)
19. Eyre, M.D., Rushton, S.P., Luff, M.L., Telfer, M.G.: Predicting the distribution of ground beetle species (Coleoptera, Carabidae) in Britain using land cover variables. J. Environ. Manag. **72**, 163–174 (2004)
20. Faulstich, M., Greiff, K.B.: Klimaschutz durch Biomasse - Ergebnisse des SRU-Sondergutachtens 2007. UWSF – Z Umweltchem Ökotox **20**, 171–179 (2008)

21. Geiger, F., Bengtsson, J., Berendse, F., Weisser, W.W., Emmerson, M., Morales, M.B., Ceryngier, P., Liira, J., Tscharntke, T., Winqvist, C., Eggers, S., Bommarco, R., Pärt, T., Bretagnolle, V., Plantegenest, M., Clement, L.W., Dennis, C., Palmer, C., Onate, J.J., Guerrero, I., Hawro, V., Aavik, T., Thies, C., Flohre, A., Hänke, S., Fischer, C., Goedhart, P. W., Inchausti, P.: Persistent negative effects of pesticides on biodiversity and biological control potential on European farmland. Basic Appl. Ecol. **11**, 97–105 (2010)
22. GenTG: Gesetz zur Regelung der Gentechnik in der Fassung vom 16. Dezember 1993 (BGBl. I S. 2066), zuletzt geändert durch Artikel 12 des Gesetzes vom 29. Juli 2009 (BGBl. I S. 2542) (1990)
23. Hance, T.: Impact of cultivation and crop husbandry practices. In: Holland, J.M. (ed.) The Agroecology of Carabid Beetles, pp. 231–250. Intercept, Andover (2002)
24. Holland, J.M.: Carabid beetles: their ecology, survival and use in agroecosystems. In: Holland, J.M. (ed.) The Agroecology of Carabid Beetles, pp. 1–40. Intercept, Andover (2002)
25. Holsten, A., Vohland, K., Cramer, W., Hochschild, V.: Ökologische Vulnerabilität von Schutzgebieten gegenüber Klimawandel - exemplarisch untersucht für Brandenburg. BfN Skripten 246 (2009)
26. Institut für Technische Thermodynamik (Deutsches Zentrum für Luft- und Raumfahrt): Ökologisch optimierter Ausbau der Nutzung erneuerbarer Energien in Deutschland (2004)
27. Jeltsch, F., Moloney, K.A., Schurr, F.M., Kochy, M., Schwager, M.: The state of plant population modelling in light of environmental change. Perspect. Plant Ecol. Evol. Syst. **9**, 171–189 (2008)
28. Kangas, J., Pukkala, T.: Operationalization of biological diversity as a decision objective in tactical forest planning. Can. J. For. Res. Revue Can. De Rech. Forestiere **26**(1), 103–111 (1996)
29. Kommission Bodenschutz beim Umweltbundesamt: Bodenschutz beim Anbau nachwachsender Rohstoffe - Empfehlungen der "Kommission Bodenschutz beim Umweltbundesamt" (2008)
30. Kratochwil, A., Schwabe, A.: Ökologie der Lebensgemeinschaften. Biozönologie. Ulmer, Stuttgart (2001)
31. Kropp, J., Holsten, A., Lissner, T., Roithmeier, O., Hattermann, F., Huang, S., Rock, J., Wechsung, F., Lüttger, A., Pompe, S., Kühn, I., Costa, L., Steinhäuser, M., Walther, C., Klaus, M., Ritchie, S., Metzger, M.: Klimawandel in Nordrhein-Westfalen - Regionale Abschätzung der Anfälligkeit ausgewählter Sektoren. Abschlussbericht des Potsdam-Instituts für Kimafolgenforschung (PIK) für das Ministerium für Umwelt und Naturschutz, Landwirtschaft und Verbraucherschutz Nordrhein-Westfalen (MUNLV) (2009)
32. Lövei, G., Sunderland, K.D.: Ecology and behavior of ground beetles (Coleoptera: Carabidae). Ann. Rev. Entomol. **41**, 231–256 (1996)
33. Luff, M.L.: Use of Carabids as environmental indicators in grasslands and cereals. Annales Zoologici Fennici, 185–195 (1996)
34. Luff, M.L.: Carabid assemblage organization and species composition. In: Holland, J.M. (ed.) The Agroecology of Carabid Beetles, pp. 41–80. Intercept, Andover (2002)
35. Möckel, S.: Kommunales Gebietsplanungsrecht - Außenverbindliche Planung für den unbesiedelten Bereich. Planungsrechtliche Steuerung der Bodennutzung für nicht besiedelte Flächen, insbesondere der land- und forstwirtschaftlichen Nutzung. In: Bundesamt für Naturschutz (ed.) Treffpunkt biologische Vielfalt, vol. 8, pp. 167–172. Bonn (2008)
36. Müller-Motzfeld, G.: Adephaga 1: Carabidae (Laufkäfer). In: Freude, H., Harde, K.W., Lohse, G.A., Klausnitzer, B. (eds.) Die Käfer Mitteleuropas, vol. 2. Spektrum, Heidelberg, Berlin (2004)
37. Nitsch, H., Osterburg, B.: Grünland – Bestandsaufnahme und Handlungsoptionen: Erkenntnisse aus dem InVeKoS-Projekt. In: Naturschutz und Landwirtschaft im Dialog "Grünland im Umbruch", Bundesamt für Naturschutz Vilm (2009)
38. Normand, S., Svenning, J.-C., Skov, F.: National and European perspectives on climate change sensitivity of the habitats directive characteristic plant species. J. Nat. Conserv. **15**, 41–53 (2007)
39. Oellers, J.: Der Einfluss von Nutzung und Klima auf Carabiden- und Collembolenzönosen in Agrarlandschaften der Jülicher Börde und der Nordeifel - Auswertung regionalisierter Muster. Diplomarbeit, RWTH, Aachen (2011)

40. Petermann, J., Balzer, S., Ellwanger, G., Schröder, E., Ssymank, A.: Klimawandel-Herausforderung für das europaweite Schutzgebietssystem Natura 2000. In: Balzer, S., Dieterich, M., Beinlich, B. (eds.) Natura 2000 und Klimaänderungen, vol. 46. Naturschutz und biologische Vielfalt. Bundesamt für Naturschutz, Bonn (2007)
41. PflSchG: Gesetz zum Schutz der Kulturpflanzen. Pflanzenschutzgesetz in der Fassung der Bekanntmachung vom 14. Mai 1998 (BGBl. I S. 971, 1527, 3512), zuletzt geändert durch Artikel 13 G vom 29. Juli (BGBI. I 2542) (1998)
42. Plachter, H., Bernotat, D., Müssner, R., Riecken, U.: Entwicklung und Festlegung von Methodenstandards im Naturschutz, vol. 70. Schriftenreihe für Landschaftspflege und Naturschutz, Bonn (2002)
43. Pott, R.: Pflanzengesellschaften Deutschlands. Ulmer, Stuttgart (1992)
44. Rainio, J., Niemela, J.: Ground beetles (Coleoptera: Carabidae) as bioindicators. Biodivers. Conserv. **12**, 487–506 (2003)
45. Roß-Nickoll, M.: Biozönologische Gradientenanalyse von Wald-, Hecken- und Parkstandorten der Stadt Aachen. Verteilungsmuster von Phyto-, Carabido- und Araneozönosen. Dissertation, Shaker Verlag, RWTH Aachen (2000)
46. Roß-Nickoll, M., Lennartz, G., Fürste, A., Mause, R., Ottermanns, R., Schäfer, S., Smolis, M., Theißen, B., Toschki, A., Ratte, H.T.: Die Arthropodenfauna von grasigen Feldrainen (off crop) und die Konsequenzen für die Bewertung der Auswirkungen von Pflanzenschutzmitteln auf den terrestrischen Bereich des Naturhaushaltes. UBA Text Berlin **10**(04), 148 (2004)
47. Roß-Nickoll, M., Schulte, C.: Vorwort Beitragsserie Biodiversität. UWSF – Z Umweltchem Ökotox **20**(2), 102–103 (2008)
48. Sachverständigenrat für Umweltfragen: Sondergutachten Sachverständigenrat für Umweltfragen: Klimaschutz durch Biomasse. Erich Schmidt Verlag, Berlin (2007)
49. Schindler, M., Schumacher, W.: Auswirkungen des Anbaus vielfältiger Fruchtfolgen auf wirbellose Tiere in der Agrarlandschaft (Literaturstudie). Schriftenreihe des Lehr- und Forschungsschwerpunktes USL, vol. 147 (2007)
50. Ssymank, A., Hauke, U., Rückriem, C., Schröder, E.: Das europäische Schutzgebietssystem Natura 2000. BfN-Handbuch zur Umsetzung der Fauna- Flora-Habitat-Richtlinie und der Vogelschutz-Richtlinie, vol. 53 (1998)
51. Thiele, H.U.: Carabid beetles in their environments. A study on habitat selection by adaptation in physiology and behaviour. Springer, Berlin (1977)
52. Thomas, C.F.G., Holland, J.M., Brown, N.J.: The spatial distribution of carabid beetles in agricultural landscapes. In: Holland, J.M. (ed.) The Agroecology of Carabid Beetles, pp. 305–344. Intercept, Andover (2002)
53. Tscharntke, T., Klein, A.M., Kruess, A., Steffan-Dewenter, I., Thies, C.: Landscape perspectives on agricultural intensification and biodiversity—ecosystem service management. Ecol. Lett. **8** (8), 857–874 (2005)
54. Umweltbundesamt: Daten zur Umwelt. http://www.umweltbundesamt-daten-zur-umwelt.de/umweltdaten/public/theme.do?nodeIdent=2419 (2009)
55. Umweltbundesamt: Daten zur Umwelt - Anteil erneuerbarer Energien am gesamten Endenergieverbrauch. http://www.umweltbundesamt-daten-zur-umwelt.de/umweltdaten/public/theme.do?nodeIdent=2850 (2011)
56. United Nations: Convention on Biological Diversity. United Nations, New York (1992)
57. USchadG: Umweltschadensgesetz vom 10. Mai 2007 (BGBl. I S. 666), zuletzt geändert durch Artikel 14 des Gesetzes vom 31. Juli 2009 (BGBl. I S. 2585) (2007)
58. UVPG: Gesetz über die Umweltverträglichkeitsprüfung in der Fassung der Bekanntmachung vom 25. Juni 2005 (BGBl. I S. 1757, 2797), zuletzt geändert durch Artikel 1 des Gesetzes vom 11. August 2009 (BGBl. I S. 2723) (1990)
59. Zieschank, R., Stickroth, H., Achtziger, R.: Seismograph für den Zustand von Natur und Landschaft. Der Indikator für Artenvielfalt. Politische Ökologie **91–92**, 58–59 (2004)

Experimental Investigation of Dissipation Element Statistics in a Jet Flow

M. Gampert, P. Schaefer and N. Peters

Abstract We present a detailed experimental investigation of conditional statistics related to dissipation elements based on the scalar field θ and its scalar dissipation rate χ. Based on high frequency two-dimensional measurements of the mass fraction of propane in a turbulent round jet discharging into surrounding air, we acquire data resolving the Kolmogorov scale in every spatial direction using a high-speed Rayleigh scattering technique and Taylor's hypothesis. The Reynolds number (based on nozzle diameter and jet exit velocity) varies between 2,000 and 6,700 and the Schmidt number between 1.2 and 1.7. The experimental results for the normalized marginal pdf $\tilde{P}(\tilde{l})$ of the length of dissipation elements follow closely the theoretical model derived by Wang and Peters (J. Fluid. Mech. 608:113–138, 2008). We also find that the mean linear distance between two extreme points of an element is of the order of the Taylor microscale λ. Furthermore, the conditional mean $\langle \Delta\theta | l \rangle$ scales with Kolmogorov's 1/3 power.

1 Introduction

The use of turbulent jet flows to improve scalar mixing is present throughout engineering applications such as chemical reactions in combustion systems, see for instance [1–4]. As the statistical properties of scalar turbulence are coupled to the underlying velocity field, it is believed that some of the basic ideas can be transferred to scalar fields. This means for instance that the assumed universal behaviour at the smallest scales of the velocity field as proposed by [5, 6] is also valid for the scalar field, see [7–9]. [10] review and compare small-scale turbulence of scalar and

M. Gampert · P. Schaefer · N. Peters (✉)
Institut für Technische Verbrennung, RWTH-Aachen University,
Templergraben 64, Aachen, Germany
e-mail: n.peters@itv.rwth-aachen.de

M. Gampert
e-mail: mgampert@itv.rwth-aachen.de

M. Klaas et al. (eds.), *Fuels From Biomass: An Interdisciplinary Approach*,
Notes on Numerical Fluid Mechanics and Multidisciplinary Design 129,
DOI 10.1007/978-3-662-45425-1_3

velocity fields, while [11] discussed the applicability of local isotropy and universality to scalar fields. The latter work further examined the scalar spectra and concluded that its scaling in grid turbulence can already be investigated at comparatively low Taylor-based Reynolds numbers of the order of $Re_\lambda = 150$ ($Re_\lambda = u\lambda/\nu$, where u denotes the longitudinal r.m.s. velocity, λ the Taylor scale and ν the kinematic viscosity), while for shear flows the Taylor-based Reynolds number has higher values of the order of $Re_\lambda \approx 1{,}000$, a finding to which we will refer later.

One of the approaches to improve the understanding of turbulent fields is to study geometrical structures using critical points of the flow field. [12] analyzed the behaviour of zero gradient points and minimal gradient surfaces in turbulent scalar fields. He argued that these points are of importance to the problem of turbulent mixing. [13] subdivided the flow field into four types of space-filling regions, characterizing them by the second invariant of the velocity derivative tensor Q and the pressure p. [14] relate the energy dissipation rate coefficient to the stagnation point structure of homogenous isotropic turbulence to prove its non-universality. [15] studied non-local geometries of eddies in scalar turbulence. Therefore, they extracted structures using curvelets and iso-contours, which they in the following characterize based on the joint probability density function of different geometrical parameters to structure them into groups with a common geometry.

Based on the extreme points of turbulent scalar fields, i.e. points of vanishing scalar gradient, [1, 16] developed the theory of dissipation elements. These elements arise as natural geometries in turbulent scalar fields, when these are analyzed by means of gradient trajectories in fields obtained by highly resolved direct numerical simulations (DNS). Starting from every grid point, trajectories along the ascending and descending gradient direction can be calculated, which inevitably end at extreme points. All points that share the same two ending points define a finite volume which is called a dissipation element. These elements are parameterized by two values: the linear length l between and the scalar difference at the extreme points. Using this theory, space-filling elements are identified, which allow the reconstruction of statistical properties of the field as a whole in terms of conditional statistics within the elements—examples of such analysis can be found in [1, 17].

From the definition it follows that the temporal evolution of dissipation elements in turbulent fields is inherently connected to the evolution of their ending points. While strain and diffusion lead to a continuous distortion of an element as a whole, the creation or annihilation of extreme points leads to their abrupt formation or disappearance. Dimensional analysis using the viscosity and the strain rate suggest that the mean linear distance l_m should be of the order of the Taylor microscale λ, see [16]. The different mechanisms have been considered in a modelled evolution equation for the probability density function (pdf) for the length distribution $P(l)$ of dissipation elements. In its normalized form $\tilde{P}(\tilde{l})$ is found to be independent of both, the Reynolds number and the type of turbulent flow. In this equation two parameters appear, cf. [1], which can be identified as a splitting respectively attachment frequency of the elements and thus indirectly correspond to the life time of extreme points, see [18–20] for further details. The normalized rate of strain $\tilde{a}$ is a third important parameter employed in [1] for the description of the pdf of the length distribution. This strain rate is defined as the difference of the velocity at the ending points, projected in direction of the connecting line between the two extreme points.

As dissipation elements have mostly been analyzed in DNS so far, a detailed experimental verification is desirable. Direct numerical simulations of homogeneous shear turbulence in connection with dissipation element analysis has revealed that a resolution of the order of the Kolmogorov scale η is needed to obtain grid independent statistics. Due to their corrugated three-dimensional geometry in combination with the required resolution, an experimental validation is therefore challenging. A first attempt was performed by [21] using tomographic PIV for three-dimensional measurements of the velocity field in a channel flow.

In the present study, we investigate the statistics of a passive scalar advected by a turbulent flow. In particular, we examine the mass fraction field of propane discharging from a turbulent free round jet into surrounding air. Therefore, we will study dissipation elements in this passive scalar field denoted by θ, which is governed by the convection-diffusion equation

$$\partial\theta/\partial t + u_i(\partial\theta/\partial x_i) = D\partial^2\theta/\partial x_i^2. \tag{1}$$

In Eq. (1), D is the binary diffusion coefficient assumed to be constant and u_i denotes the velocity component in i-direction, while repeated indices imply summation. In addition, a gradient quantity, namely the scalar dissipation rate χ is considered. The latter is for a passive scalar θ defined as

$$\chi = 2D(\partial\theta/\partial x_i)^2. \tag{2}$$

A wide range of experimental investigations of such a scalar field can be found in the literature. Three-dimensional data, however, is limited as often single- or multi-point measurements in combination with Taylor's hypothesis are conducted, see for instance [22, 23], which is of limited use in the context of dissipation element analysis. The development of advanced laser optical techniques with a high pulse energy at a high repetition rate has facilitated the experimental investigation of spatially three-dimensional conserved scalar quantities. In such measurements, the three-dimensional information is found either by imaging in parallel, spatially distinct two-dimensional planes or via a sweeping of a single two-dimensional laser sheet in sheet normal direction, see for instance [24] for an overview. For the present purpose, however, where gradient trajectories are to be determined, both approaches are impractical.

In Sect. 2, we therefore present a method, which combines a high-speed Rayleigh scattering technique with Taylor's hypothesis to resolve the Kolmogorov scale η in all three spatial directions at Reynolds numbers Re_λ up to 106. In this Sect. 3, we discuss the experimental results in terms of extremal points in scalar fields [12] and an experimental validation of dissipation element theory [1], before the paper is concluded in Sect. 4.

2 Experimental Arrangement

In the course of this section we present the experimental arrangement and give some background information regarding Rayleigh scattering and Taylor's hypothesis. We conclude with the description of the post-processing before the analysis of the experimental results.

2.1 Experimental Arrangement

In the present study, a turbulent round propane jet discharging from a nozzle with a diameter d = 6 mm into surrounding air has been chosen as the core of the experimental set-up. The scalar field, i.e. the mass fraction of propane, is visualized via Rayleigh scattering of a diode pumped double cavity Nd:YLF laser (Litron Lasers LDY303HE-PIV) at the molecules. The laser emits frequency-doubled light at a wavelength of 527 nm, has a pulse energy of 2×22.5 mJ with a pulse width of 150 ns at 1 kHz and can operate at up to 10 kHz. To account for energy fluctuations, the signal is corrected on a shot by shot basis by a 12 bit energy monitor (LaVision Online Energy Monitor).

Laser-Rayleigh scattering is used to determine the instantaneous mass fraction of the binary mixture of jet and reservoir gas in a small focal plane within the turbulent core of the jet around the centre line. Laser-Rayleigh scattering has been used and documented in many previous studies, see for instance [25, 26], and is therefore only described briefly here. The technique makes use of the fact that gas molecules elastically scatter photons, and that different molecules have different Rayleigh-scattering cross-sections. In the present study for instance, the cross-section of propane is roughly thirteen times higher than the one of the surrounding air. The Rayleigh scattering light intensity in the perpendicular direction to the light source from a binary gas mixture is related in a linear manner to the concentration of the gas exiting from the jet. Hence, the two end points $\theta_1 = 0$ and $\theta_1 = 1$ of this linear relation are recorded for calibration purposes, before the conversion from signal to concentration is simply accomplished by linear interpolation.

For the illumination of a two-dimensional plane, a sheet optic for thin, collimated sheets of 130 μm diameter and 25 mm height is installed behind laser and energy monitor, thereby illuminating a plane perpendicular to the jet centerline, see Fig. 1 for an overview of the full experimental set-up. The resulting signal is recorded with a 12 bit LaVision high speed CMOS-camera HighspeedStar6 with a full resolution up to 5.4 kHz and 8 GB internal memory in combination with a two step high speed intensified relay optic (LaVision HighSpeed IRO). This image intensifier is an electronic shutter device with a maximal repetition rate of 2 MHz and an extremely variable exposure time. In contrast to a standard CMOS or CCD, which usually has an exposure time in the *ms* range, the IRO can be operated in the *ns* range, thereby allowing time resolved analysis of shortest light pulses.

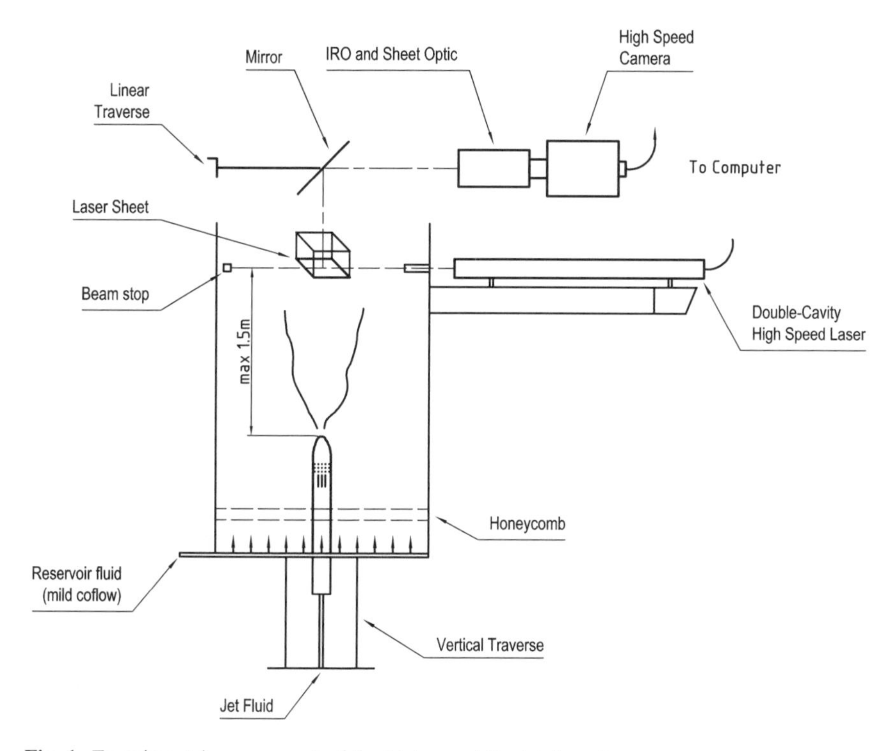

Fig. 1 Experimental arrangement of the high-speed Rayleigh system

In addition, this IRO has an extremely reduced vignetting to maximize the area recorded by the camera. For further information regarding camera and IRO sensitivity and signal as well as noise considerations see [27].

In order to observe the Rayleigh signal without interaction between the optical arrangement and the turbulent flow, a mirror is installed in some distance to the laser sheet, which has a thin coating of enhanced aluminium reflecting above 95 % of the incoming light at a wavelength of 527 nm, thereby minimizing signal losses. To protect the propane jet from exterior influences such as dust particles, a mild coflow of clean, dry air discharges from a surrounding tube with a diameter of 150 mm and a length of 450 mm. In the latter a honeycomb is installed in the lower third of the tube to guarantee a uniform velocity profile.

2.2 Data Post-processing

Based on the raw images recorded by the camera, several corrections of the data have to be applied before a proper analysis can be performed. In a first step, noise stemming from dark current and background are subtracted before the intensity of each image is corrected. This is done on a shot-by-shot basis to compensate for fluctuations of the laser energy using the energy monitor data as well as for an inhomogeneous illumination within the laser sheet. The size of the images is then reduced from $1{,}024^2$ to 850^2 pixels to remove areas without any signal due to vignetting between IRO and camera. Afterwards, a Mie-filter consisting of a mixed intensity threshold and particle size approach is applied to remove undesirable effects originating from dust particles in the test region. The signal of a dust particle is captured on several images, once it enters the area of interest, due to the relatively low jet velocity in combination with the high-speed recording. Based on these corrected images, the signals corresponding to pure air and pure propane respectively are calibrated and used to convert the recorded photon counts to propane mass fraction.

In a next step, the recorded time series of the plane at a fixed downstream location x/d is transformed into a spatial signal in streamwise direction with $\Delta x = U \cdot \Delta t$ based on Taylor's hypothesis, see [28]. Hence, we obtain a frozen three-dimensional concentration field. This approximation estimates the spatial derivative in the streamwise x-direction from the local instantaneous value of the time derivative from a single-point or planar measurement, when the required three-dimensional multipoint measurements are impractical or unavailable. In the limit of low turbulence intensities, the motion of gradients relative to the local mean flow can be approximated as one of pure convection. Due to the importance of two-point statistics and spatial gradient quantities in turbulence, it is common to use Taylor's hypothesis to estimate spatial derivatives.

Among the most widespread uses of Taylor's hypothesis is the estimation of dissipation rates, see Eq. (2), in turbulent shear flows, cf. [29], though the scalar dissipation can be measured directly at one point in contrast to ε. As in the present case time series measurements within a plane are performed, a direct evaluation is only possible for the cross-stream spatial derivative components in y- and z-direction. For a calculation of χ, consequently a mixed spatio-temporal scalar dissipation approximation has to be formed by combining the available spatial derivatives with the time derivative, yielding

$$\chi = 2D\left[\left(-\frac{1}{U}\frac{\partial\theta}{\partial t}\right)^2 + \left(\frac{\partial\theta}{\partial y}\right)^2 + \left(\frac{\partial\theta}{\partial z}\right)^2\right]. \tag{3}$$

The camera can resolve a plane of $1{,}024^2$ pixels at a frequency of 5 kHz. This bounds the jet exit velocity U_0 to a value, at which the resolution in x-direction remains below the Kolmogorov scale at the various downstream positions x/d. The resulting experimental parameters are shown in Table 1, where the series of measurements in the upper part are performed at a fixed jet exit velocity. For those in the lower part of the table, however, the highest possible velocity U_j has been used, where η is still assumed to be resolved. The Taylor based Reynolds number Re_λ stems from $Re_\lambda = 1.3\sqrt{Re_0}$ (with $Re_0 = U_0 d/\nu$), cf. [30]. The mean concentration $\langle\Theta\rangle$ on the center line is given as well as the local Schmidt number Sc, calculated with the kinematic viscosities of propane ($\nu_{C_3H_8} = 4.6 \times 10^{-6}\,\mathrm{m}^2/\mathrm{s}^{-1}$) and air

Table 1 Characteristics of the propane jet at centerline locations from x/d = 15−40

Downstream position x/d	15	20	30	40
Jet exit velocity U_j (m/s)	2.15	2.15	2.15	2.15
Mean concentration $\langle\Theta\rangle$	0.24	0.19	0.14	0.11
Scalar Taylor scale λ (mm)	0.43	0.43	0.32	0.38
Kolmogorov scale η (mm)	0.12	0.15	0.21	0.25
Batchelor scale η_B (mm)	0.10	0.13	0.17	0.20
Schmidt number Sc	1.45	1.52	1.59	1.63
Reynolds number Re_0	2,800	2,800	2,800	2,800
Reynolds number Re_λ	69	69	69	69
Resolution $\Delta x/\eta$	0.91	0.65	0.34	0.22
Resolution $\Delta z/\eta$	0.13	0.10	0.07	0.06
Downstream position x/d	10	20	30	40
Jet exit velocity U_j (m/s)	1.55	2.75	4.0	5.1
Mean concentration $\langle\Theta\rangle$	0.38	0.18	0.12	0.08
Scalar Taylor scale λ (mm)	0.82	0.35	0.30	0.35
Kolmogorov scale η(mm)	0.12	0.12	0.12	0.12
Batchelor scale η_B(mm)	0.11	0.10	0.09	0.09
Schmidt number Sc	1.28	1.54	1.6	1.66
Reynolds number Re_0	2,020	3,590	5,220	6,650
Reynolds number Re_λ	58	78	94	106
Resolution $\Delta x/\eta$	0.96	0.97	0.98	0.97
Resolution $\Delta z/\eta$	0.13	0.12	0.12	0.12

($\nu_{air} = 16.04 \times 10^{-6}\,\mathrm{m}^2/\mathrm{s}^{-1}$) at 300 K, cf. [31], and the molecular diffusion coefficient of propane into air ($D_{C_3H_8-air} = 9.09 \times 10^{-6}\,\mathrm{m}^2/\mathrm{s}^{-1}$). Following [32], we assume a constant molecular diffusivity D as variations remain within a few percent independent of the exact composition of the binary gas mixture. Therefore, we calculate the Schmidt number in our test volume using the viscosity on the center line ν_c. Based on this definition, the Schmidt number increases along the center line from 0.51 (with the viscosity of pure propane at the nozzle orifice) to 1.76 (with the viscosity of air far downstream). An approximation formula taken from [33] is used to estimate ε, based on which the Kolmogorov scale η ($= (\nu^3/\varepsilon)^{1/4}$) and the Batchelor scale η_B ($=Sc^{-1/2}\eta$) have been calculated. Though the latter quantities are not calculated from direct measurements, and therefore have to be treated with caution, they will be used in the following to interpret the data. Note that in the following the measurements at the same downstream positions will be referred to as for instance $x/d = 30.1$ and $x/d = 30.2$, where (1) always corresponds to the lower jet exit velocity (and consequently lower Reynolds number) data and (2) to the higher one given in Table 1.

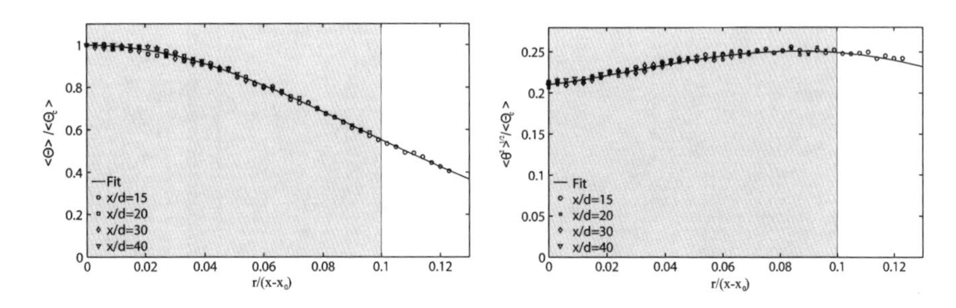

Fig. 2 Radial profiles for mean $\langle \Theta \rangle / \langle \Theta_c \rangle$ and r.m.s. value $\langle \theta^2 \rangle^{1/2} / \langle \Theta_c \rangle$ at x/d = 15, 20.1, 30.1 and 40.1

Furthermore, values for the scalar Taylor length scale λ in streamwise direction defined by

$$\lambda = \left(\frac{\langle \theta^2 \rangle}{\langle (\partial \theta / \partial x)^2 \rangle} \right)^{1/2} \tag{4}$$

are listed in Table 1 to give an impression of the relative size of the recorded area, which extends over a square with 16 mm side length. Finally, a three-dimensional diffusion and a spectral cut-off filter are applied in the boxes. High-frequency noise was thus removed by isotropic truncation applying the 2/3 rule, so that one obtains a three-dimensional concentration field at various downstream positions. Based on these, we then calculate the scalar dissipation rate, which is computed using a highly accurate and numerically robust compact scheme. Note, that the analysis which will be performed in the following, has been restricted in radial direction to $\tilde{r} < 0.1$. Here, the non-dimensional similarity coordinate is defined by $\tilde{r} = r/(x - x_0)$, x_0 denotes the virtual origin of the jet and has been found to be $x_0/d = -1.75$. The radial limitation is illustrated in Fig. 2, were we show the radial profiles of the mean and the r.m.s. value of the concentration at the four downstream positions with constant nozzle exit velocity. Furthermore, it is indicated to which extent the data is used for the analysis of the experimental results. Thereby, the entrainment zone in which external intermittency plays a fundamental role, see for instance [34], is essentially excluded.

3 Analysis of the Experimental Results

In the course of this section, we will analyze the experimental results with respect to the distribution of extremal points in the measured scalar fields and the scaling of their distance. Then we perform a dissipation element analysis and seek detailed experimental validation of the theory and investigate the orientation of dissipation elements in the jet flow.

3.1 Investigation of Extremal Points in Scalar Fields

In Sect. 1 it has been described briefly that dissipation elements are defined by the spatial region containing all points from which gradient trajectories reach the same two extrema as ending points. In a first step, we therefore examine the distribution of maxima and minima (red and blue points respectively) in the measurement volume. Figure 3 depicts these points in only a smaller part of the box to facilitate visibility of the structure. Overall, we observe approximately an equal number of maximum and minimum points. For each dissipation element the two ending points have been connected by a straight green line.

[1, 16, 20] investigated the scaling of the mean linear distance l_m between two extreme points using DNS of various turbulent flows and scalar fields. The authors conclude that the mean linear separation length of extreme points is of the order of the Taylor scale of the velocity field defined as

$$\lambda_u = \sqrt{10\frac{k}{\varepsilon}\nu}, \tag{5}$$

in which k denotes the turbulent kinetic energy. Based on their analysis, it can also be concluded that only extreme points with a large separation length are subject to secondary splitting, while for small separations diffusive processes become dominant, which leads to a mutual annihilation of the extreme points. For the general case of arbitrary Schmidt numbers, [12] concluded that the separation distance of extreme points scales as

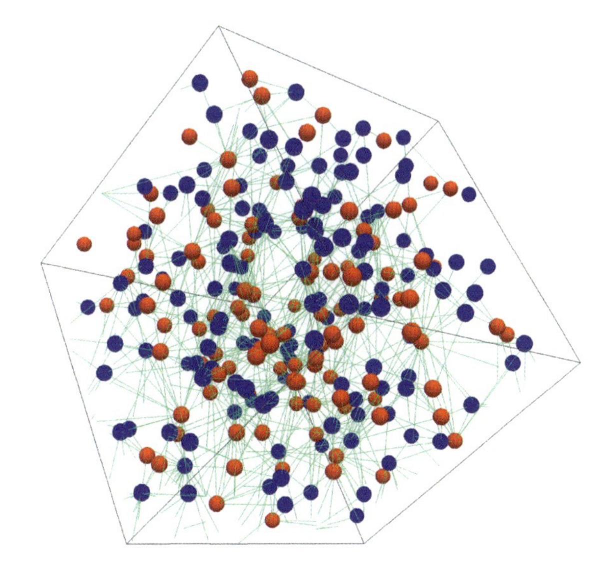

Fig. 3 (*top*) Distribution of extremal points in a section of the box. *Blue points* are minima and *red points* are mixima, which are connected by straight *green lines* for each dissipation element

$$l_m \propto \left(\frac{D}{\gamma}\right)^{\frac{1}{2}}, \tag{6}$$

where γ denotes the root mean square rate of strain, a scaling which coincides with the scaling proportional to the Taylor length found by [1] for unity Schmidt numbers. Figure 4 shows the ratio of the mean linear distance to the Taylor microscale over the nozzle distance for the different measurements and the fields of θ and χ. The afore discussed proportionality is clearly illustrated as $l_m/\lambda \approx 1$ for θ independent of the measurement location x/d. The same observation is valid for the χ field, where the constant value of the ratio lies at around $l_m/\lambda \approx 0.6$. As will be discussed below, this lower value is no surprise due to the shorter distance between two extreme points in the field of the scalar dissipation rate. This scaling of the mean distance between extreme points with the Taylor microscale is also found for other critical points: [14] recently showed that the average distance between neighbouring stagnation points is proportional to λ using the generalized Rice theorem.

Furthermore, this scaling of the mean linear distance between maximum and minimum points can be related to the number density of extreme points via $N_{ex} \propto l_m^{-3}$. Clearly this scaling is also valid in the fields of θ and χ as depicted in Fig. 5. Combining the results displayed in Figs. 4 and 5 we can further state $N_{ex} \propto \lambda^{-3}$, a finding which is of particular interest, as a similar scaling has been obtained by [35] for the number density n_s of stagnation points. They concluded $n_s \approx C_s L^{-3}(L/\eta)^{D_s}$, where C_s is a dimensionless number and $D_s = 2$ can be interpreted as a fractal dimension. Introducing the relation between η, λ and L, one

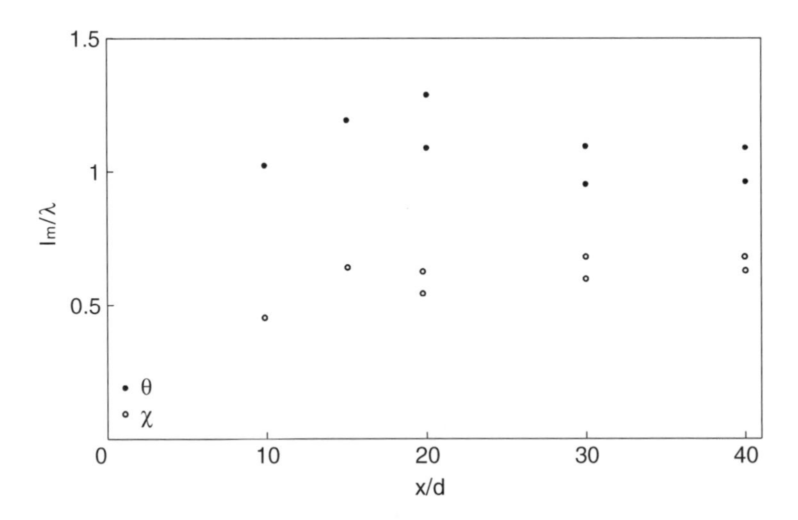

Fig. 4 The ratio between the mean linear distance to the Taylor microscale l_m/λ for the mass fraction field denoted by θ and the scalar dissipation field denoted by χ as a function of the downstream position x/d

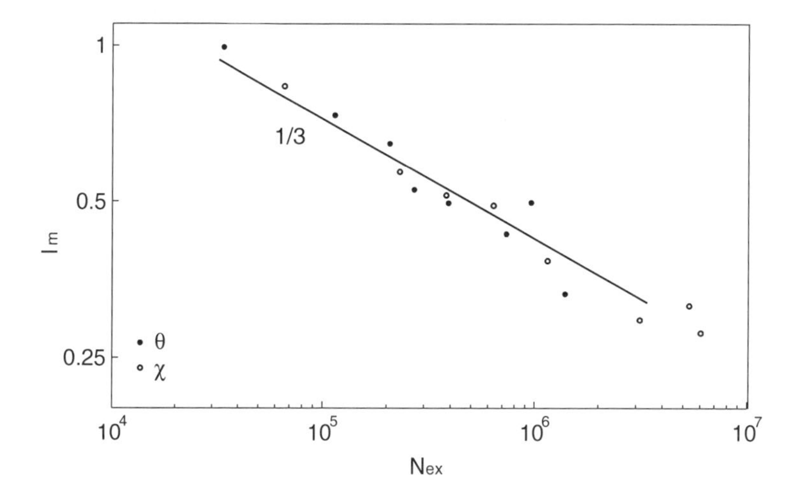

Fig. 5 The scaling of the mean linear distance l_m with the number of extreme points N_{ex}

obtains $n_s \propto \lambda^{-3}$, which equals the scaling of the number density of extreme points in a scalar field as obtained in the present experimental data. Finally, that many more extremal points are found in the χ-field as compared to the θ-field, indicates the well-known highly intermittent structure of the χ-field, see for instance [10, 23].

3.2 Dissipation Element Analysis

The motivation for dissipation elements is the reconstruction of the entire scalar field by means of an adequate parametrization of the geometric and scalar properties of the elements. Different examples for the elements obtained from various measurements of the θ-field are displayed in Fig. 6. One can observe that experimentally obtained dissipation elements have the same convoluted, irregular and corrugated shape already known from DNS. As has been discussed in the introduction, the elements are space-filling so that neighbouring ones are strongly intertwisted and make a clearly defined parameterization necessary.

The corresponding joint probability density function $P(l, \Delta\theta)$ is expected to contain most of the information needed for a statistical reconstruction. Based on a trajectory search algorithm, the field of the propane mass fraction has been analyzed for the different measurements and the resulting jpdf for x/d = 30.1 is shown in Fig. 7 (to relate the values of l and $\Delta\theta$ given in this figure to other characteristic flow quantities see Table 1). In this illustration, different physical effects can be identified. Besides a distinct maximum, one observes a decrease at the origin, corresponding to the annihilation of small elements due to molecular diffusion.

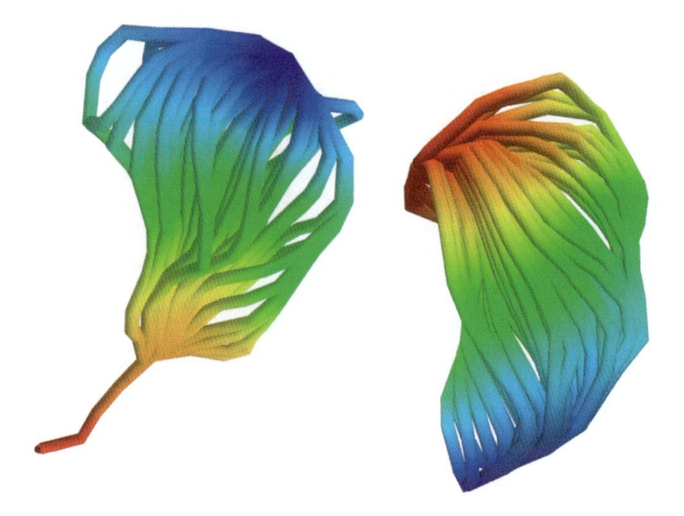

Fig. 6 Examples of dissipation elements calculated from the concentration field θ. The scalar value increases from the minimum (*blue*) to the maximum (*red*) point

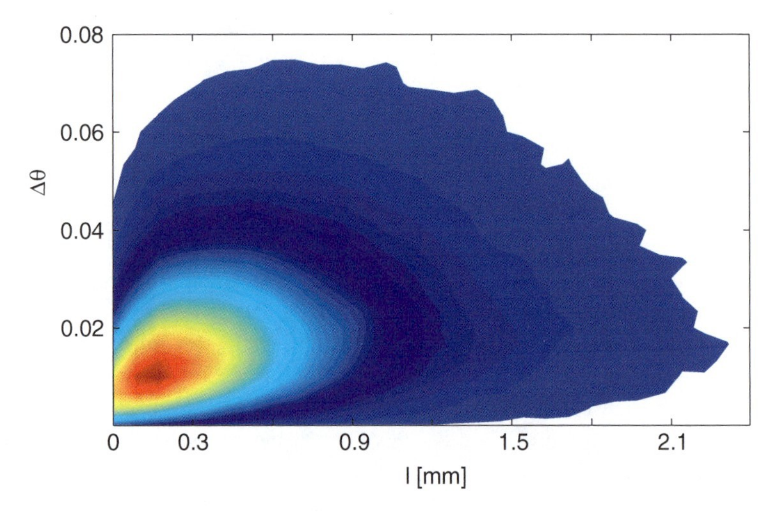

Fig. 7 Joint probability density function $P(l, \Delta\theta)$ for dissipation elements in the θ-field from measurements at x/d = 30.1

The marginal pdf $P(l)$ of the linear length is defined by

$$P(l) = \int_{0}^{\infty} P(\Delta\theta, l) d\Delta\theta. \tag{7}$$

For this pdf in its normalized form $\tilde{P}(\tilde{l})$, with $\tilde{P} = Pl_m$ and $\tilde{l} = l/l_m$, a model equation can be derived, cf. [1], which can be solved numerically yielding the steady solution, which is depicted by the solid curves in Figs. 8 and 9.

The latter show the results for the normalized pdf of the length distribution $\tilde{P}(\tilde{l})$ obtained for the fields of θ and χ at the different downstream positions. In general, one observes a very good qualitative agreement of the shape of the experimental results with the solution of the theoretically derived model. This is solved using $D_e = 0.6$ for θ as this has been determined to be the optimal value for passive scalar fields, cf. [1]. However, for χ an optimal agreement is obtained using $D_e = 1.5$. Slight differences can be identified for x/d = 30.1 and 40.2, where the model marginally underpredicts the maximal value of the pdf. The linear increase at the origin as well as the exponential tail follow closely the predicted solution, see especially the log-in insets. For the measurements at x/d = 10−20, the location and the overall shape of the pdf are tilted slightly to the left, resulting in a small deviation from the model, though the branches left and right from the maximum qualitatively agree nicely. However, deviations in the exponential tails (see the slopes of the pdf in the insets) of the pdf are hard to interpret due to the limited number of sample points for large elements. Nevertheless, Figs. 8 and 9 illustrate that $\tilde{P}(\tilde{l})$ seems not to be a function of the Reynolds number as the values of Re_λ vary from 58 to 106.

In a next step, we will analyze the conditional mean of $\Delta\theta$ conditioned on the length of the respective dissipation element, defined by

$$\langle \Delta\theta | l \rangle = \int_{0}^{\infty} \Delta\theta P(\Delta\theta | l)\, d\Delta\theta, \tag{8}$$

where

$$P(\Delta\theta | l) = P(l, \Delta\theta)/P(l). \tag{9}$$

In unconditioned statistics the first moment is equal to zero. For statistics based on gradient trajectories however, this is not the case, as the scalar value increases monotonically along a trajectory from the minimum to the maximum point. Consequently, we will study the conditional first order moment to examine its scaling. [1, 16, 17] investigated this conditional difference based on various scalar fields from DNS—for instance, a passive scalar ϕ or the instantaneous dissipation ε, which were found to scale with Kolmogorov's 1/3 power law. For the different measurement locations and conditions presented in Sect. 2, the results for $\langle \Delta\theta | l \rangle$ are

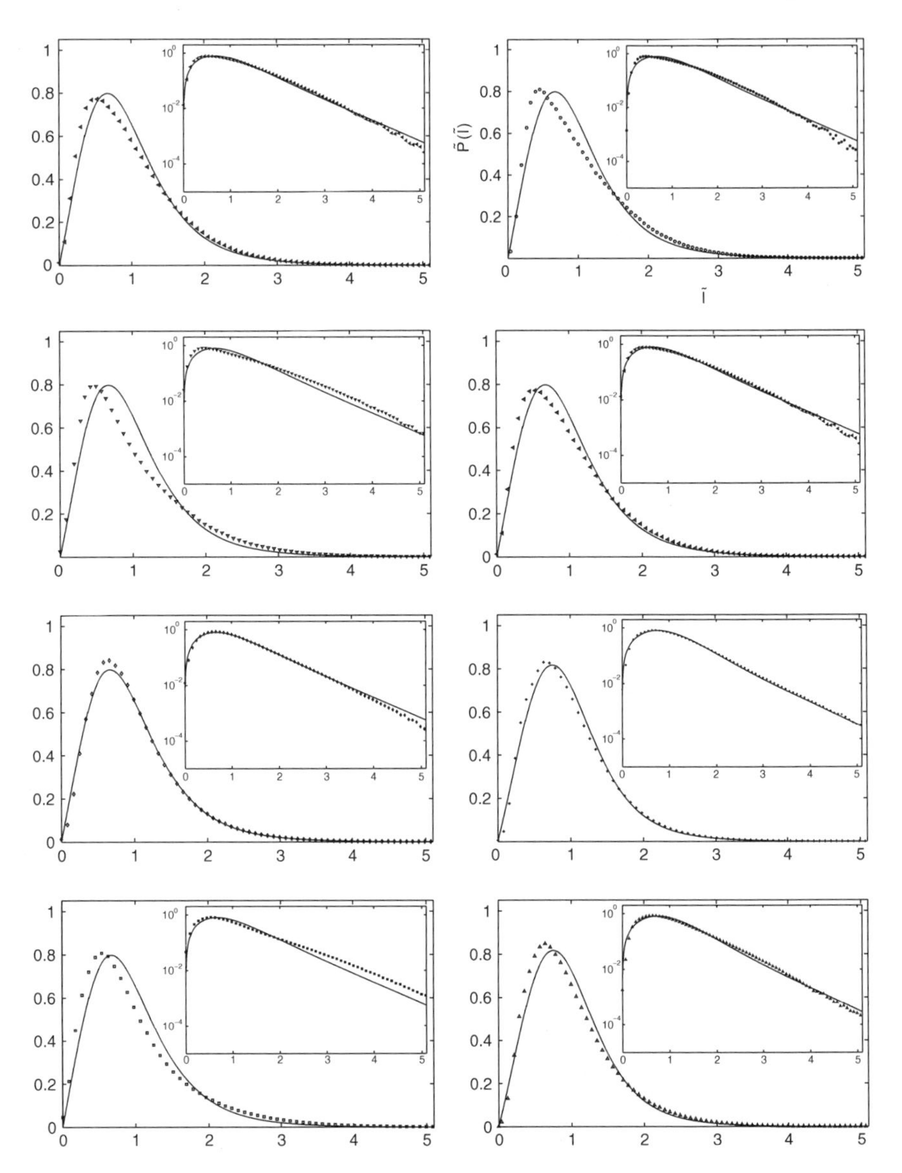

Fig. 8 Marginal pdf $\tilde{P}(\tilde{l})$ for dissipation elements in the θ-field from measurements between x/d = 10–40

shown in Fig. 10 in compensated form (Note: for a better graphical illustration all data apart from x/d = 40.2 have been multiplied with multiples of five to distinguish between the different datasets).

The data obtained at the different measurement locations included in Fig. 10, show for small lengths a viscous scaling, followed by a scaling of $\langle \Delta\theta | l \rangle$ with 1/3 as indicated by the solid line, which is more or less accurate for the different data.

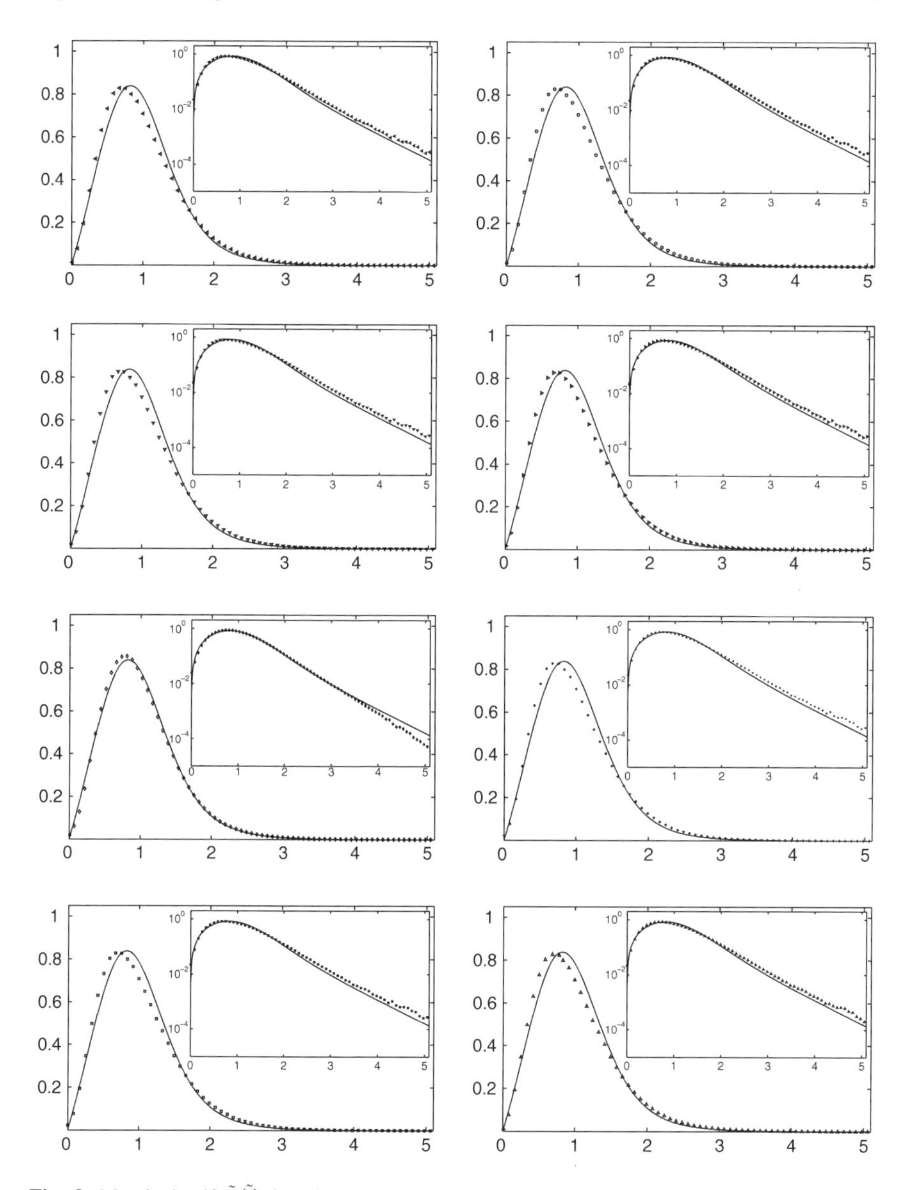

Fig. 9 Marginal pdf $\tilde{P}(\tilde{l})$ for dissipation elements in the χ-field from measurements between x/d = 10–40

The best agreement and widest scaling region of almost one decade is obtained with increasing Reynolds number and nozzle distance. Referring to [11, 36], we also want to stress at this point the fact that a scaling following the one suggested by Kolmogorov is more or less accurate for all of the shown data even at Taylor-based Reynolds numbers below sixty and at distances $l \approx 3\lambda \approx 10\eta$. A similar

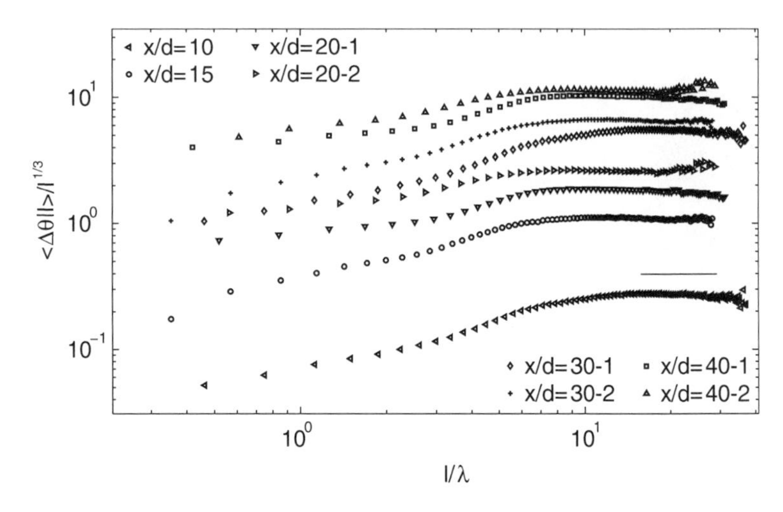

Fig. 10 Conditional mean $\langle \Delta\theta|l\rangle$ for x/d = 10–40

observation has been reported by [37], who investigated the velocity field of DNS of low-Reynolds number flows. They concluded that the asymptotic state of turbulence is attained for the velocity gradients at far lower Reynolds numbers than those required for the identification of the inertial range.

4 Conclusion

We have presented a detailed experimental analysis of dissipation elements in the scalar concentration field θ and the scalar dissipation rate χ. To this end, we have conducted three-dimensional measurements of the mass fraction of propane discharging into quiescent air from a turbulent round jet using a high-speed Rayleigh scattering technique in combination with Taylor's hypothesis. This procedure allows to acquire data resolving the Kolmogorov scale in every spatial direction at nozzle based Reynolds numbers between 2,000 and 6,700 and Schmidt numbers between 1.2 and 1.7.

In a first step, we have examined the distribution of extremal points in the scalar fields and have shown that the mean linear distance between a maximum and a minimum scales as $l_m \propto N_{ex}^{1/3}$. Here, N_{ex} denotes the overall number density of extreme points and is proportional to the Taylor micro-scale as $l_m/\lambda \approx 1$ at all measurement positions and conditions. Afterwards, we have shown that the experimental results for the normalized marginal pdf $\tilde{P}(\tilde{l})$ of the length of dissipation elements follows closely the theoretical model. Furthermore, the conditional mean $\langle \Delta\theta|l\rangle$ already scales with Kolmogorov's 1/3 for separation distances of $\mathcal{O}(\lambda)$ though $\mathcal{O}(Re_\lambda) = 10^1 - 10^2$.

Acknowledgments This work was funded by the NRW-Research School "BrenaRo" and the Cluster of Excellence "Tailor-Made Fuels from Biomass", which is funded by the Excellence Initiative of the German federal state governments to promote science and research at German universities.

References

1. Wang, L., Peters, N.: Length scale distribution functions and conditional means for various fields in turbulence. J. Fluid Mech. **608**, 113–138 (2008)
2. Warhaft, Z.: Passive scalars in turbulent flows. Annu. Rev. Fluid Mech. **32**, 203–240 (2000)
3. Peters, N.: Turbulent Combustion. Cambridge University Press, Cambridge (2000)
4. Dimotakis, P.E.: Turbulent mixing. Ann. Rev. Fluid Mech. **37**, 329–356 (2005)
5. Kolmogorov, A.N.: The local structure of turbulence in an incompressible viscous fluid for very large Reynolds numbers. Dokl. Akad. Nauk SSSR **30**, 301–305 (1941)
6. Kolmogorov, A.N.: Dissipation of energy under locally isotropic turbulence. Dokl. Akad. Nauk SSSR **32**, 16–18 (1941)
7. Obukhov, A.R.: The structure of the temperature field in a turbulent flow. Jzv. Akad. Nauk. SSSR, Ser. Geographic. Geophys. **13**, 58–69 (1949)
8. Corrsin, S.: On the spectrum of isotropic temperature field in isotropic turbulence. J. Appl. Phys. **22**, 469–473 (1951)
9. Batchelor, G.K., Proudman, I.: The large scale structure of homogeneous turbulence. Phil. Trans. Roy. Soc. Lond. A **248**, 369–405 (1956)
10. Sreenivasan, K.R., Antonia, R.A.: The phenomenology of small-scale turbulence. Annu. Rev. Fluid Mech. **29**, 435–472 (1997)
11. Sreenivasan, K.R.: On local isotropy of passive scalars in turbulent shear flows. Proc. R. Soc. London A **434**, 165–182 (1991)
12. Gibson, C.H.: Fine structure of scalar fields mixed by turbulence i. zero gradient points and minimal gradient surfaces. Phys. Fluids **11**, 2305–2315 (1968)
13. Wray, A.A., Hunt, J.C.R.: Algorithms for classification of turbulent structures. In Moffat, H.K., Tsinober, A. (eds.) Topological Fluid Mechanics, pp. 95–104. Cambridge University Press,Cambridge (1990)
14. Goto, S., Vassilicos, J.C.: The dissipation rate coefficient is not universal and depends on the internal stagnation point structure. Phys. Fluids **21**, 035104 (2009)
15. Bermejo-Moreno, I., Pullin, D.I.: On the non-local geometry of turbulence. J. Fluid Mech. **603**, 101–135 (2008)
16. Wang, L., Peters, N.: The length scale distribution function of the distance between extremal points in passive scalar turbulence. J. Fluid Mech. **554**, 457–475 (2006)
17. Schaefer, P., Gampert, M., Goebbert, J.H., Wang, L., Peters, N.: Testing of different model equations for the mean dissipation using Kolmogorov flows. Flow Turbul. Combust. **85**, 225–243 (2010)
18. Schaefer, P., Gampert, M., Wang, L., Peters, N.: Fast and slow changes of the length of gradient trajectories in homogenous shear turbulence. In: Eckhardt, B. (ed.) Advances in Turbulence XII, pp. 565–572. Springer, Berlin (2009)
19. Schaefer, P., Gampert, M., Gauding, M., Peters, N., Treviño, C.: The secondary splitting of zero gradient points in a turbulent scalar field. J. Eng. Math. (2011). doi:10.1007/s10665-011-9452-x
20. Gampert, M., Goebbert, J.H., Schaefer, P., Gauding, M., Peters, N., Aldudak, F., Oberlack, M.: Extensive strain along gradient trajectories in the turbulent kinetic energy field. New J. Phys. **13**, 043012 (2011)

21. Schaefer, L., Dierksheide, U., Klaas, M., Schroeder, W.: Investigation of dissipation elements in a fully developed turbulent channel flow by tomographic particle-image velocimetry. Phys. Fluids **23**, 035106 (2010)
22. Antonia, R.A., Hopfinger, E.J., Gagne, Y., Anselmet, F.: Temperature structure functions in turbulent shear flows. Phys. Rev. A **30**, 2704–2707 (1984)
23. Mydlarski, L., Warhaft, Z.: Passive scalar statistics in high-Péclet-number grid turbulence. J. Fluid Mech. **358**, 135–175 (1998)
24. Su, L.K., Clemens, N.T.: Planar measurements of the full three-dimensional scalar dissipation rate in gas-phase turbulent flows. Exp. Fluids **27**, 507–521 (1999)
25. Dowling, D.R., Dimotakis, P.E.: Similarity of the concentration field of gas-phase turbulent jets. J. Fluid Mech. **218**, 109–141 (1990)
26. Su, L.K., Clemens, N.T.: The structure of fine-scale scalar mixing in gas-phase planar turbulent jets. J. Fluid Mech. **488**, 1–29 (2003)
27. Weber, V., Bruebach, J., Gordon, R.L., Dreizler, A.: Pixel-based characterisation of cmos high-speed camera systems. Appl. Phys. B. 2011 doi:10.1007/s00340-011-4443-1
28. Taylor, G.I.: The spectrum of turbulence. Proc. R. Soc. London Ser. A **164**, 476 (1938)
29. Antonia, R.A., Sreenivasan, K.R.: Log-normality of temperature dissipation in a turbulent boundary layer. Phys. Fluids **20**, 1800–1804 (1977)
30. Peinke, J., Renner, C., Friedrich, R.: Experimental indications for markov properties of small-scale turbulence. J. Fluid Mech. **433**:383–409 (2001)
31. Lide, D.: Handbook of Chemistry and Physics, 88 edn. CRC Press, Boca Raton (2007–2008)
32. Hirschfelder, J.O., Curtiss, C.F., Bird, R.B.: Molecular theory of gases and liquids. Wiley, New York (1966)
33. Friehe, C.A., van Atta, C.W., Gibson, C.H.: Jet turbulence dissipation rate measurements and correlations. In: AGARD Turbulent Shear Flows, CP-93:18.1–18.7 (1971)
34. Mellado, J.P., Wang, L., Peters, N.: Gradient trajectory analysis of a scalar field with internal intermittency. J. Fluid Mech. **626**, 333–365 (2009)
35. Dávila, J., Vassilicos, J.C.: Richardson's pair diffusion and the stagnation point structure of turbulence. Phys. Rev. Lett. **91**, 14 (2003)
36. Sreenivasan, K.R.: The passive scalar spectrum and the Obukhov-Corrsin constant. Phys. Fluids **8**, 189–196 (1996)
37. Schumacher, J., Sreenivasan, K.R., Yakhot, V.: Asymptotic exponents from low-reynolds-number flows. New J. Phys. **9**, 89 (2007)

The Cellulolytic System of Cyst Nematodes

Dirk Heesel, Ulrich Commandeur and Rainer Fischer

Abstract The growing energy consumption and the dwindling resources of fossil energy carriers make the exploitation of renewable energy from biomass a key technology for the maintenance of modern industrial societies and lifestyles. Because of the conflict of demand between crops used for first generation biofuels (made from starch) with food production, the development of second generation biofuels from lignocellulose is recognized as preferable system. The recalcitrance of lignocellulose to enzymatic hydrolysis makes the utilization of this renewable feedstock difficult and cost intensive, due to high enzyme loads, pretreatment processes, and the low reactant/product concentrations in lignocellulose hydrolysates. The study outline presented here aims to contribute to the economic efficiency of second generation biofuel production by exploring a cellulolytic system consisting of a minimal set of cellulases, which we expect to optimize for high feedstock loading. Such systems are provided by nature in the form of proteins secreted into the saliva of plant parasitic nematodes of the genera *Meloidogyne*, *Heterodera* and *Globodera*. These cellulolytic systems typically consist of between two and six cellulases, pectinases and expansin-like proteins. In this article, we review information about root cyst nematodes, their cellulolytic system and provide an experimental outline for achieving insights into the cellulolytic system of the root cyst nematode *Globodera tabacum solanacearum*.

1 Introduction

The most important fuel from biomass in current production is bioethanol [9]. This first generation fuel is produced by the microbial fermentation of glucose produced by enzymatic hydrolysis of corn starch. The energy balance for this production

D. Heesel (✉) · U. Commandeur · R. Fischer
Institute for Molecular Biotechnology, RWTH Aachen University, Worringer Weg 1, 52074 Aachen, Germany
e-mail: dirk.heesel@rwth-aachen.de

M. Klaas et al. (eds.), *Fuels From Biomass: An Interdisciplinary Approach*,
Notes on Numerical Fluid Mechanics and Multidisciplinary Design 129,
DOI 10.1007/978-3-662-45425-1_4

process is only slightly positive [7]. Because of the insufficient gain in energy, and because of the conflict between using corn in this manner or as food stocks, an alternative to first generation biofuels is desirable.

Another possible feed stock for biofuel production is lignocellulose, the building material of plant cell walls. Lignocellulose is a composite material consisting of different sugar polymers, which can be hydrolyzed either by acid catalysis [4] or enzymatically [30] into sugar monomers. Both hydrolytic processes are inefficient under the current process conditions. Microbial cellulolytic systems from numerous fungi, recently reviewed in [11, 37], and from bacteria [3] are well characterized, and almost always composed of a mixture of endoglucanases, exoglucanases, beta-glucosidases and accessory proteins. In contrast to these well known systems, cellulases derived from animal sources, particularly insect cellulases [14], have only recently come into focus and little is known of their specific compositions [47]. In particular, the cellulolytic systems of plant parasitic nematodes remain poorly characterized [8, 18, 33, 46].

2 Lignocellulose

The composition of plant cell wall material is well characterized (Popper 2008). However, the exact structure is not completely understood and varies between different plants, plant tissues, adjacent cells and different parts of plant cell walls (Knox 2008).

The main polymer in plant cell walls is cellulose. The fraction of cellulose in the wood of angiosperms is 40–50 %, along with a hemicellulose portion of 25–30 % and lignin of 20–25 %.

The building block of cellulose is cellobiose, a β-1,4-linked glucose dimer. Cellulose is synthesized by cellulose synthase, a protein complex located in the plasma membrane of plant cells. It was long known that the cellulose synthase complex is formed of six particles and it was assumed that one particle is formed by 6 cellulose synthases [20]. Recent studies found that it is more probable, that one particle is made up from three cellulose synthases and that a complex consists of 18 cellulose synthases. Each cellulose synthase produces a cellulose chain, with the cellulose chains of one complex forming a microfibril. Because of the β-1,4-glycosydic links of the anhydro-glucopyranoses in the cellobiose building blocks, each building block can form 2 intermolecular hydrogen bonds on each side of the building block in the 020 plane. Hydroxyl groups of one chain or different chains therefore build a rigid hydrogen bond network giving cellulose a crystalline structure [6].

As mentioned above there are other sugar polymers in plant cell walls which build a network with cellulose, one important example of this are the xyloglucans. Xyloglucan is a category of hemicelluloses with a backbone of β-1,4-linked glucose residues which are branched with α-1,4-xylose residues. Further sugar residues can be linked to the xylose residues giving rise to a huge variety of xyloglucans [21, 23, 38].

3 Cellulases

Cellulases depolymerize the cellulose part of lignocellulose. However, many cellulases have a broader substrate spectrum, i.e. a number of cellulases are additionally active against β-glucan or, less commonly, xylose [2]. There are two general types of cellulases, categorized as exocellulases or endocellulases [28]. Exocellulases degrade cellulose from either its reducing end or its non reducing end, while endocellulases attack internal β-1,4-glycosidic links in amorphous regions of cellulose fibrils, generating new ends for exocellulases [13]. There are also reports of an additional type of cellulase which initially hydrolyzes an internal glycosidic bond, thus acting like an endocellulase, and then cleaves adjacent bonds releasing cellobiose from the glucose chain, by which it acts like an exocellulase [24]. All three types of cellulases have to loosen a stretch of a single polymer from the crystalline fibril to fit it in their active centers. Most cellulases are multi domain proteins consisting of a catalytic domain and one or several cellulose binding domains (CBM). Some bacterial cellulases additionally contain fibronectin like domains which are thought to enlarge the accessible surface area [50, 52]. Alternatively, there are also cellulases consisting of only a catalytic domain [51].

Most cellulases hydrolyze cellulose to cellobiose or cellotriose. These products of cellulose hydrolysis are further degraded to glucose by β-glucosidases. As cellulose is accompanied by hemicelluloses and lignin, cellulolytic systems are normally comprised of cellulases, hemicellulases and lignin degrading enzymes, such as laccases.

4 The Role of Cellulases in Infection of Plants by Cyst Nematodes

Nematodes of the genera *Globodera* and *Heterodera* are known as cyst nematodes. The stage two juveniles (J2) of these plant parasitic nematodes invade roots by intracellular movement [36, 46]. Therefore, host cell walls in the migratory path have to be locally destroyed by the nematodes. This is achieved by a combination of mechanical force [48] and the secretion of endogenous cellulases [46]. It was shown that these J2 male and female juveniles secrete cellulases and pectinase from their subventral esophageal gland into the saliva [19]. These enzymes are then injected through a stylet into the host plants' root cortex cells [43]. While male cyst nematodes produce these enzymes during later life stages, the females, which remain all their life at the syncytial feeding site, stop the expression of cellulases at juvenile stage three of growth [18].

Cellulases and pectinases are recognized to be pathogenicity factors which play a crucial role in intracellular migration. When J2 juveniles reach the vascular cylinder of the host root they start to establish a feeding site. Those feeding sites are syncytia (polynuclear cells), formed by the nematode-induced fusion of adjacent endodermal and pericycle cells near the xylem vessels [26]. It was observed, by differential interference contrast microscopy, that the J2 larvae pump secretory granules from their dorsal esophageal gland into a cell which becomes the origin of the developing syncytia [48]. The fusion of endodermal and pericycle cells requires partial cell wall dissolution. Although there is no direct evidence that cyst nematode cellulases are involved in this process, there is evidence that host cellulases are upregulated during the development of syncytia [19, 27]. Additionally, there is evidence of an upregulated pectin acetylesterase in both the syncytia and the surrounding tissue, which is induced by *Heterodera schachtii* larvae [45].

The endogenous cellulases produced by nematodes (accompanied with the mechanical force) act effectively to destroy root cell walls in the migratory path. This is of interest because these cellulolytic systems are comprised of only two [18] to six [15] cellulases, depending on the nematode species. These systems thus degrade lignocellulosic biomass with significantly fewer components than what is typically found in microbial cellulolytic systems. Therefore, the recombinant production of these enzyme sets could be a promising and cost effective alternative to the recombinant production of microbial cellulolytic systems.

5 The Cellulolytic System of *Globodera tabacum*

In clade 12 of the phylum *Nematoda* [22] endoglucanases belonging to glycosyl hydrolase family 5 (GH5) are ubiquitous [41], although this family is not restricted to this clade. Plant parasitic nematodes belonging to the genera *Meloidogyne*, *Hirschmanniella* and *Pratylenchus* and the genera *Globodera* and *Heterodera* belonging to the superfamily *Hoplolaimoidea* are thought to have inherited GH5 genes from early *Pratylenchidae*, which acquired a GH5 gene by horizontal gene transfer [41]. Hence, the coding sequences of GH5 endoglucanases in nematodes are related to some bacterial cellulases and are similarly assembled, i.e. containing a GH5 catalytic domain and a family 2 carbohydrate-binding module (CBM2) [31].

Two endoglucanases are secreted by *Globodera tabacum* [18], known as endoglucanase1 (*Gts*Eng1) and endoglucanase2 (*Gts*Eng2), both belonging to glycosyl hydrolase family 5 (Fig. 1). *Gts*Eng1 (uniprot accession number Q9U6M5) is assembled of a catalytic domain and a type two CBM (CBM2), while *Gts*Eng2 (uniprot accession number Q9U6M4) is assembled of a catalytic domain, which has 92 % amino acid sequence identity with the catalytic domain of *Gts*Eng1, and a C-terminal repetitive sequence known as an AKP domain. To our best knowledge, the function of this C-terminal domain is so far unresolved.

There are no exoglucanases reported for *Globodera tabacum*, or for other cyst nematodes. The only cell wall degrading enzymes which have been identified, other than *Gts*Eng1 and *Gts*Eng2, are two pectate lyases [29, 39] giving rise to the question of whether nematode cellulases have a broad substrate spectrum or if activities against several hemicelluloses are missing in the cellulolytic system of *Globodera tabacum*. There is one non-catalytic accessory protein verified for *Globodera tabacum*, called ExpansinB3 (uniprot accession number G9FYR3).

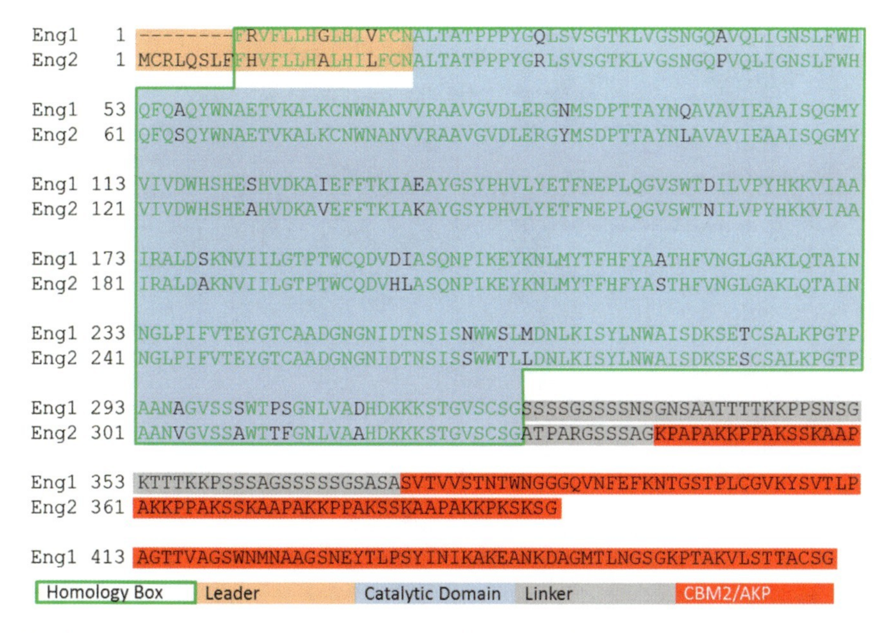

Fig. 1 Alignment of the tobacco cyst nematode *Globodera tabacum solanacearum* endoglucanases, *Gts*Eng1 and *Gts*Eng2. *Green frame* region of high homology between GtsEng1 and GtsEng2; *orange highlighted region* leader peptide; *blue highlighted region* catalytic domain (GH5); *grey highlighted region* linker; *red highlighted region* C-terminal domains (CBM2/AKP)

Expansin-like proteins are known to be present in a range of cellulolytic organisms. Expansins and expansin-like proteins contain a RlpA-like double-psi β-barrel domain (Fig. 2), which is a homolog of a domain from rare lipoprotein A, consisting of parallel and antiparallel β-sheets linked by two psi loops [49]. These proteins act to disrupt the hydrogen bond network of cellulose. This leads to changes in cellulose topography, a decrease in its crystallinity and a reduction in its recalcitrance to enzymatic hydrolysis [10, 25]. The most prominent example of an expansin is *Tr*Swollenin, which is part of the secretome of the soft rot ascomycete *Trichoderma reesei* [42].

Because of the small number of enzymes in the cellulolytic system of *Globodera tabacum*, and because of its established capability to degrade the cell walls of root cortex cells, it can be seen as a minimal set of cellulases for plant cell wall degradation.

6 Aims of the Project

The aim of this research project is to characterize the cellulolytic activity of *Gts*Eng2. The project will include heterologous expression of *Gts*Eng2, determination of its substrate spectrum and the investigation of the optimal reaction

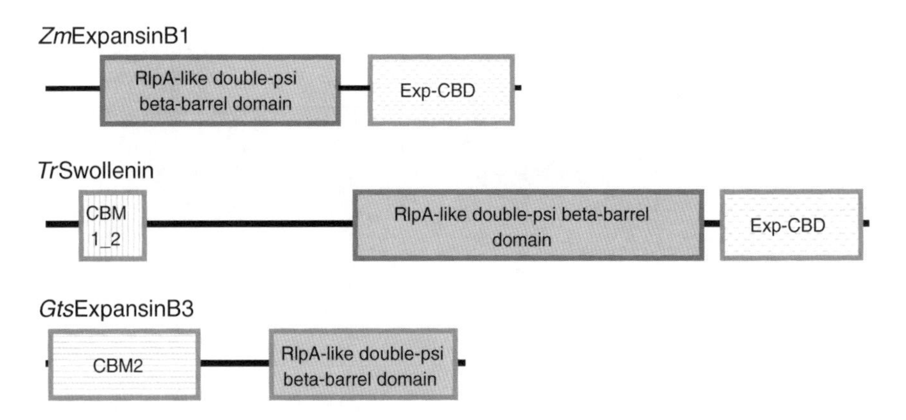

Fig. 2 Expansin and expansin-like proteins. *ZmExpansinB1* ExpansinB1 of maize (*Mw* 29 kDa; 169 amino acids); *TrSwollenin* Expansin-like protein of *T. reesei* (*Mw* 51 kDa; 487 amino acids), *GtsExpansiB3* expansin-like protein of *G. tabacum solanacearum* (*Mw* 27 kDa; 263 amino acids); *Exp-CBD* expansin-type cellulose binding domain; CBM1_2, fungal carbohydrate binding module; *CBM2* type-2 carbohydrate binding module

conditions in terms of the ideal temperature, pH and buffer composition for *Gts*Eng2 activity and stability.

As well as establishing the range of activity for this protein, the project will provide insights into the function of the AKP-domain. Furthermore, the effect of the expansin-like protein *Gts*ExpansinB3 on the hydrolysis of different cellulosic substrates by *Gts*Eng2 will be investigated, as will the effect of enzymes not verified for *Globodera tabacum* but of potential synergistic benefit, e.g. laccases, on the activity of *Gts*Eng2 to degrade plant biomass.

7 Methods

The hydrolysis of cellulosic substrates can be monitored by a range of methods. In the project described here, the following methods will be applied: determination of soluble reducing sugars by the PAHBAH assay [35], estimation of the degree of polymerization using gel permeation chromatography (GPC) and qualitative estimation of small hydrolysis products using thin layer chromatography (TLC).

Additionally, changes in the topology or crystallinity of cellulose can be monitored by scanning electron microscopy (SEM) or by X-ray diffraction. Finally, we aim to develop an assay to identify substrates with specific topologies by designing fluorescent protein probes assembled from one or more CBMs and a fluorescent protein, such as GFP or mCherry.

7.1 Heterologous Expression

As the extraction of secreted proteins from nematodes is not a feasible approach for protein characterization and impossible for large scale purposes, the endoglucanase *Gts*Eng2 and the accessory protein *Gts*ExpansinB3 will be heterologously expressed. Therefore, we plan to test *Escherichia coli* and *Kluyveromyces lactis* as heterologous protein expression systems. The first, *E. coli*, is a gram negative bacterial expression system and is the most frequently used protein expression organism in molecular biotechnology. Many expression strains and vectors of *E. coli* have been generated and are commercially available. Unfortunately, one drawback of protein expression in *E. coli* is that is displays poor secretion of many enzymes. In contrast, *K. lactis* is a unicellular eukaryotic expression system, which is known to feature high levels of protein secretion. The coding sequences for our genes of interest will be synthesized and subsequently ligated into commercially available expression vectors pET22 and pKLAC1, for use in the two systems respectively.

The *E. coli* expression vector pET22 is an episomal vector which confers resistance to ampicillin, which is important for selection of transformants and vector maintenance [44]. The expression cassette consists of a T7-promoter sequence followed by the leader peptide coding sequence from the pectate lyaseB of *Erwinia carotovora* and a multiple cloning site in which the coding sequence of interest can be inserted. Downstream of the multiple cloning site is a His-tag coding sequence and a terminator sequence. Expression vectors with the T7 promoter can only be used in expression strains which encode a T7-polymerase within their genomic DNA. Therefore we will employ BL21 (DE3) strains for enzyme expression [44].

The *K. lactis* expression vector pKLAC1 can be cloned in *E. coli* as an episomal vector. The expression cassette is designed for integration into the *K. lactis* host

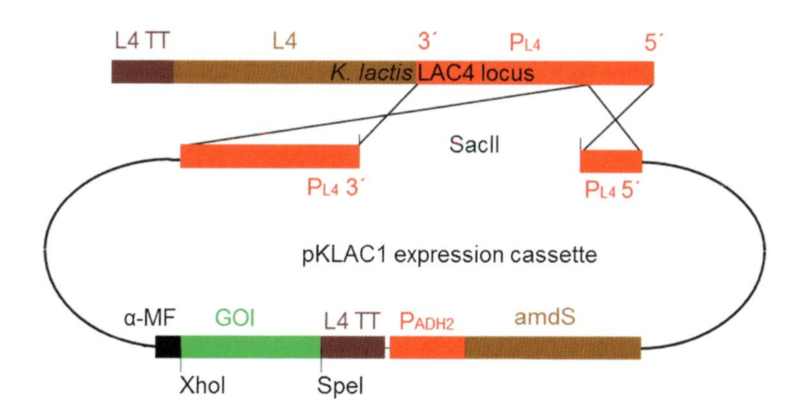

Fig. 3 Integration of linearized pKLAC1 into the *K. lactis* host genome. The expression cassette of pKLAC1 is integrated into the host genome by a double recombination event. *PL4* Lac4 promoter; *α-MF* α-mating factor; *L4 TT* Lac4 terminator of transcription; *PADH2* alcoholdehydrogenase2 promoter; *andS* acetamidase; *GOI* gene of interest

genome leading to stable transformants. Therefore, pKLAC1 contains only one resistance marker for cloning and propagation in *E. coli.* The expression cassette of pKLAC1 consists of the *K. lactis* lactase promoter PLAC4-PBI, the coding sequence for the *K. lactis* α-mating factor secretion domain, multiple cloning sites and the LAC4 terminator sequence (Fig. 3). This expression cassette can integrate several times into the host genome [40]. The more copies a transformant carries in his genome the higher the expression may be.

7.2 Reducing Sugar Assay

The PAHBAH-assay is a colorimetric test which is based on the reaction of p-hydroxybenzoic acid hydrazide in a basic environment with reducing sugars. During this reaction, an anionic form of the hydrazide is formed which can be quantified by adsorption measurement at 410 nm [35]. This method can be utilized after enzymatic digestion of various substrates to determine the amount of reducing sugars released, and thus can determine volumetric activities of enzyme preparations and specific activities of enzymes on different cellulosic substrates.

7.3 GPC Analysis

Gel permeation chromatography (GPC) is a technique which separates substances of different size by measuring the time needed to flow through a porous resin-filled column. Small substances, which can access all the pores, flow through a bigger share of the column volume than substances which, due to their size, can access only the bigger pores or no pores at all. Therefore, substances with a high molecular weight will be eluted from the column prior to substances with small molecular weight or, as in this application, molecules with a high degree of polymerization will be eluted prior to molecules with a low degree of polymerization. The solute concentrations in the eluate stream can than be analyzed by light refraction.

For GPC analysis the polymers to be analyzed have to be in solution. This is simple for water soluble substrates and reaction products, but more difficult for water insoluble polymers, such as Avicel or α-cellulose, which have to be dissolved in an organic solvent prior to analysis. It has been shown that Avicel and α-cellulose can be dissolved in a mixture of dimethyl formamide and 10–20 % (v/v) 1-ethyl-3-methylimidazolium acetate (EMIM-acetate). GPC can be employed for the detection, quantification and visualization of depolymerized cellulosic substrates [12] as demonstrated for carboxy-methyl cellulose (CMC, Fig. 4).

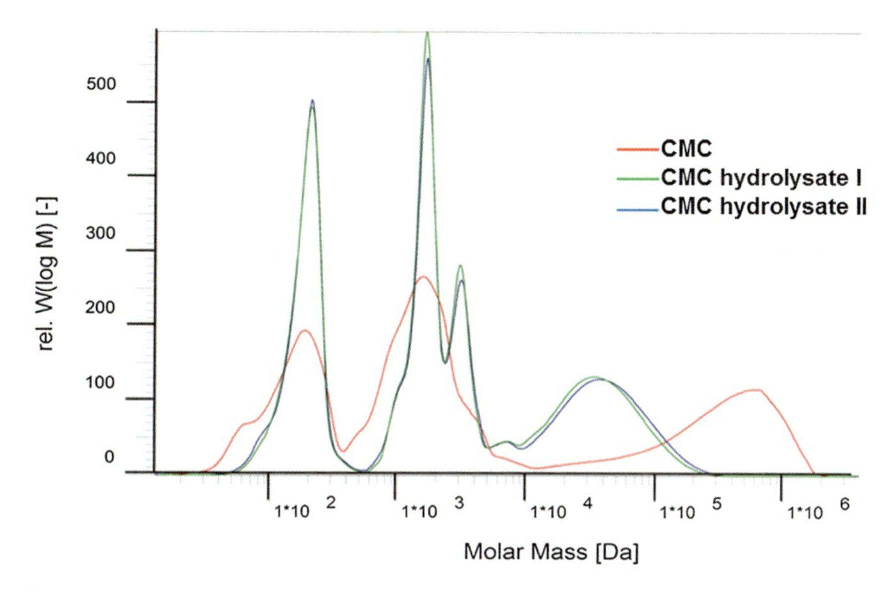

Fig. 4 Gel permeation chromatograph of the soluble, model substrate carboxymethyl cellulose (CMC) and of two independently produced CMC hydrolysates

7.4 Estimation of CBM Binding Specificities and of Substrate Characteristics Using Fluorescent Protein Probes

CBMs are domains found in carbohydrate active enzymes which facilitate the binding of an enzyme to its substrate [17, 32]. Different CBMs have different binding specificities. For example, members of the CBM families CBM1 and CBM3 are reported to bind to the hydrophobic 110 face of insoluble crystalline cellulose [34], while members of CBM4 bind specifically to amorphous cellulose [1]. Members of the CBM family CBM2 may show specificity to either crystalline cellulose [16] or xylan [5]. The two subtypes are referred as CBM2a and CBM2b, respectively.

For characterizing the interaction of the AKP domain of *Gts*Eng2 and the CBM2 of *Gts*ExpansinB3 with cellulose samples, fusion proteins of mCherry or GFP with the AKP domain and the CBM2 of *Gts*ExpansinB3 will be recombinantly expressed and used to record adsorption isotherms on different substrates. Comparison of the adsorption capacity of the AKP domain with those of other CBM-fusion proteins will help clarify their role in substrate binding.

This system will also be used to analyze the effects of *Gts*ExpansinB3 on cellulose surface structure by comparing untreated and *Gts*ExpansinB3-treated samples. A change in adsorption capacity (Fig. 5) of a substrate for fluorescent protein probes with different specificities will document the expected changes in substrate characteristics upon *Gts*ExpansinB3 treatment.

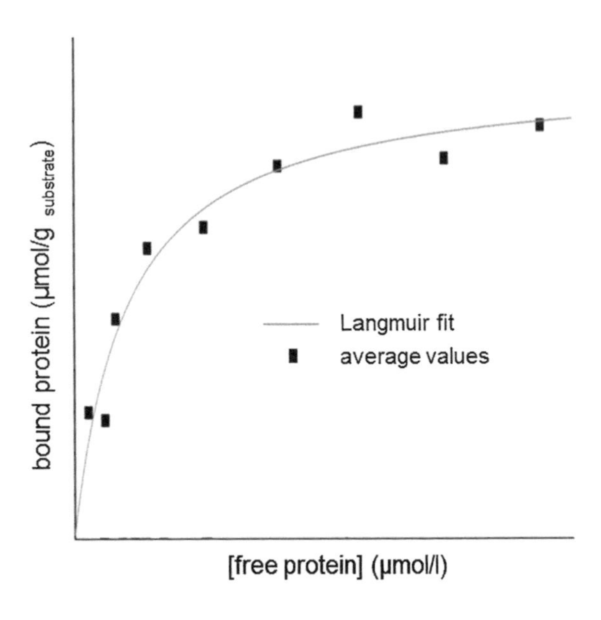

Fig. 5 A Langmuir adsorption isotherm. The isotherm was recorded from the adsorption of a CBM-mCherry fusion protein to crystalline cellulose (Avicel)

References

1. Abou, H.M., Nordberg, K.E., Bartonek-Roxa, E., Raghothama, S., Simpson, P., Gilbert, H., Williamson, M., Holst, O.: Carbohydrate-binding modules from a thermostable *Rhodothermus marinus* xylanase: cloning, expression and binding studies. Biochem. J. **345**, 53–60 (2000)
2. Asha, B.M., Revathi, M., Yadav, A., Sakthivel, N.: Purification and characterization of a thermophilic cellulase from a novel cellulolytic strain, *Paenibacillus barcinonensis*. J. Microbiol. Biotechnol. **22**(11), 1501–1509 (2012)
3. Bhalla, A., Bansal, N., Kumar, S., Bischoff, K.M., Sani, R.K.: Improved lignocellulose conversion to biofuels with thermophilic bacteria and thermostable enzymes. Bioresour. Technol. **128**, 751–759 (2013)
4. Binder, J. B., Raines, R. T.: Fermentable sugars by chemical hydrolysis of biomass. Proc. Nat. Acad. Sci. **107**(10), 4516–4521 (2010)
5. Black, G., Hazlewood, G., Millward-Sadler, S., Laurie, J., Gilbert, H.: A modular xylanase containing a novel non-catalytic xylan-specific binding domain. Biochem. J. **307**, 191–195 (1995)
6. Blackwell, Ga: The structure of native cellulose. Biopolymers **13**, 1975–2001 (1974)
7. Brehmer, B., Bals, B., Sanders, J., Dale, B.: Improving the corn-ethanol industry: studying protein separation techniques to obtain higher value-added product options for distillers grains. Biotechnol. Bioeng. **101**(1), 49–61 (2008)
8. Béra-Maillet, C., Arthaud, L., Abad, P., Rosso, M.-N.: Biochemical characterization of MI-ENG1, a family 5 endoglucanase secreted by the root-knot nematode *Meloidogyne incognita*. Eur. J. Biochem. **267**(11), 3255–3263 (2000)
9. Carroll, A., Somerville, C.: Cellulosic biofuels. Annu. Rev. Plant Biol. **60**, 165–182 (2009)
10. Cosgrove, D.J.: Loosening of plant cell walls by expansins. Nature **407**(6802), 321–326 (2000)
11. Dashtban, M., Schraft, H., Qin, W.: Fungal bioconversion of lignocellulosic residues; opportunities and perspectives. Int. J. Biol. Sci **5**(6), 578 (2009)

12. Engel, P., Hein, L., Spiess, A.C.: Derivatization-free gel permeation chromatography elucidates enzymatic cellulose hydrolysis. Biotechnol. Biofuels **5**(1), 77 (2012)
13. Eriksson, K.-E., Pettersson, B.: Extracellular enzyme system utilized by the fungus *Chrysosporium lignorum* for the breakdown of cellulose. In: Biodeterioration of Materials, vol. 2, pp. 116–120. Applied Science Publishers Ltd, London (1972)
14. Fischer, R., Ostafe, R., Twyman, R.M.: Cellulases from insects. In: Vilcinskas A Yellow Biotechnology II, vol. 136, pp. 51–64. Springer, Berlin (2013)
15. Gao, B., Allen, R., Maier, T., Davis, E.L., Baum, T.J., Hussey, R.S.: The parasitome of the phytonematode *Heterodera glycines*. Mol. Plant Microbe Interact. **16**(8), 720–726 (2003)
16. Gilkes, N.R., Jervis, E., Henrissat, B., Tekant, B., Miller, R.C., Warren, R.A., Kilburn, D.G.: The adsorption of a bacterial cellulase and its two isolated domains to crystalline cellulose. J. Biol. Chem. **267**(10), 6743–6749 (1992)
17. Gilkes, N., Warren, R., Miller, R., Kilburn, D.G.: Precise excision of the cellulose binding domains from two *Cellulomonas fimi* cellulases by a homologous protease and the effect on catalysis. J. Biol. Chem. **263**(21), 10401–10407 (1988)
18. Goellner, M., Smant, G., De Boer, J.M., Baum, T.J., Davis, E.L.: Isolation of beta-1,4-endoglucanase genes from *Globodera tabacum* and their expression during parasitism. J Nematol **32**(2), 154–165 (2000)
19. Goellner, M., Wang, X., Davis, E.L.: Endo-β-1,4-glucanase expression in compatible plant–nematode interactions. The Plant Cell Online **13**(10), 2241–2255 (2001)
20. Herth, W.: Arrays of plasma-membrane “rosettes” involved in cellulose microfibril formation of Spirogyra. Planta **159**(4), 347–356 (1983)
21. Hoffman, M., Jia, Z., Peña, M.J., Cash, M., Harper, A., Blackburn II, A.R., Darvill, A., York, W.S.: Structural analysis of xyloglucans in the primary cell walls of plants in the subclass *Asteridae*. Carbohydr. Res. **340**(11), 1826–1840 (2005)
22. Holterman, M., van der Wurff, A., van den Elsen, S., van Megen, H., Bongers, T., Holovachov, O., Bakker, J., Helder, J.: Phylum-wide analysis of SSU rDNA reveals deep phylogenetic relationships among nematodes and accelerated evolution toward crown clades. Mol. Biol. Evol. **23**(9), 1792–1800 (2006)
23. Hsieh, Y.S., Harris, P.J.: Xyloglucans of monocotyledons have diverse structures. Mol. Plant **2** (5), 943–965 (2009)
24. Irwin, D.C., Spezio, M., Walker, L.P., Wilson, D.B.: Activity studies of eight purified cellulases: specificity, synergism, and binding domain effects. Biotechnol. Bioeng. **42**(8), 1002–1013 (1993)
25. Jaeger, G., Girfoglio, M., Dollo, F., Rinaldi, R., Bongard, H., Commandeur, U., Fischer, R., Spiess, A.C., Buechs, J.: How recombinant swollenin from *Kluyveromyces lactis* affects cellulosic substrates and accelerates their hydrolysis. Biotechnol. Biofuels **4**(1), 33 (2011)
26. Jones, M., Northcote, D.: Nematode-induced syncytium-a multinucleate transfer cell. J. Cell Sci. **10**(809), 1 (1972)
27. Karczmarek, A., Fudali, S., Lichocka, M., Sobczak, M., Kurek, W., Janakowski, S., Roosien, J., Golinowski, W., Bakker, J., Goverse, A., Helder, J.: Expression of two functionally distinct plant endo-β-1,4-glucanases is essential for the compatible interaction between potato cyst nematode and its hosts. Mol. Plant Microbe Interact. **21**(6), 791–798 (2008)
28. King, K., Smibert, R.: Distinctive properties of β-glucosidases and related enzymes derived from a commercial *Aspergillus niger* cellulase. Appl. Microbiol. **11**(4), 315–319 (1963)
29. Kudla, U., Milac, A.-L., Qin, L., Overmars, H., Roze, E., Holterman, M., Petrescu, A.-J., Goverse, A., Bakker, J., Helder, J., Smant, G.: Structural and functional characterization of a novel, host penetration-related pectate lyase from the potato cyst nematode *Globodera rostochiensis*. Mol. Plant Pathol. **8**(3), 293–305 (2007)
30. Kumar, R., Singh, S., Singh, O.: Bioconversion of lignocellulosic biomass: biochemical and molecular perspectives. J. Ind. Microbiol. Biotechnol. **35**(5), 377–391 (2008)
31. Kyndt, T., Haegeman, A., Gheysen, G.: Evolution of GHF5 endoglucanase gene structure in plant-parasitic nematodes: no evidence for an early domain shuffling event. BMC Evol. Biol. **8** (1), 305 (2008)

32. Langsford, M., Gilkes, N., Singh, B., Moser, B., Warren, R., Kilburn, D.: Glycosylation of bacterial cellulases prevents proteolytic cleavage between functional domains. FEBS Lett. **225** (1), 163–167 (1987)
33. Ledger, T.N., Jaubert, S., Bosselut, N., Abad, P., Rosso, M.-N.: Characterization of a new β-1,4-endoglucanase gene from the root-knot nematode *meloidogyne incognita* and evolutionary scheme for phytonematode family 5 glycosyl hydrolases. Gene **382**, 121–128 (2006)
34. Lehtiö, J., Sugiyama, J., Gustavsson, M., Fransson, L., Linder, M., Teeri, T.: The binding specificity and affinity determinants of family 1 and family 3 cellulose binding modules. Proc. Natl. Acad. Sci. U.S.A. **100**(2), 484–489 (2003)
35. Lever, M.: A new reaction for colorimetric determination of carbohydrates. Anal. Biochem. **47** (1), 273–279 (1972)
36. Mankau, R., Linford, M.B.: Host-parasite relationships of the clover cyst nematode. *Heterodera trifolii* Goffart. Illinois Agric. Exp. Stn. Bul. **667**, 33–36 (1960)
37. Peterson, R., Nevalainen, H.: *Trichoderma reesei* RUT-C30–thirty years of strain improvement. Microbiology **158**(1), 58–68 (2012)
38. Peña, M.J., Darvill, A.G., Eberhard, S., York, W.S., O'Neill, M.A.: Moss and liverwort xyloglucans contain galacturonic acid and are structurally distinct from the xyloglucans synthesized by hornworts and vascular plants*. Glycobiology **18**(11), 891–904 (2008)
39. Popeijus, H., Overmars, H., Jones, J., Blok, V., Goverse, A., Helder, J., Schots, A., Bakker, J., Smant, G.: Degradation of plant cell walls by a nematode. Nature **406**(6791), 36–37 (2000)
40. Read, J.D., Colussi, P.A., Ganatra, M.B., Taron, C.H.: Acetamide selection of *Kluyveromyces lactis* cells transformed with an integrative vector leads to high-frequency formation of multicopy strains. Appl. Environ. Microbiol. **73**(16), 5088–5096 (2007)
41. Rybarczyk-Mydłowska, K., Maboreke, H.R., van Megen, H., van den Elsen, S., Mooyman, P., Smant, G., Bakker, J., Helder, J.: Rather than by direct acquisition via lateral gene transfer, GHF5 cellulases were passed on from early Pratylenchidae to root-knot and cyst nematodes. BMC Evol. Biol. **12**(1), 221 (2012)
42. Saloheimo, M., Paloheimo, M., Hakola, S., Pere, J., Swanson, B., Nyyssönen, E., Bhatia, A., Ward, M., Penttilä, M.: Swollenin, a *Trichoderma reesei* protein with sequence similarity to the plant expansins, exhibits disruption activity on cellulosic materials. Eur. J. Biochem. **269** (17), 4202–4211 (2002)
43. Smant, G., Goverse, A., Stokkermans, J.P., De Boer, J.M., Pomp, H., Zilverentant, J.F., Overmars, H.A., Helder, J., Schots, A., Bakker, J.: Potato root diffusate-induced secretion of soluble, basic proteins originating from the subventral esophageal glands of potato cyst nematodes. Phytopathology **87**(8), 839–845 (1997)
44. Studier, F.W., Moffatt, B.A.: Use of bacteriophage T7 RNA polymerase to direct selective high-level expression of cloned genes. J. Mol. Biol. **189**(1), 113–130 (1986)
45. Vercauteren, I., de Almeida Engler, J., De Groodt, R., Gheysen, G.: An *Arabidopsis thaliana* pectin acetylesterase gene is upregulated in nematode feeding sites induced by root-knot and cyst nematodes. Mol. Plant Microbe Interact. **15**(4), 404–407 (2002)
46. Wang, X., Meyers, D., Yan, Y., Baum, T., Smant, G., Hussey, R., Davis, E.: *In planta* localization of a β-1,4-endoglucanase secreted by *Heterodera glycines*. Mol. Plant Microbe Interact. **12**(1), 64–67 (1999)
47. Watanabe, H., Tokuda, G.: Cellulolytic systems in insects. Annu. Rev. Entomol. **55**(1), 609–632 (2010)
48. Wyss, U., Zunke, U.: Observations on the behaviour of second stage juveniles of *Hetero* inside host roots. Revue Nematol **9**(2), 153–165 (1986)
49. Yennawar, N.H., Li, L.-C., Dudzinski, D.M., Tabuchi, A., Cosgrove, D.J.: Crystal structure and activities of EXPB1 (Zea m 1), a β-expansin and group-1 pollen allergen from maize. Proc. Natl. Acad. Sci. U.S.A. **103**(40), 14664–14671 (2006)
50. Zhou, W., Irwin, D.C., Escovar-Kousen, J., Wilson, D.B.: Kinetic studies of *Thermobifida fusca* Cel9A active site mutant enzymes. Biochemistry **43**(30), 9655–9663 (2004)

51. Zhou, X., Kovaleva, E.S., Wu-Scharf, D., Campbell, J.H., Buchman, G.W., Boucias, D.G., Scharf, M.E.: Production and characterization of a recombinant beta-1,4-endoglucanase (glycohydrolase family 9) from the termite *Reticulitermes flavipes*. Arch. Insect Biochem. Physiol. **74**(3), 147–162 (2010)
52. Zverlov, V., Mahr, S., Riedel, K., Bronnenmeier, K.: Properties and gene structure of a bifunctional cellulolytic enzyme (CelA) from the extreme thermophile '*Anaerocellum thermophilum*' with separate glycosyl hydrolase family 9 and 48 catalytic domains. Microbiology **144**(Pt 2), 457–465 (1998)

New Pathways for the Valorization of Fatty Acid Esters

Thomas Hermanns, Jürgen Klankermayer and Walter Leitner

Abstract Biomass represents a major renewable carbon source and thus is increasingly important as a feedstock for the synthesis of fuels and chemicals. To achieve sustainable conversion of biogenic resources in the future, it is important to develop new selective catalytic reaction pathways. Olefin metathesis is one important tool in organic synthesis to convert unsaturated compounds while maintaining their functionalities. While classical heterogeneous catalysts are not capable of the transformation of such compounds, homogeneous ruthenium based catalysts seem suitable because of their high tolerance towards functional groups. In the following article we describe our investigations for the valorization of oleic acid methyl esters. Beyond the use as bio fuel there is an interest in adding the compounds value by conversion to base and fine chemicals. Therefore sustainable reaction paths have to be developed, in which innovative reaction media like supercritical fluids and ionic liquids are used for the establishment of a multiphase reaction system.

1 Introduction

The limitation of the fossil resources crude oil, natural gas and coal and the consequences for energy and feedstock supply are in recent discussion by politics, industry and research. Also the emissions of greenhouse gases like CO_2, produced by burning those resources, have to be drastically reduced to avoid its influence on the climatic change. The worldwide consumption of energy in 2008 was met by fossil resources by 81 % [1]. In the traffic sector the main demand is met by liquid fuels derived from crude oil. On a long term regenerative alternatives for gasoline

T. Hermanns · J. Klankermayer · W. Leitner (✉)
Institute for Technical and Macromolecular Chemistry,
RWTH Aachen University, Aachen, Germany
e-mail: leitner@itmc.rwth-aachen.de

M. Klaas et al. (eds.), *Fuels From Biomass: An Interdisciplinary Approach*,
Notes on Numerical Fluid Mechanics and Multidisciplinary Design 129,
DOI 10.1007/978-3-662-45425-1_5

and diesel have to be found. Fuels from biomass can make a contribution, since biomass is a major renewable source of organic carbon.

In addition most of the items of daily use are made from petrochemicals or in their production process been in contact with petrochemicals. The valorization of Biomass to base or fine chemicals has particular potential, since those compounds are equipped with functional groups by nature, which otherwise would be introduced expensively by refinery and synthesis. At the same time those functional groups are the biggest challenge in specific chemical conversions and utilization of those compounds.

1.1 Conversion of Biomass to Fuels

The early efforts of biomass as fuel conversion produced the so called biofuels of the first generation. Those are ethanol from fermentation of starch and sugar and biodiesel from fats and oils. To produce biodiesel the triglycerides from oil-bearing crops are transesterificated with methanol, yielding fatty acid methyl esters with 16 to 20 carbon atoms. The structural similarity of the product mixture is high enough to use those esters as alternative diesel fuel (Fig. 1).

The processes to produce biofuels of the first generation are well established and in worldwide use. However those biofuels do not constitute a satisfactory solution to the energy problems, since the cultivation of crops for fuel production are in direct opposition to cultivation of food and feed crops.

But the usage of fats and oil is also possible via the transformation to base or fine chemicals while maintaining the given functionalities. Among the unsaturated fatty acid esters oleic acid methyl ester (OME) is the most abundant compound. Therefore and because of its good availability this ester is the ideal model compound for investigations on fatty acid ester valorization.

1.2 Unconventional Reaction Media

Next to using biomass to replace fossil resources the sustainability of a chemical reaction and the whole process play a crucial role. Improvements in energy consumption of chemical processes, in the choice of solvents and the general influence

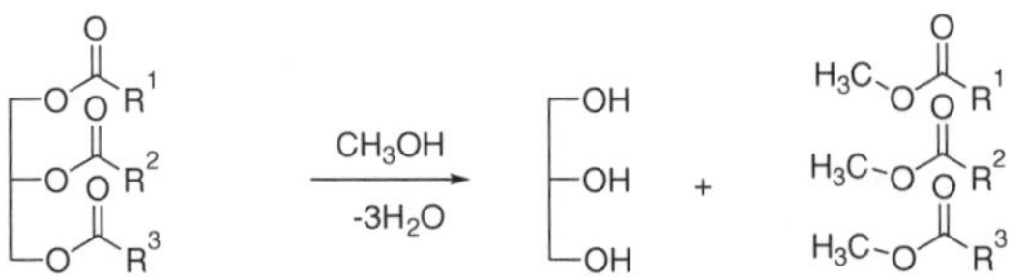

CH_3OH / $-3H_2O$; OH, OH, OH; H_3C; R^1, R^2, R^3

Fig. 1 Transesterification of plant oils to fatty acid methyl esters (R_1, R_2, R_3 = C_{16}–C_{20})

on environment and personal health are essential. Conventional organic solvents are often hazardous to health and environment and are used in large amounts resulting in major waste streams in industrial processes. Separation processes are a good example for this: the isolation of a product or the separation of a homogeneous catalyst from the reaction mixture are usually realized by distillation or extraction, demanding significant amounts of energy or solvents.

Potential for improvement lies in utilizing the physical properties of the reaction mixture, the solvents and the catalyst in a way that avoids downstream separation. One approach is multiphase systems, in which the catalyst is dissolved in a separate phase. This concept is for example used for the hydroformylation during the Ruhrchemie/Rhône-Poulenc-process or for the oligomerization of olefins in the SHOP-Process. In recent concepts unconventional media like supercritical fluids and ionic liquids are taken into consideration for multiphase processes.

Substances that are heated above their critical temperature and pressurized above their critical pressure are in the so called supercritical state. In this state the border between liquid and gas vanishes and the fluid combines the properties of both states: the high miscibility and low mass transfer limitations of a gas and the high solvent capacity and the heat capacity of a liquid. Above all the compressibility near the critical point is highly increased resulting in a high tunability of the fluid by minor changes in pressure and temperature. Especially the solvent capacity correlates with the fluid's density, which is easily altered by pressure or temperature.

CO_2 presents an attractive choice for a supercritical fluid solvent. With a critical temperature of 31 °C and a critical pressure of 73.8 bar [2] the supercritical state is achieved under rather mild conditions (Fig. 2). Furthermore it is non-toxic, abundant and cheap. Supercritical CO_2 (*sc*CO_2) has been successfully used for homogeneous catalysis in the laboratory and for industrial purification of natural compounds and for the decaffeination of coffee beans [3–6].

Ionic liquids (IL) are compounds, which—much like conventional salts—consist of ions but because of their molecular structure feature melting points below 100 °C. Numerous compounds are liquid at room temperature and thus are called

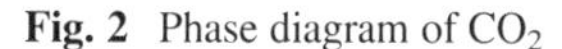
Fig. 2 Phase diagram of CO_2

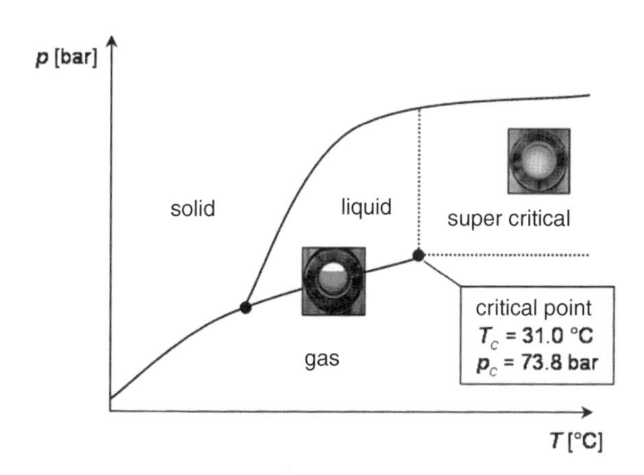

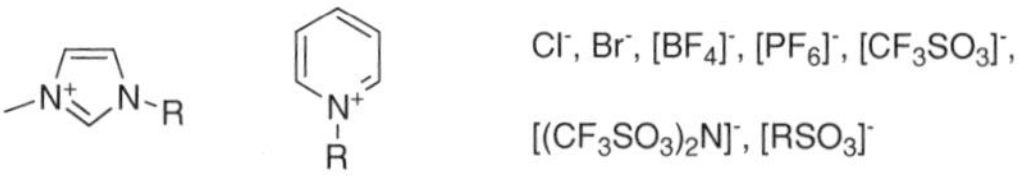

Cl^-, Br^-, $[BF_4]^-$, $[PF_6]^-$, $[CF_3SO_3]^-$,

$[(CF_3SO_3)_2N]^-$, $[RSO_3]^-$

Fig. 3 Common IL cations: 1-alkyl-3-methylimidazolium, *N*-alkylpyridinium, common IL anions: chloride, bromide, tetrafluoroborate, hexafluorophosphate, triflate, bistriflimide, alkylsulfonate

room temperature ionic liquids (RTIL) [7]. Organic cations with a high steric demand and a broad distribution of charge are commonly used for ILs, while the anions can be either simple halides or more complex organic anions (Fig. 3). The properties of ionic liquids like their melting point, their viscosity and reactivity can be nearly arbitrarily tuned by modifications of the ions and their composition. Usually the vapor pressure is negligibly low, so they are commonly described as non-volatile at all. Thus they can provide a possibility to dissolve and immobilize a catalyst, while the substrates or the products are dissolved in a non-polar, volatile phase (e.g. *sc*CO_2) [8–10].

A recent overview about possible applications of ionic liquids in catalysis is given by D. Zhao and V.I. Pârvulescu in their review articles [11, 12]. Ionic liquids cannot be regarded as "green" solvent as such, since their production is usually complex and their environmental hazards and their toxicities are often not explored yet. But they offer the possibilities to establish alternative process strategies suitable to save energy and to minimize waste streams.

1.3 Olefin Metathesis

Metathesis is the exchange of molecule parts between two reactants. In the case of olefin metathesis the substituents of a C=C-double bound are exchanged (Fig. 4). Olefin metathesis was described in 1964 to happen at heterogeneous tungsten- and cobalt-molybdenum catalysts [13]. An early industrial application was the *Phillips Triolefin Process* in the 1970s, which in recent times is used in reverse as the *olefin conversion technology* (OCT) process to produce propene from ethylene and *n*-butene.

The most important industrial application using olefin metathesis is the *Shell higher olefins* (SHOP) process in which one step is the metathesis of olefins with an inner double bond with ethylene to α-olefins with a terminal double bond [14].

Fig. 4 General reaction scheme of olefin metathesis

The first homogeneous catalysts in use were metal halides like WCl_6 or the Tebbe-reagent $Cp_2TiCl_2/MeAl_3$ with mediocre activity and stability [15, 16]. The development of well-defined highly active homogeneous catalysts in the 1990ies helped olefin metathesis to become one of the most prominent tools in synthetic chemistry. While homogeneous tungsten- and molybdenum complexes (so called Schrock-catalysts) [17, 18] are highly active, the ruthenium based complexes exhibit a relatively high tolerance towards functional groups in the substrates. The so called Grubbs-catalysts are built from a Ru-benzylidene structure with tricyclohexyl phosphine ligands in the first generation and *N*-heterocarbene ligands in the later generation (**1** and **2**, Fig. 5) [16, 19].

A.H. Hoveyda modified the benzylidene-architecture by introducing a bidentate ligand at the metal centre (**5**) [20]. The chelate effect stabilizes the complexes better than before. A substitution of the phosphine ligand by a *N*-heterocarbene (**6**) yields a phosphorous free catalyst with an excellent stability towards functional groups and yet sufficient activity [21].

Depending on the substrates different types of olefin metathesis can be distinguished. If two molecules of the same kind react with each other the reaction is called self-metathesis (Fig. 6a). If the reaction takes place with two different substrate molecules, it is called cross metathesis (Fig. 6b). With ethylene as a reaction partner the cross metathesis yields two α-olefins and is called ethenolysis. During the ring closing metathesis two double bounds of the same molecule react and form a cyclic product (Fig. 6c). The ring closing of diethyl diallylmalonate to the corresponding cyclopentene is one of the major benchmark reactions to test the activity of catalysts for ring closing metathesis. The reverse reaction of opening a cyclic olefin is usually followed by the polymerization of the resulting diolefins with two terminal double bonds (Fig. 6d). The so called ring opening metathesis polymerization (ROMP) is used for the commercial polymerization of norbornene and cyclooctene [3, 22].

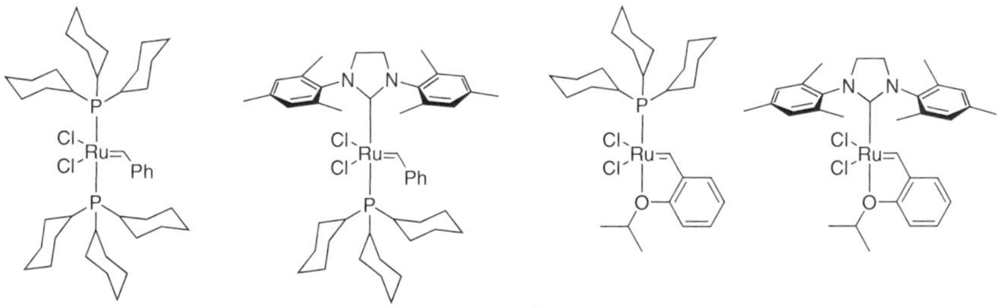

Fig. 5 Homogene Metathesekatalysatoren vom Grubbs-Typ 1. Generation (**1**), Grubbs-Typ 2. Generation (**2**), Hoveyda-Typ 1. Generation (**3**) und Hoveyda-Typ 2. Generation (**4**)

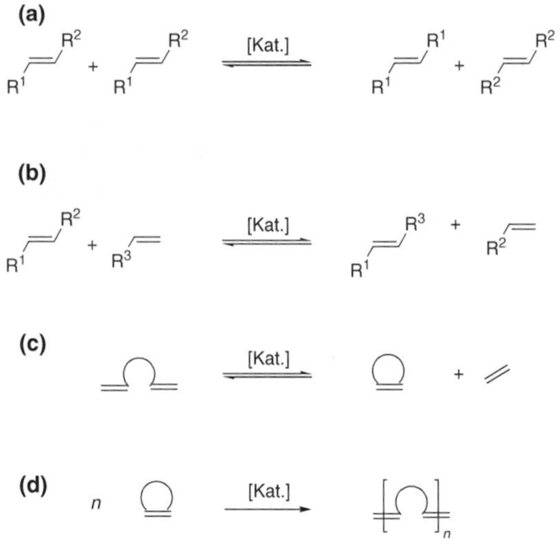

Fig. 6 Different types of olefin metathesis: self-metathesis (**a**), cross metathesis (**b**), ring closing metathesis (RCM) (**c**), ring opening metathesis polymerization (ROMP) (**d**)

2 Valorization of Oleic Acid Methyl Ester

2.1 Metathesis of Fatty Acid Esters

The metathesis of unsaturated fatty acid esters and in particular of oleic acid methyl ester yields in products covering a range of industrial application possibilities, either directly used or as an intermediate building block.

The self-metathesis of OME (**5**) gives 9-octadecene (**6**) and 1,18-dimethyl-9-octadecenedioate (**7**) (Fig. 7).

Olefin **6** can be dimerized and afterwards transformed to 10,11-dioctyleicosane (**9**) via hydrogenation. **9** is a star shaped hydrocarbon and a high performance lubricant for automotive applications [23] (Fig. 8).

The diester **7** that yields from the self-metathesis of OME offers to ester functions for further derivatization. Via a Dieckmann condensation a cyclic product can be formed. The resulting 17-membered ring (**10**) is the intermediate for the synthesis of *Z*-civetone (**11**), a fragrant highly valuable to the perfume industry, of which the classical lab synthesis is very complex (Fig. 9) [24].

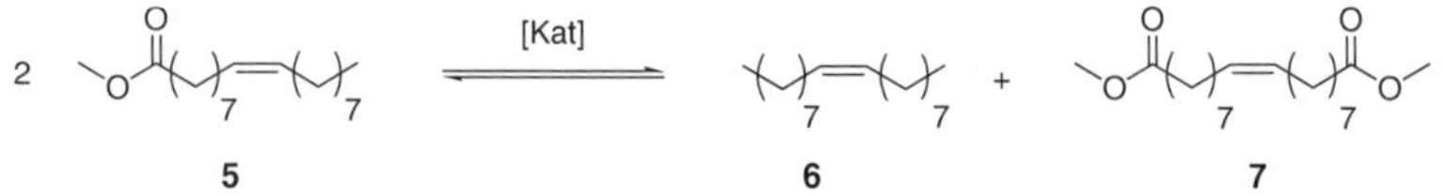

Fig. 7 Self-metathesis of OME (**5**)

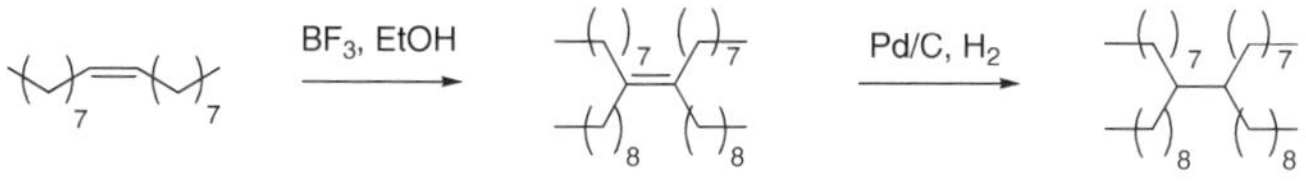

Fig. 8 Synthesis of 10,11-dioctyleicosane (9) form ester (6) [23]

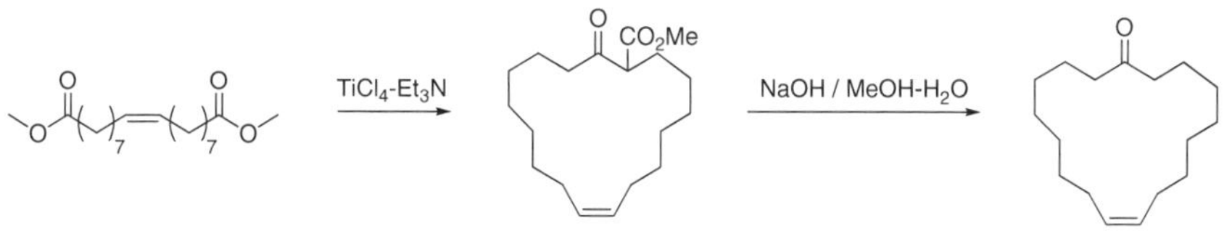

Fig. 9 Synthesis of Z-civetone (11) from diester 7 [24]

Furthermore long chained α,ω-diester like 7 can undergo the polycondensation with diols, to yield new types of polyesters (Fig. 10) [25]. By changing from diols to diamines the production yields the analogous polyamides.

Besides the self-metathesis the ethenolysis of OME represents an important reaction providing products with shorter chain length and terminal double bonds: 1-decene (12) und 9-decenoic acid methyl ester (13) (Fig. 11).

The resulting products 12 and 13 are used as base chemicals in polymer, lubricant and surfactant production. For example 12 can be oligomerized to poly-α-olefins, a class of sulfur and phosphorous free synthetic lubricants with high thermal stability [26]. The ester 13 can be used as a base chemical in copolymerization reactions with ethylene for the production of methyl ester functionalized polyethylenes [27]. Alternatively 13 can be condensed at the ester group yielding a diene with terminal double bounds that can undergo acyclic diene metathesis (ADMET), this means self-metathesis of two dienes yielding a longer chained olefin and ethylene and consequentially a new type of polymer [25]. The use of metathesis technology for the

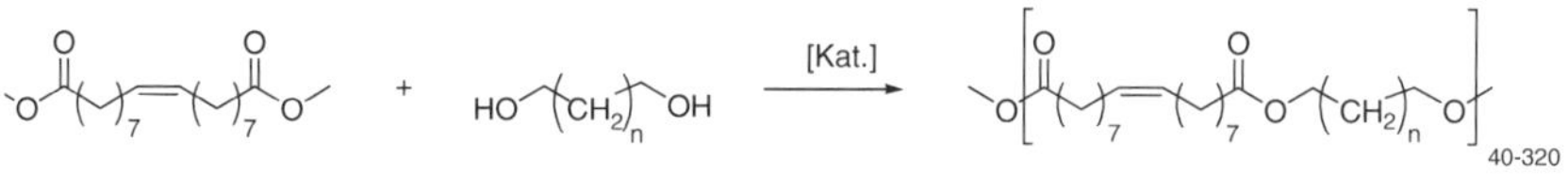

Fig. 10 Conversion of diester 7 with diols to polyesters,. cat: $Ca(ac)_2/Sb_2O_3$ or $Ti(O\text{-}n\text{-}Bu)_4$ [25]

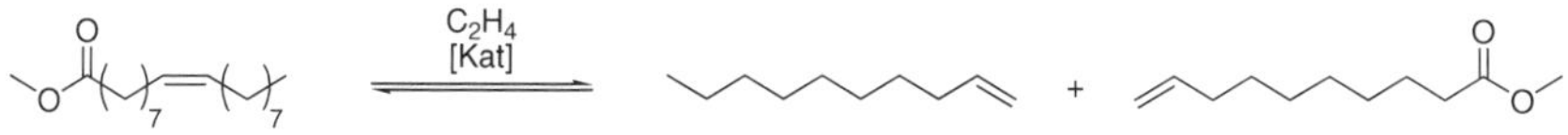

Fig. 11 Ethenolysis of OME (5)

valorization of natural oils has been industrially implemented in the Elevance Biorefinery Process [http://www.elevance.com/technology/biorefinery].

2.2 Reaction Concept

In the previous section the possibilities of valorization of OME by olefin metathesis were shown. Since the common heterogeneous catalysts for metathesis are not stable against oxygen containing functional groups, they are not suitable of the conversion of OME. Therefore ruthenium based catalysts of the Grubbs or Hoveyda type have to be used for this kind of transformation [28].

The major challenge in homogeneous catalysis is the separation of the catalyst from the products or the reaction mixture for the success of a process. Either for economic reasons, to avoid loss of expensive catalyst, or for health reasons in life sciences to gain heavy metal free products. A novel approach is the immobilization of the catalyst in an IL as stationary phase and the dissolution of the substrate in a second, mobile phase [29].

In their work about the ethenolysis of OME in ionic liquids C. Thurier et al. found, that ruthenium based catalysts of the Hoveyda type dissolved in alkyl imidazolium ILs reached conversions up to 95 % (with **3**) and up to 53 % (with **4**). The advantage of the immobilization of the catalyst in IL was shown in this work: it was possible to recycle the catalytic active phase up to three times in batchwise reactions [30].

The concept of immobilizing a catalyst to achieve the recyclability of a reaction can be refined to a continuous flow application, by choosing a suitable solvent for the mobile phase, to deliver the substrate to the catalyst followed by the removal of the product away from the catalytic active phase (Fig. 12). As described in Sect. 1.2 the properties of a supercritical fluid make compressed CO_2 a suitable candidate for a solvent. The low polarity of CO_2 guarantees a sufficient immiscibility for phase separation. It is also to be expected, that the resulting olefin **6** exhibits a higher solubility in the non-polar solvent than the esters, resulting in selectivity for one of the products of the self-metathesis. A first example of OME metathesis in IL/ CO_2 systems was reported recently by the group of Cole-Hamilton [31].

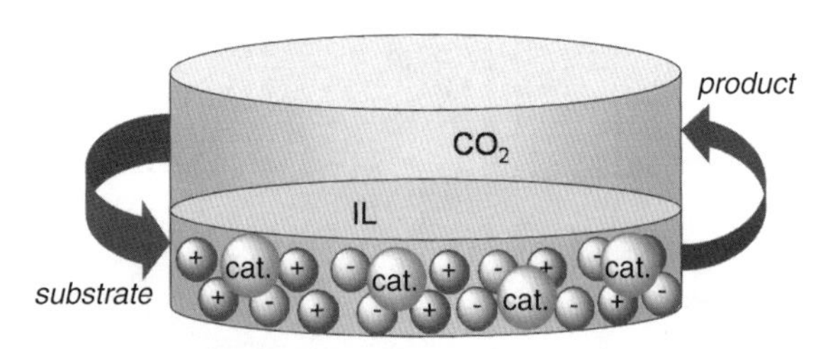

Fig. 12 Two-phase reaction conept: the catalyst is immobilized in the IL phase, while substrate and product are transported in and out in the mobile CO_2-phase

2.3 Investigations Towards Reactivity in IL/CO_2 Mixtures

In collaboration with the group of C. Bruneau [30], the self-metathesis and the ethenolysis of OME were investigated in an IL/CO_2 two phasic system. As catalysts complexes **3** and **4** were chosen for the reactions. The influence of compressed CO_2 on the reaction system was investigated by batch wise reactions at different temperatures and different CO_2 pressures. For the immobilizing IL phase the commercially available 1,3-dimethylimdiazolium bistriflimide [mmim][NTf$_2$] (see Fig. 13) was used.

The two-phasic system was established by 1 mL of IL in which catalyst was dissolved with a loading of 1 mol% (relative to the OME loading) and addition of 0.4 mL of OME, forming its own phase. The reaction times were 2 h each. Since self-metathesis is an equilibrium reaction the maximum conversion without the removal of the products will be 50 %.

As shown in Fig. 14 complex **3** shows just medium conversions of up to 23 % at 25 and 40 °C, which can be increased to 36 % by increasing the reaction temperature to 60 °C. At a CO_2 pressure of p = 80 bar, which means, temperature and pressure are beyond the critical point, the conversion drops most likely because of a separation of the catalyst from the substrate.

Complex **4** on the other hand showed under all investigated temperatures full conversion of 50 %. Furthermore there is no negative effect of supercritical CO_2 on the reaction as can be seen in Fig. 15. Complex **4** was investigated further in different ionic liquids.

Since we found full conversion over all conditions in the reaction time of 2 h the complex was tested for his activity at shorter reaction times. It was found, that at room temperature full conversion is reached after 30 min. At elevated temperature of 40 °C the reaction was complete after 5 min (see Fig. 16).

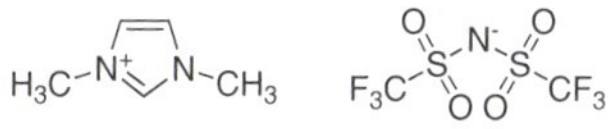

Fig. 13 1,3-dimethylimidazoilium bistriflimid [mmim][NTf$_2$], mp: 26 °C, η = 31 mPas [32]

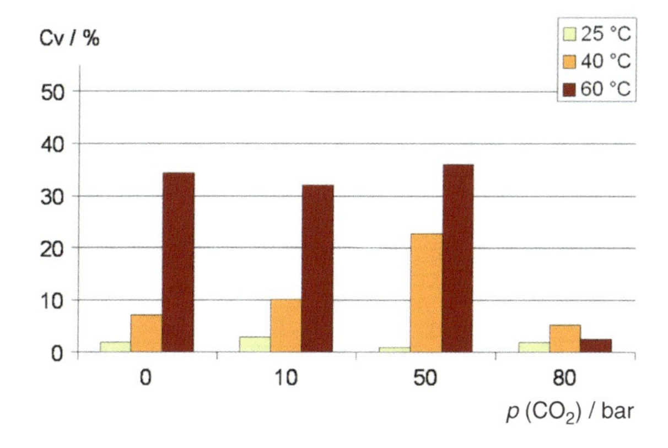

Fig. 14 Self-metathesis of OME catalyzed by **3** at different temperatures and CO_2 pressures

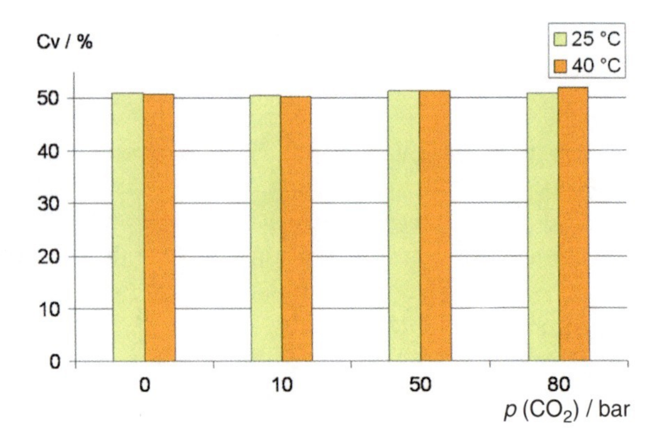

Fig. 15 Self-metathesis of OME catalyzed by **4** at different temperatures and CO_2 pressures

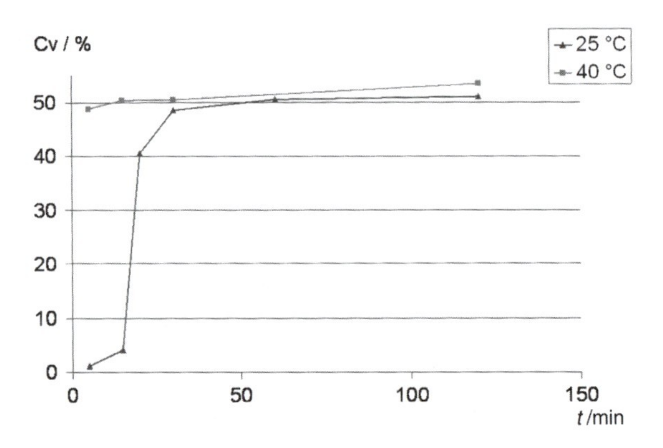

Fig. 16 Activity of complex **4**: self-metathesis of 0.4 mL OME with 1 mol% catalyst loading in 1 mL IL

The recyclability of the reaction system was investigated with a choice of different commercially available imidazolium ILs (see Fig. 17). At the cations the alkyl chain length was varied. Variation in the anions was from bistriflimide [NTf_2] to tetracyanoborate [$B(CN)_4$] to tris(perfluoroethyl)trifluorophosphate [fap]. After dissolution of the catalyst in 1 mL of IL OME (0.4 mL) was added and after a reaction time of 15 min removed. For recycling of the catalytic active phase, again 0.4 mL of OME were added, and so on. The reactions were done in the presence of CO_2 in supercritical conditions ($T = 40$ °C, $p = 80$ bar). The NTf_2 based ILs showed the best results in the repetitive batch reaction. In 1,3-dimethyl imidazolium NTf_2 a set of three reactions were possible in which the conversion was in the range of 50–40 %. In 1-butyl-3-methylimidazolium NTf_2 the conversion decreased more rapidly to below 20 % in the third batch. In 1-methyl-3-octylimidazolium NTf_2 it appears that the reaction system has a longer activation period, since the first batch only reached 10 % conversion before increasing to nearly full conversion. This might

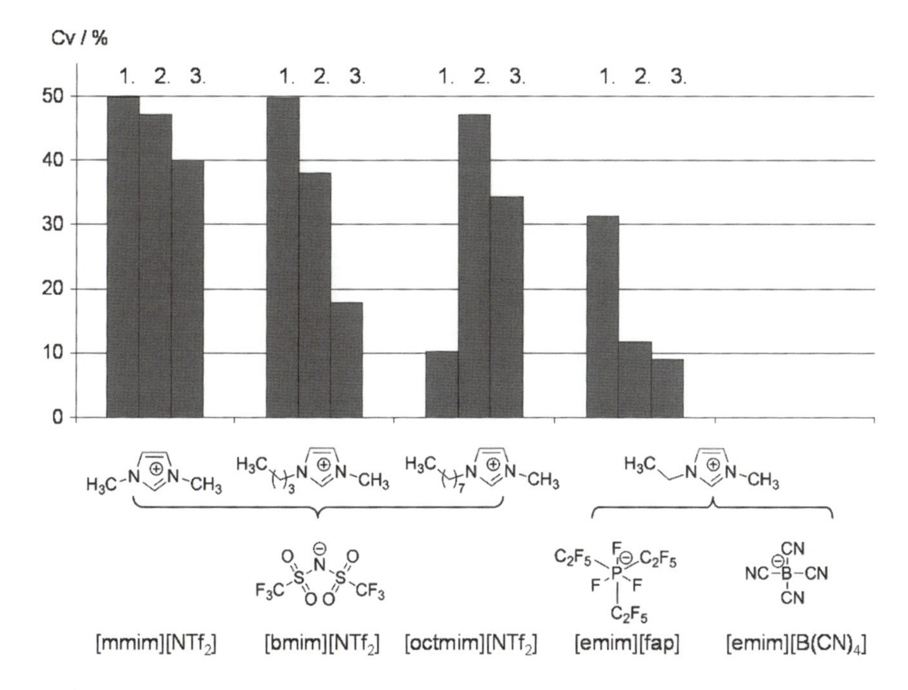

Fig. 17 Batch-wise repeated self metathesis of OME in different IL

likely be induced by the significantly higher viscosity of this IL. In general there is a decreasing trend in catalytic activity in the repetitive batch experiments. This happens most likely due to loss of catalyst during the removal of the reactants phase between batches.

Differently from the NTf_2 based ILs the reactivity in the fap-IL is lower from the beginning. The first batch reached around 30 % conversion decreasing rapidly. Using the $B(CN)_4$-IL no conversion could be obtained. The catalyst-IL solution turned immediately from originally green to dark brown, indicating that the catalyst is being decomposed in the IL.

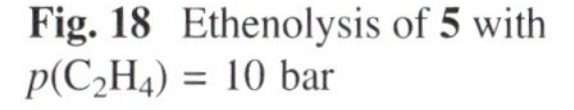

Fig. 18 Ethenolysis of **5** with $p(C_2H_4)$ = 10 bar

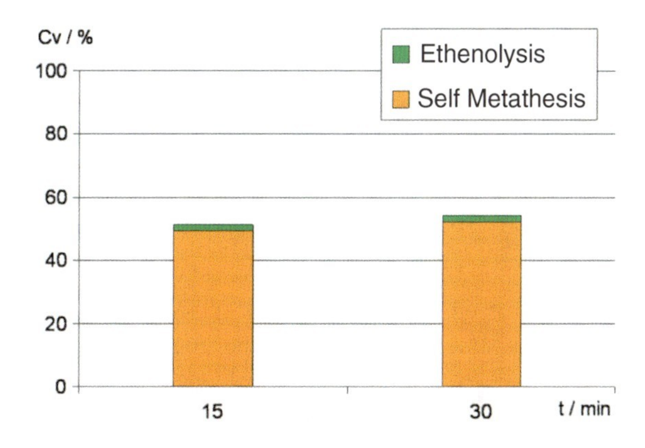

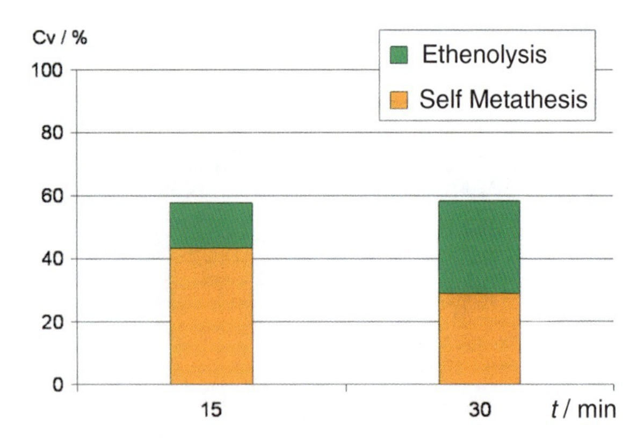

Fig. 19 Ethenolysis of **5** with $p(C_2H_4) = 50$ bar

The ethenolysis of OME was investigated with comparable conditions in ionic liquid [mmim][NTf_2] and in presence of $scCO_2$. Figures 18 and 19 show the conversions of self-metathesis and ethenolysis as a function of reaction time and ethylene pressure.

At a low ethylene pressure of $p = 10$ bar, conversions of 50 % (at 15 min) and 54 % at 30 min reaction time were obtained. As shown in Fig. 18 in the proportional bars only 2 % each of the conversions were ethenolysis. Apparently the pressure of ethylene is too low to sufficiently repress the competing self-metathesis reaction. At an increased pressure of ethylene of $p = 50$ bar the conversions could be increased up to 58 %, of which 14 % (at 15 min) and 29 % (at 30 min) were ethenolysis. The selectivity towards ethenolysis for the latter is then 50 %. Those results comply with those of C. Thurier et al. [30], who found during their investigations a competition between self-metathesis and ethenolysis when catalyst **4** was applied.

3 Conclusions

The addition of value to fatty acid methyl ester will be an important process in future, to accomplish the general change from diminishing fossil resources to renewable feedstock. The metathesis of unsaturated fatty acid methyl esters with homogeneous ruthenium based catalysts provides an attractive approach for this goal. To tackle the procedural challenge of catalyst removal from the product, the use of a multiphase system containing ionic liquids is of great interest.

Our investigations for the self-metathesis of OME in a two phase system showed, that a recyclability of the catalytic active phase is possible. Furthermore we found, that with the proper catalyst the reaction is not influenced by supercritical CO_2. This fact provides an opportunity for the transfer to a continuously run reaction system, in which $scCO_2$ will act as a solvent and mobile phase. At the same time we expect from the system to allow for product selectivity towards the olefin, because of different solubilities of the esters and olefins in the stationary as well as in the mobile phase.

To optimize the reaction system it will be necessary to find a catalyst-IL combination that provides a high solubility for the catalyst, thus minimizing catalyst leaching into the mobile phase. Furthermore a better suited catalyst for ethenolysis will be needed. Since an ethenolysis selective catalyst could be expected to show less selectivity for self-metathesis, the use of two catalysts in one system that can be switched from on reaction to the other by adding ethylene is thinkable.

References

1. Yearly updated statistics of the International Energy Agency, Key World Energy Statistics. http://www.iea.org/statistics
2. Angus, S., Armstrong, B., de Reuck, K.M.: International thermodynamic tables of the fluid state: carbon dioxide. Pergamon Press, Oxford, IUPAC (1976)
3. Fürstner, A., Koch, D., Langemann, K., Leitner, W., Six, C.: Olefin metathesis in compressed carbon dioxide. Angew. Chem. Int. Ed. **36**, 2466–2469 (1997). doi:10.1002/anie.199724661
4. Koch, D., Leitner, W.: Rhodium-catalyzed hydroformylation in supercritical carbon dioxide. J. Am. Chem. Soc. **120**, 13398–13404 (1998). doi:10.1021/ja980729w
5. Kainz, S., Brinkmann, A., Leitner, W., Pfaltz, A.: Iridium-catalyzed enantioselective hydrogenation of imines in supercritical carbon dioxide. J. Am. Chem. Soc. **121**, 6421 (1999). doi:10.1021/ja984309i
6. Franciò, G., Wittmann, K., Leitner, W.: Highly efficient enantioselective catalysis in supercritical carbon dioxide using the perfluoroalkyl-substituted ligand (R, S)-3-H2F6-BINAPHOS. J. Organomet. Chem. **621**, 130–142 (2001). doi:10.1016/S0022-328X(00)00778-6
7. Wasserscheid, P., Keim, W.: Ionic liquids-new "solutions" for transition metal catalysis. Angew. Chem. Int. Ed. **39**, 3772–3789 (2000). doi:10.1002/1521-3773(20001103)39:21<3772:AID-ANIE3772>3.0.CO;2-5
8. Blanchard, L.A., Hancu, D., Beckman, E.J., Brennecke, J.F.: Green processing using ionic liquids and CO2. Nature **399**, 28–29 (1999). doi:10.1038/19887
9. Bösmann, A., Franciò, G., Janssen, E., et al.: Activation, tuning, and immobilization of homogeneous catalysts in an ionic liquid/compressed co_2 continuous-flow system. Angew. Chem. Int. Ed. **40**, 2697–2699 (2001). doi:10.1002/1521-3757(20010716)113:14<2769:AID-ANGE2769>3.0.CO;2-M
10. Solinas, M., Pfaltz, A., Cozzi, P.G., Leitner, W.: Enantioselective hydrogenation of imines in ionic liquid/carbon dioxide media. J. Am. Chem. Soc. **126**, 16142–16147 (2004). doi:10.1021/ja046129g
11. Zhao, D.: Ionic liquids: applications in catalysis. Catal. Today **74**, 157–189 (2002). doi:10.1016/S0920-5861(01)00541-7
12. Pârvulescu, V.I., Hardacre, C.: Catalysis in ionic liquids. Chem. Rev. **107**, 2615–2665 (2007). doi:10.1021/cr050948h
13. Banks, R.L., Bailey, G.C.: Olefin disproportionation. A new catalytic process. Ind. Eng. Chem. Prod. Res. Dev. **3**, 170–173 (1964). doi:10.1021/i360011a002
14. Mol, J.: Industrial applications of olefin metathesis. J. Mol. Cat. A **213**, 39–45 (2004). doi:10.1016/j.molcata.2003.10.049
15. Tebbe, F.N., Parshall, G.W., Reddy, G.S.: Olefin homologation with titanium methylene compounds. J. Am. Chem. Soc. **100**, 3611–3613 (1978). doi:10.1021/ja00479a061

16. Grubbs, R.H.: Olefin metathesis. Tetrahedron **60**, 7117–7140 (2004). doi:10.1016/j.tet.2004.05.124
17. Schrock, R.R.: Living ring-opening metathesis polymerization catalyzed by well-characterized transition-metal alkylidene complexes. Acc. Chem. Res. **23**, 158–165 (1990). doi:10.1021/ar00173a007
18. Schrock, R.R., Murdzek, J.S., Bazan, G.C., et al.: Synthesis of molybdenum imido alkylidene complexes and some reactions involving acyclic olefins. J. Am. Chem. Soc. **112**, 3875–3886 (1990). doi:10.1021/ja00166a023
19. Trnka, T.M., Grubbs, R.H.: The development of L 2 X 2 RuCHR olefin metathesis catalysts: an organometallic success story. Acc. Chem. Res. **34**, 18–29 (2001). doi:10.1021/ar000114f
20. Kingsbury, J.S., Harrity, J.P.A., Bonitatebus, P.J., Hoveyda, A.H.: A recyclable Ru-based metathesis catalyst. J. Am. Chem. Soc. **121**, 791–799 (1999). doi:10.1021/ja983222u
21. Garber, S.B., Kingsbury, J.S., Gray, B.L., Hoveyda, A.H.: Efficient and recyclable monomeric and dendritic ru-based metathesis catalysts. J. Am. Chem. Soc. **122**, 8168–8179 (2000). doi:10.1021/ja001179g
22. Grubbs, R.H.: Handbook of metathesis, 1st edn. Wiley, Weinheim (2003)
23. Choo, Y.-M., Ooi, K.-E., Ooi, I.-H., Tan, D.D.H.: Synthesis of a palm-based star-shaped hydrocarbon via oleate metathesis. J. Am. Oil Chem. Soc. **73**, 333–336 (1996). doi:10.1007/BF02523427
24. Tanabe, Y., Makita, A., Funakoshi, S., et al.: Practical synthesis of (Z)-Civetone utilizing Ti-Dieckmann condensation. Adv. Synth. Catal. **344**, 507–510 (2002). doi:10.1002/1615-4169(200207)344:5
25. Warwel, S., Brüse, F., Demes, C., et al.: Polymers and surfactants on the basis of renewable resources. Chemosphere **43**, 39–48 (2001). doi:10.1016/S0045-6535(00)00322-2
26. Yadav, G.D., Doshi, N.S.: Development of a green process for poly-a-olefin based lubricants. Green Chem. **4**, 528–540 (2002). doi:10.1039/b206081g
27. Warwel, S., Wiege, B., Fehling, E., Kunz, M.: Copolymerization of ethylene with ω-unsaturated fatty acid methyl esters using a cationic palladium complex. Macromol. Chem. Phys. **202**, 849–855 (2001). doi:10.1002/1521-3935(20010301)202:6<849:AID-MACP849>3.0.CO;2-8
28. Dinger, M.B., Mol, J.C.: High turnover numbers with ruthenium-based metathesis catalysts. Adv. Synth. Catal. **344**, 671–677 (2002). doi:10.1002/1615-4169(200208)344:6/7<671:AID-ADSC671>3.0.CO;2-G
29. Hintermair, U., Franciò, G., Leitner, W.: Continuous flow organometallic catalysis: new wind in old sails. Chem. Commun. **47**, 3691–3701 (2011). doi:10.1039/c0cc04958a
30. Thurier, C., Fischmeister, C., Bruneau, C., et al.: Ethenolysis of methyl oleate in room-temperature ionic liquids. ChemSusChem **1**, 118–122 (2008). doi:10.1002/cssc.200700002
31. Duque, R., Öchsnet, E., Clavier, H., Caijo, F., Nolan, S.P., Manduit, M., Cole-Hamilton, D.J., Gree Chemistry **13**, 1187–1195 (2011). doi: 10.1039/C1GC15048K
32. Tokuda H, Tsuzuki S, Susan M.A.B.H., et al.: How ionic are room-temperature ionic liquids? An indicator of the physicochemical properties. J. Phys. Chem. B **110**, 19593–19600 (2006). doi: 10.1021/jp064159v

Soluble Organocalcium Compounds for the Activation and Conversion of Carbon Dioxide and Heteroaromatic Substrates

Phillip Jochmann, Thomas P. Spaniol and Jun Okuda

Abstract The effective activation of (hetero)aromatic compounds is of particular interest for the production of tailor made compounds that can serve as key intermediates in the development of alternative combustion fuels. As a sustainable alternative for late transition metals, organocalcium complexes are studied in the context of activation of carbon dioxide and aromatic N- and O-heterocycles. Highly regioselective C–H bond activation and carbometalation reactions have been observed for conversions with pyridine derivatives. Rapid insertion of CO_2 into calcium carbon bonds of the obtained products is observed. Furan derivatives are found more inert and the formation of polymeric products is described. Slow isomerization of 2,5-dihydrofuran (2,5-DHF) to 2,3-dihydrofuran (2,3-DHF) is reported.

1 Introduction

The limited availability of fossil energy resources and the steadily increasing demand for energy call for sustainable alternatives (Fig. 1). In order to replace long-known energy sources, such as nuclear power, scientific disciplines have to work together. In this context, chemistry plays a key role for the understanding and development of energy conversion processes, which can subsequently be exploited by engineers [1].

Today, chemistry already makes an important contribution to energetic processes and its relevance will steadily increase in the coming decades, envisioning sustainable and broadly applicable alternatives for the interconversion of regenerative energy. Hence, the near future bears immense challenges for chemists in conjunction with other scientific fields.

P. Jochmann · T.P. Spaniol · J. Okuda (✉)
Institute of Inorganic Chemistry, RWTH Aachen University, Landoltweg 1, 52056 Aachen, Germany
e-mail: jun.okuda@ac.rwth-aachen.de; sekr.okuda@ac.rwth-aachen.de

M. Klaas et al. (eds.), *Fuels From Biomass: An Interdisciplinary Approach*,
Notes on Numerical Fluid Mechanics and Multidisciplinary Design 129,
DOI 10.1007/978-3-662-45425-1_6

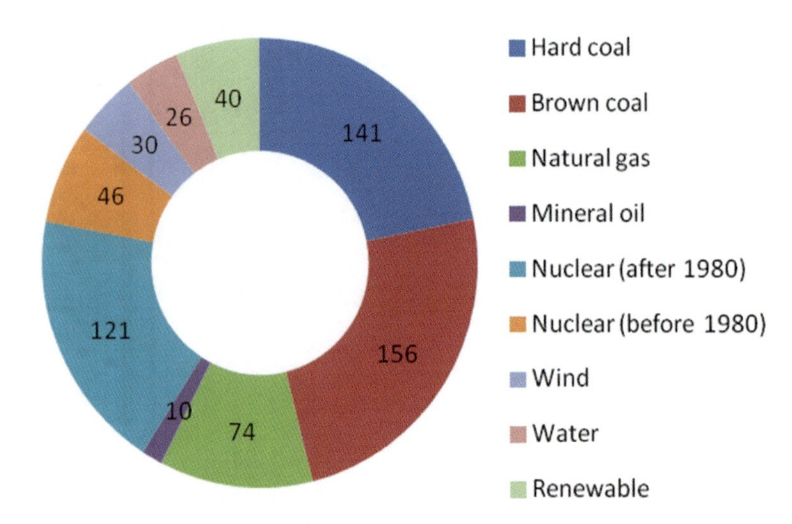

Fig. 1 Energy sources for electrical power in Germany (until 2008) in TWh. Reproduced from Ref. [2]

These efforts may not be seen as short- or mid-term aims, but are rather part of an ongoing development, which may improve energy efficiency for the future.

For the above purposes it is a prerequisite to fully understand basic processes in great detail in order to steer them in the desired direction. Fundamental research has to be carried out carefully to not only develop resilient alternatives for fossil energy resources, but also to optimize the degree of efficiency of existing processes and to quantify long-term economical and ecological consequences of these strategies.

Existing concepts for sequestering renewable and available energy carriers include (a) biomass conversion, (b) CO_2 utilization, and (c) light harvesting (typically for the generation of H_2). The latter strategy is a successfully developing field, which greatly contributes to the generation of energy storage chemicals and materials. The present work will focus on methods (a) and to a somewhat lesser extent on (b). Approaches in homogeneous reaction systems in the liquid phase will be highlighted in the context of biomass conversion (a) and CO_2 utilization (b). Heterogeneous strategies involving solid state reagents are not within the scope of this work.

Developments in chemical biomass conversion concentrate on the dehydration and degradation of glucose and cellobiose into furan derivatives like 2-HMF and subsequent transformations to finally yield high energy combustion fuels. Catalysts that have been proven suitable for such transformations are mineral acids, organic acids, and metal salts [3, 4]. All reactions reported have been carried out under extremely polar reaction conditions and common solvents are dimethylformamide (DMF), dimethylacetamide (DMA), ionic liquids (ILs), and supercritical CO_2 [4, 5]. Organometallic catalysts have been designed and reported only recently [6–8]. The use of organometallic reagents is challenging as biomass-derived substrates usually bear high oxygen contents and sources of H_2O. These can irreversibly bind to the metal center resulting in deactivation. In other words, the generation of thermodynamically too stable metal-oxo or -amido species has to be prevented in catalytic applications.

Versatile strategies have been developed in the recent years to use CO_2 as carbon feedstock [9, 10]. This circumvents the “mineralization” of CO_2 underground, a technique that has not been studied sufficiently to foresee long-term consequences. Instead, carbon dioxide might conveniently be converted into valuable chemical products, that is, linear and cyclic organic carbonates, polymers by co-polymerization of CO_2, urea and urethane derivatives, carboxylic acids, esters, lactones, isocyanates, formic acid, methanol, and related compounds [10–15]. The latter molecules are discussed to serve as fuels or as starting materials for the generation of higher hydrocarbons [16]. A broad application of organometallic catalysts has been emphasized for the fixation and transformation of CO_2 [17, 18].

Newly developed calcium-based reagents have been chosen as reagents for stoichiometric and catalytic transformations of biomass-related molecules. The catalytic potential of this relatively young class of compounds has just begun unfolding [19]. In comparison to traditional catalysts of the late transition metals, calcium-based reagents combine the following advantages: (i) Calcium is the third most abundant metal (by mass) in the earth crust and is, therefore, cheap and readily available. (ii) In contrast to many heavy metal compounds, organocalcium species are non-toxic and environmentally benign. If no other arguments (e.g., optical properties or machining) speak against, no additional efforts have to be undertaken to remove catalyst traces from the final product. (iii) Calcium is considered a light metal. This may play an additional role for logistic considerations as the cost-value ratio decreases compared to reagents of the heavy metals.

Bis(allyl)calcium (**1**) was recently reported and was probed in reactions with pyridine and furan derivatives which serve as aromatic model compounds [20–22]. Because little is known about the fundamental properties of such organocalcium complexes, a detailed analysis of the reactivity towards heteroaromatic substrates has been undertaken. The presented findings will be discussed in the context of activation of biomass-derived intermediates.

2 Results and Discussion

The isolation and full characterization of bis(allyl)calcium ($[Ca(C_3H_5)_2]$, (**1**)) was reported in 2009 [20]. Complex **1** was obtained in quantitative yield from salt metathesis of CaI_2 with the potassium precursor $[K(C_3H_5)]$ in tetrahydrofuran (THF) as the solvent. The synthetic protocol consists of simply mixing both reagents, filtration (removal of precipitated KI) and removal of all volatiles under reduced pressure. This procedure was shown to be generally applicable in the synthesis of other 1-alkenyl complexes of calcium, that is bis(1-butenyl)calcium (**2**), bis(2-methylallyl)calcium (**3**), and bis(1-hexenyl)calcium (**4**) (Fig. 2). All potassium precursors can be prepared by deprotonation of the corresponding α-olefin with KO*t*Bu/Li*n*Bu.

KO*t*Bu / Li*n*Bu, pentane, - 78 °C to 25 °C → n CaI_2, THF, 25 °C → $Ca(THF)_x$

	K salt	n	Product
propene: R^1 = H, R^2 = H	97%	n = 0.50	> 99% (**1**, x = 0)
1-butene: R^1 = H, R^2 = CH_3	77%	n = 0.38	94% (**2**, x = 0.15-0.25)
isobutene: R^1 = CH_3, R^2 = H	84%	n = 0.50	94% (**3**, x = 0.25-0.75)
1-hexene: R^1 = H, R^2 = C_3H_7	79%	n = 0.50	99% (**4**, x = 0.15-0.25)

Fig. 2 Preparation of allylic calcium compounds (**1–4**) of propene, *n*-butene (1), isobutene (2), and *n*-hexene

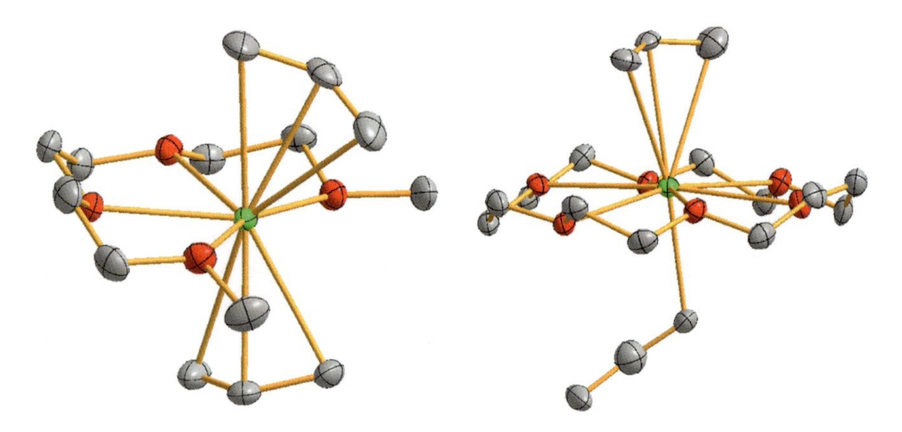

Fig. 3 Molecular structures of adducts of **1** with triglyme (*left*) [20] and 18-crown-6 (*right*) [25] in the solid state. Thermal *ellipsoids* are shown at the 50 % probability level. Hydrogen atoms omitted for clarity. *Green* calcium, *red* oxygen, *grey* carbon

Neat **1** was found to be stable up to ca. 225 °C, as established by DSC. All isolated bis(1-alkenyl)calcium complexes are stable in THF solution and realize η^3-coordinated ligands, as was evident from NMR analysis. This is not surprising regarding the ionic radius of Ca^{2+} (1.00 Å for a coordination number of six) [23] and the pronounced tendency of calcium to coordinate delocalized anions. For unsymmetrically substituted allyl ligands (i.e., in **2** and **4**) a fluxional behavior concerning the *endo* and *exo* isomers was reported [24].

The parent compound **1** was crystallized after addition of stoichiometric amounts of triethylenglycol dimethylether (triglyme) and the solid state structure was determined by X-ray diffraction (Fig. 3, left).

Although ether cleavage reactions have frequently been observed for organo-calcium complexes, **1** was found to be stable even in refluxing THF and no degradation of triglyme was detected. However, when **1** was treated with 18-crown-6, the Brønsted basicity of the $(C_3H_5)^-$ anions became obvious. Whereas at low temperatures, the 18-crown-6 adduct of **1** could be isolated (Fig. 3, right), a half time of $t_{1/2}$ = 2.5 h was determined for the degradation of the crown ether at ambient temperature in [D_8]THF solution [25]. Degradation of the crown ether is initiated by formal deprotonation of a methylene group and concomitant release of

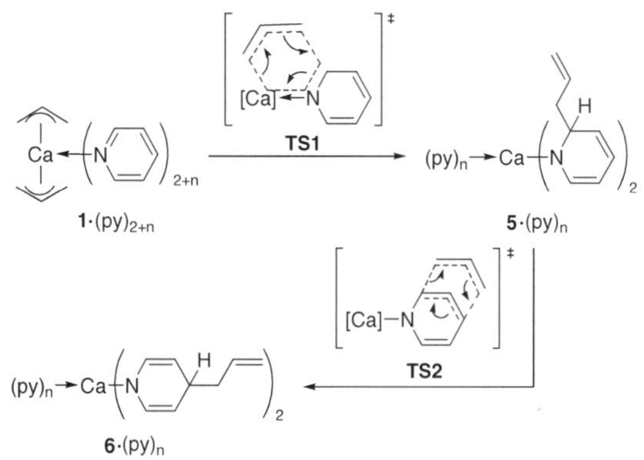

Fig. 4 Proposed mechanism for the insertion of pyridine into the calcium allyl bond

propene. Final products from ether cleavage are vinyloxide terminated alcoholates of different chain lengths.

Reaction of compound **1** with pyridine (py) resulted in the rapid and quantitative formation of the 1,4-insertion product **6** (Fig. 4) [21]. The overall mechanism was deduced from NMR analysis. The reaction is initiated by coordination of pyridine at the calcium center to give complex **1**•$(py)_{2+n}$. Attack on the *ortho* position by the nucleophilic allyl group results in the formation of the 2-allylated product **5**•$(py)_n$ via a six-membered, metalacyclic transition state **TS1**. Intermediate **5** displays a half time of $t_{1/2} = 10$ min (25 °C). The final 1,4-insertion product **6**•$(py)_n$ is formed by a rate-determining Cope rearrangement. In this second, six-membered transition state **TS2**, a lack of conformational flexibility of both the allyl and the pyridine ring fragment disfavors the 1,3-rearrangement. This sequence of allylic rearrangements is analogous to Claisen and subsequent Cope rearrangement observed for *ortho* disubstituted allyloxybenzenes to give 4-allylcyclohexa-2,5-dienones [26–28].

Product **6**•$(py)_4$ was isolated in near quantitative yield and fully characterized (for the molecular solid state structure see Fig. 5). A range of other pyridine derivatives was shown to undergo the same carbometalation reaction with **1**. These include 3,5-lutidine, 2,2′-bipyridine, acridine, quinoline, and isoquinoline [22]. All allylated products were isolated in high to quantitative yields.

However, a different pattern was found for reactions of **1** with pyridine derivatives with methyl substituents in *ortho* or *para* positions. The reaction of **1** with stoichiometric amounts of 2-picoline led to the formation of the expected carbometalation product **7** (Fig. 6). Compound **7** was subsequently transformed to yield the C–H bond activation product **8** with concomitant release of propene. This product was isolated as the mono(THF) adduct **8**•(THF).

The unexpected formation of **8** from **7** was studied by in situ ^{1}H NMR spectroscopy and the reaction was found to follow first order kinetics. This allows for the assumption of an intramolecular mechanism involving cleavage of a C–C bond, C–H bond activation and formation of a new Ca–C bond. Elimination of gaseous propene facilitates the reaction. The formation of C–H bond activation products was also observed for 4-picoline and 2,6-lutidine [22]. Yields of the obtained products were moderate to high.

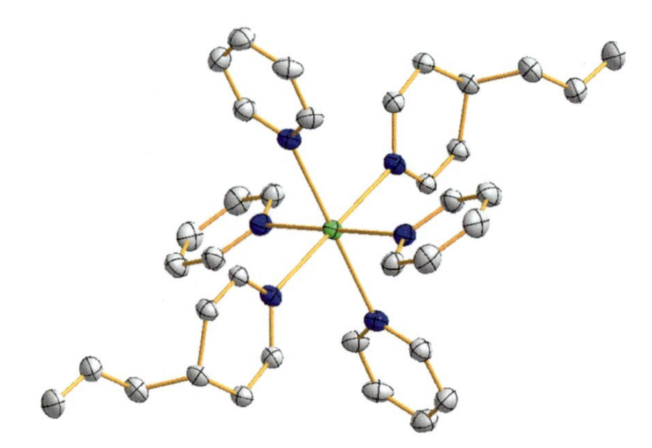

Fig. 5 Molecular structures of adducts of **6**•$(py)_4$ in the solid state [21]. Thermal ellipsoids are shown at the 50 % probability level. Hydrogen atoms omitted for clarity. *Green* calcium, *blue* nitrogen, *grey* carbon

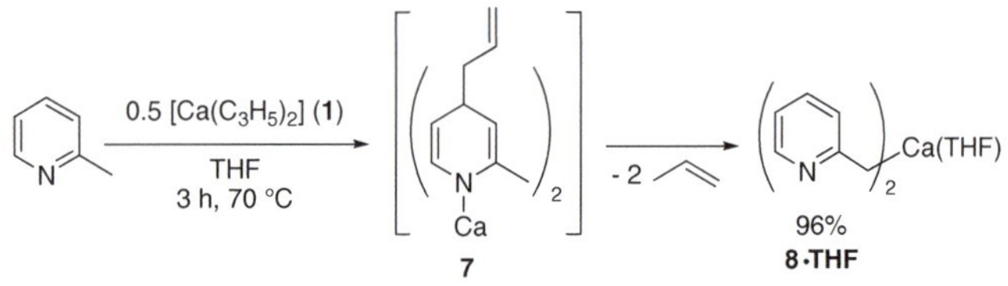

0.5 $[Ca(C_3H_5)_2]$ (**1**)
THF
3 h, 70 °C
N
Ca
2
7
- 2
Ca(THF)
2
96%
8·THF

Fig. 6 Reaction of 2-picoline with bis(allyl)calcium (**1**)

0.5 $[Ca(C_3H_5)_2]$ (**1**)
THF
7 d, 25 °C
N
Ca
2
9
+
N
Ca
2
10
- 2
$Ca(THF)_2$
2
> 99%
11·(THF)$_2$

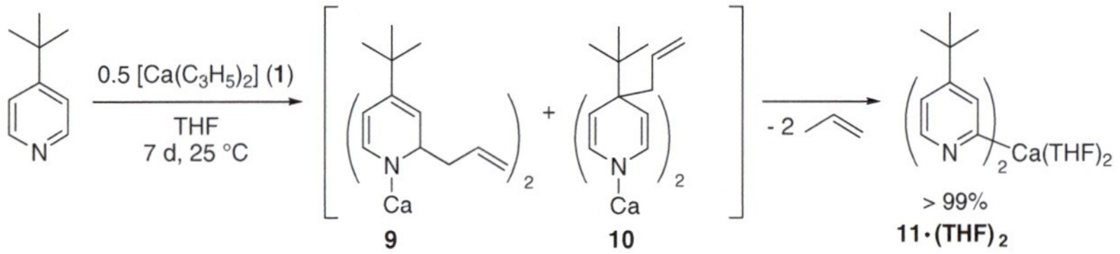

Fig. 7 Reaction of 4-*tert*-butylpyridine with bis(allyl)calcium (**1**)

The reaction of **1** with 4-*tert*-butylpyridine was studied to evaluate the steric influence of a bulky *tert*-butyl group at the *para* position of the pyridine ring. As shown by in situ NMR experiments, the 2- and 4-carbometalated intermediates **9** and **10** were formed in the presence of the starting compounds (Fig. 7). This equilibrium mixture resulted in the quantitative formation of the thermodynamically favored *ortho* metalation product bis(4-*tert*-butylpyridin-2-yl)calcium (**11**), isolated as dark purple bis(THF) adduct in quantitative yield. Steric shielding of the 4-position of pyridine still allowed for the formation of the 2- and 4-allylated intermediates. These intermediates showed an increased life-time in the reaction mixture. However, the 4-allylated carbometalation product **10** reacted to give the ring-metalation product **11** with concomitant release of propene. This is in contrast to pyridine derivatives with vacant 4-positions and isolation of the corresponding 4-allylated products.

In summary, two reaction types were observed for reactions of **1** with pyridine derivatives: (i) carbometalation to yield allylated and dearomatized N-heterocycles and (ii) metalated C–H bond activation products with intact aromaticity. Both activation of C–H(sp^3) and C–H(sp^2) bonds was observed.

The reactivity of complexes **1**, **8**•(THF), and **11**•$(THF)_2$ towards carbon dioxide was probed by treating THF solutions of the corresponding calcium compounds with CO_2 (1 bar). In each case, the formation of the corresponding calcium carboxylate was observed either in situ by ^{1}H NMR spectroscopy in [D_8]THF or after the dried calcium salt was redissolved in D_2O. As this is a proof of principle for CO_2 incorporation, isolation of the reaction products in high purity was not attempted. **1** showed the highest activity for insertion of CO_2 (1 bar) to yield calcium but-3-enoate [20]. Quantitative conversion was observed after ca. 10 min. **8**•(THF), and **11**•$(THF)_2$ were converted to calcium 2-(pyridin-2-yl)acetate and calcium 4-(*tert*-butyl)picolinate, respectively. The latter two reactions afforded 5–30 % of the free acids, what is most likely the result of hydrolysis.

After well-defined transformations of pyridines with **1** were observed, the reactivity towards furan and derivatives was studied.

The reaction of bis(allyl)calcium (**1**) with 1–3 equivalents of furan led to complete conversion of **1** during 24 h, as evident from ^{1}H NMR spectroscopy. However, complete conversion of furan was not observed. The reaction mixture turned from yellow to red to black and formation of a dark precipitate was observed at late stages (Fig. 8).

In situ ^{1}H NMR spectra displayed resonances, which were attributed to small amounts of 2-calciated furan amongst other unidentified products. ^{1}H NMR spectra recorded at higher conversions proved inconclusive, partly because of the inhomogeneous reaction mixture. Multiple broad resonances in the olefinic and aromatic region suggest the formation of polymeric products. To investigate the composition of the dark precipitate, it was treated with Et_2O and deuterium oxide. Extraction with Et_2O and subsequent GC/MS analysis of the ethereal phase revealed the absence of any soluble species. Similarly, extraction with D_2O and subsequent NMR analysis proved the precipitate insoluble in water. These observations are in line with the assumption of polymeric products. Due to the observation that even understoichiometric amounts of furan were not fully converted (Fig. 8, n = 1), a multiple activation of furan and a partly ring-opened and branched polymer product

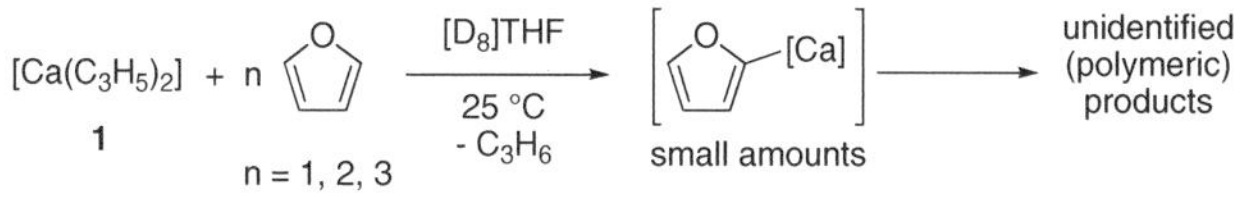

$[Ca(C_3H_5)_2]$ + n furan (n = 1, 2, 3) —[D_8]THF, 25 °C, - C_3H_6→ [2-[Ca]-furan] (small amounts) → unidentified (polymeric) products

Fig. 8 Reactions of **1** with furan

is proposed. Related 2-furyl calcium hydride was described to be formed from cocondensation of furan and calcium metal in the presence of THF, but characterization was performed only after functionalization with Me_3SiCl [29–32]. A strong competition between solvent donors and the furan oxygen atom has been described. This might be the reason for the slow reaction rate in the activation of furan, compared to the activation of pyridine by bis(allyl)calcium.

Reactions of bis(allyl)calcium (**1**) with 2-methylfuran and 2,5-dimethylfuran showed slow formation of propene, as evident from ^{1}H NMR spectra. A conversion of ca. 3 and 5 %, with respect to **1**, was observed after 24 h and 5 days, respectively. Additional heating to 55 °C for 24 h led to ca. 9 % conversion for both reactions. No product resonances were observed by NMR spectroscopy. During the course of the reaction, the mixture turned black and a precipitate began to form.

It is proposed that similar reactions take place for 2-methyl- and 2,5-dimethylfuran. A significant decrease in the reaction rate was observed for the dimethyl derivative. This is due to substituted α positions and inhibition of C–H activation of the furan ring. An insoluble, polymeric product is proposed to be formed in reactions of **1** with 2-methyl- and 2,5-dimethylfuran.

Reaction of bis(allyl)calcium (**1**) with 2,5-dihydrofuran showed slow formation of propene, as evident from ^{1}H NMR spectra. A conversion of ca. 45 %, with respect to **1**, was observed after 11 days. After 30 days all **1** was converted and a defined signal pattern was observed together with broad signals in the aliphatic and olefinic region of the proton NMR spectrum. These resonances match those observed for 4-substituted 2,3-dihydrofurans and reach their maximum concentration after a reaction time of 45 days [33, 34]. Synthetical approaches towards substituted 2,3-dihydrofurans commonly use derivatives of 4-hydroxybutyraldehyde in cyclization reactions [35]. Isomerization reactions from 2,5- towards 2,3-dihydrofurans usually employ harsh conditions and/or late transition metal catalysts [36–40]. An early protocol utilizing KO*t*Bu for the isomerization of 2,5-dihydrofuran suggests a basic mechanism that may also account for **1** [41].

In an attempt to yield catalytic carbocalciation of carbon carbon double bonds, reactions of bis(allyl)calcium with $[ZrCp_2Cl_2]$ were investigated. Similar procedures have been reported to catalytically produce carbomagnesiated products [42–45].

When $[D_8]$THF solutions of **1** and $[ZrCp_2Cl_2]$ were combined, a colorless precipitate, most likely of calcium dichloride, was formed within 1 min (Fig. 9). ^{1}H NMR spectra of the filtered solutions displayed coupling patterns, which were assigned to $[Zr(\eta^5\text{–}Cp)_2(\eta^3\text{–}C_3H_5)_2]$. Diallyldicyclopentadienyl zirconium has been described before and shows a dynamic behavior of the allyl ligands at ambient

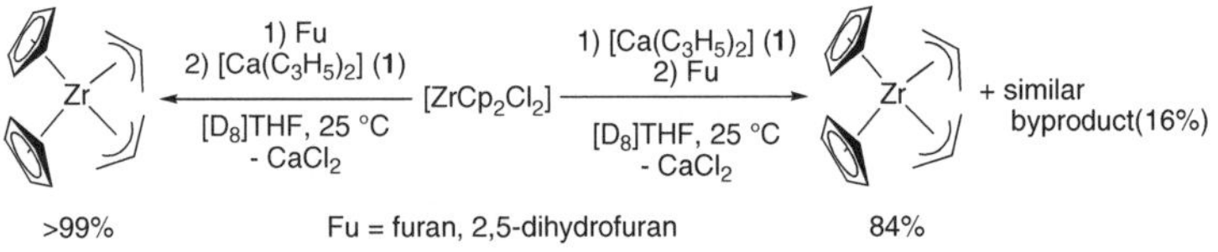

1) Fu
2) $[Ca(C_3H_5)_2]$ (1)
$[D_8]$THF, 25 °C
- $CaCl_2$
$[ZrCp_2Cl_2]$
1) $[Ca(C_3H_5)_2]$ (1)
2) Fu
$[D_8]$THF, 25 °C
- $CaCl_2$
Zr
Zr
+ similar byproduct(16%)
>99%
Fu = furan, 2,5-dihydrofuran
84%

Fig. 9 Reactions of $[ZrCp_2Cl_2]$ with bis(allyl)calcium (**1**) and furan temperature [46–49]. Whereas [Zr(η^5–Cp)$_2$(η^3–C_3H_5)$_2$] was identified as the only product from the reaction of [$ZrCp_2Cl_2$] with **1** (quantitative yield, in situ), the addition of a furan derivative prior to the addition of **1** produced a second product in ca. 16 % yield. This minor product shows an integral ratio of ca. (η^5–Cp)/(η^3–C_3H_5) = 3/2. Although this compound has not been reported, it might be speculated that [Zr(η^5–Cp)$_3$(η^3–C_3H_5)$_2$][Ca(THF)$_n$] was formed. No attempts were undertaken to identify this species.

Regardless of whether furan or 2,5-dihydrofuran were added before or after the addition of **1**, no conversion of the cyclic ethers was observed within 1 week.

3 Conclusion

The preparation and isolation of 1-alkenyl calcium complexes was achieved. All products were isolated in high to quantitative yields and fully characterized. The parent compound bis(allyl)calcium (**1**) showed quantitative conversion upon reaction with CO_2 to give calcium but-3-enoate. This reaction proceeds under mild conditions (25 °C, 1 bar CO_2).

Reactions of aromatic N-heterocycles with bis(allyl)calcium **1** give two categories of products: (i) carbometalated and therefore dearomatized, and (ii) C–H bond activated products with retained aromaticity. The reaction course is predefined by the substitution pattern of the pyridine skeleton. Allyl ligands are smoothly transferred to the electron deficient carbon atoms in α or γ position to the nitrogen atom. Here, aromaticity is sacrificed for the formation of new C–C and Ca–N bonds. When methyl substituents are present in the α or γ position, the allyl ligand acts as a base to generate propene and to form new Ca–C bonds. Carbometalated intermediates were frequently observed and the intramolecular C(sp^3)–H bond activation is accompanied by rearomatization. A C(sp^2)–H bond activation is achieved by steric shielding of the γ position, as was demonstrated by the reaction of 4-*tert*-butylpyridine. This leads to ring metalation with concomitant release of propene. The transformations of carbometalated products giving C–H bond activated products may hint at a reversibility of the initial insertion step. Indication for a reversible pyridine functionalization by organoruthenium catalysts has recently been reported [50, 51]. The above results constitute a general procedure for the regioselective production of allylated and calciated pyridines and quinolines. Due to predominantly quantitative yields, side product formation or contamination of the products is minimal. Similar to CO_2 insertion in the calcium carbon bond in **1**, CO_2 reacts with calciated pyridines to afford the corresponding calcium carboxylates.

The conversion of furan and derivatives proves difficult. Reactions of **1** with furan resulted in sluggish and inhomogeneous mixtures, which most likely consist of partly ring-opened and branched polymers. C–H bond activation in α position seems to be a key step, as methylation of the 2- and 5-positions in furan dramatically decreased the reaction rate. Clean albeit incomplete conversion of 2,5-dihydrofuran led to the formation of a 2,3-dihydrofuran derivative as the product of isomerization. Catalytic conversion of furan and 2,5-dihydrofuran proved unsuccessful, due to rapid elimination of calcium chloride and the formation of allyl cyclopentadienyl zirconium compounds. The latter showed no reactivity towards furan or 2,5-dihydrofuran.

Organocalcium complexes are shown to be effective in the activation of heteroaromatic substrates. Different modes of bond cleavage and formation are realized, strongly depending on the nature of the heterocycle employed. Clean conversion of furans remains an issue. It is proposed that activation proceeds via the α positions, but ill-defined product species complicate the application of the calcium reagent studied for the production of furans as combustion fuels.

Furthermore, the calcium reagents reported here cannot be used in common solvents for the conversion of biomass-derived substrates. As many of the above-mentioned solvents like DMA or DMF contain activated double bonds, the development of catalysts that can be handled in such media seems desirable.

In the context of furan conversion, it is proposed to investigate less Brønsted basic calcium reagents that bear calcium oxygen or calcium nitrogen bonds. As calcium carbon bonds are extremely reactive towards activated double bonds (as in furans), the utilization of calcium alkoxides might result in a more controlled reactivity. By this, the observed polymerization of furans might be circumvented and a more selective ring opening or furyl coupling is more likely.

In addition to other reported calcium amide compounds the present work gives access to a large variety of so far unknown calcium pyridyl complexes which can be probed in controlled transformations of furans.

4 Experimental Section

General remarks: Experimental details for all published compounds and procedures can be found in the corresponding references (i.e., Refs. [20–22, 24, 25]). All operations were performed under an inert atmosphere of argon using standard Schlenk line or glove box techniques. $[D_8]$THF was distilled under argon from sodium/benzophenone ketyl prior to use. D_2O was used as purchased. THF and pentane were purified using a MB SPS-800 solvent purification system. NMR spectra were recorded on a Bruker DRX 400 spectrometer (^{1}H, 400.1 MHz) at 25 °C unless otherwise stated. Chemical shifts for ^{1}H were referenced internally using the residual solvent resonances and are reported relative to tetramethylsilane.

General procedure for the reaction of bis(allyl)calcium with furan, 2-methyl furan and 2,5-dimethyl furan. To a solution of bis(allyl)calcium (23 mg, 0.19 mmol) in $[D_8]$THF (0.25 mL), a solution of the furan derivative (furan: 13 mg, 0.19 mmol; 26 mg, 0.38 mmol; 39 mg, 0.56 mmol; 2-methylfuran: 30 mg, 0.37 mmol; 2,5-dimethylfuran: 36 mg, 0.37 mmol) in $[D_8]$THF (0.25 mL) was added. In each case, a color change to red was observed within few hours. Extended reaction times led to an inhomogeneous dark brown/black reaction mixture. Analysis by NMR or GC/MS proved unsuccessful, as the products were not soluble in THF, Et_2O, and D_2O. During the course of the reaction, 2-furyl calcium species were identified: ^{1}H NMR (400 MHz, [D_8]THF, 25 °C): δ = 6.12 (dd, $^3J_{HH}$ = 1.5, 2.8 Hz, 1H, 4-CH), 6.51 (d, $^3J_{HH}$ = 2.8 Hz, 1H, 3-CH), 7.66 (d, $^3J_{HH}$ = 1.5 Hz, 1H, 5-CH).

Reaction of bis(allyl)calcium with 2,5-dihydrofuran. To a solution of bis (allyl)calcium (15 mg, 0.12 mmol) in [D_8]THF (0.3 mL), a solution of 2,5-dihydrofuran (19 mg, 0.27 mmol) in [D_8]THF (0.3 mL) was added. After 45 days, a sharp signal pattern was observed in the ^{1}H NMR spectrum of the dark reaction mixture. This was identified as a 4-substituted 2,3-dihydrofuran: ^{1}H NMR (400 MHz, [D_8]THF, 25 °C): δ = 2.53 (tt, $^3J_{HH}$ = 9.6 Hz, $^4J_{HH}$ = 2.3 Hz, 2H, 3-CH_2), 4.20 (t, $^3J_{HH}$ = 9.8 Hz, 2H, 2-CH_2), 6.29 (q, $^4J_{HH}$ = 2.4 Hz, 1H, 5-CH). It is proposed, that the 4-position was calciated.

Reaction of bis(allyl)calcium with [$ZrCp_2Cl_2$] and furan or 2,5-dihydrofuran. (a) To a solution of [$ZrCp_2Cl_2$] (25 mg, 0.09 mmol) in [D_8]THF (0.3 mL), a solution of bis(allyl)calcium (10 mg, 0.08 mmol) in [D_8]THF (0.3 mL) was added. A colorless precipitate and a color change from yellow to red/brown were observed. The ^{1}H NMR spectrum of the reaction solution showed two signal patterns (ratio = 10/1.6). The one with higher concentration was assigned to [Zr(η^5–Cp)$_2$(η^3–C_3H_5)$_2$]: ^{1}H NMR (400 MHz, [D_8]THF, 25 °C): δ = 2.82 (d, $^3J_{HH}$ = 11.5 Hz, 8H, CH_2), 5.52 (s, 10H, C_5H_5), 5.72 (quint, $^3J_{HH}$ = 11.5 Hz, 2H, CH). This corresponds well to NMR data reported for allylzirconium complexes [46–49]. The second signal pattern was observed at lower field: 3.14 (d, $^3J_{HH}$ = 11.0 Hz, CH_2), 5.98 (quint, $^3J_{HH}$ = 11.0 Hz, CH), 6.24 (s, C_5H_5). Addition of furan or 2,5-dihydrofuran to the reaction mixture did not lead to any reaction within 1 day.

(b) A solution of $ZrCp_2Cl_2$ (25 mg, 0.09 mmol) in [D_8]THF (0.5 mL) was added to furan (14 mg, 0.21 mmol) or 2,5-dihydrofuran (14 mg, 0.20 mmol). No reaction was observed by NMR spectroscopy. Bis(allyl)calcium (10 mg, 0.08 mmol) was added to the reaction mixture and a colorless precipitate was formed. ^{1}H NMR spectra indicated the coexistence of [Zr(η^5–Cp)$_2$(η^3–C_3H_5)$_2$] and the furan derivative, respectively. No reaction was observed within 1 day.

References

1. Schlögl, R.: The role of chemistry in the energy challenge. ChemSusChem **3**, 209–222 (2010). doi:10.1002/cssc.200900183
2. Schlögl, R.: Chemistry's role in regenerative energy. Angew. Chem. Int. Ed. **50**, 6424–6426 (2011). doi:10.1002/anie.201103415
3. Kuster, B.F.M.: 5-Hydroxymethylfurfural (HMF). A Review Focussing on its ManufactureStarch – Stärke **42**, 314–321 (1990). doi:10.1002/star.19900420808
4. Zakrzewska, M.E., Bogel-Łukasik, E., Bogel-Łukasik, R.: Ionic liquid-mediated formation of 5-hydroxymethylfurfural—a promising biomass-derived building block. Chem. Rev. **111**, 397–417 (2010). doi:10.1021/cr100171a

5. Ståhlberg, T., Fu, W., Woodley, J.M., Riisager, A.: Synthesis of 5-(hydroxymethyl)furfural in ionic liquids: paving the way to renewable chemicals. ChemSusChem **4**, 451–458 (2011). doi:10.1002/cssc.201000374
6. Geilen, F.M.A., vom Stein, T., Engendahl, B., Winterle, S., Liauw, M.A., Klankermayer, J., Leitner, W.: Highly selective decarbonylation of 5-(hydroxymethyl)furfural in the presence of compressed carbon dioxide. Angew. Chem. Int. Ed. **50**, 6831–6834 (2011). doi:10.1002/anie.201007582
7. Geilen, F.M.A., Engendahl, B., Harwardt, A., Marquardt, W., Klankermayer, J., Leitner, W.: Selective and flexible transformation of biomass-derived platform chemicals by a multifunctional catalytic system. Angew. Chem. Int. Ed. **49,** 5510–5514 (2010). doi:10.1002/anie.201002060
8. Nichols, J.M., Bishop, L.M., Bergman, R.G., Ellman, J.A.: Catalytic C–O bond cleavage of 2-Aryloxy-1-arylethanols and its application to the depolymerization of lignin-related polymers. J. Am. Chem. Soc. **132**, 12554–12555 (2010). doi:10.1021/ja106101f
9. Peters, M., Köhler, B., Kuckshinrichs, W., Leitner, W., Markewitz, P., Müller, T.E.: Chemical technologies for exploiting and recycling carbon dioxide into the value chain. ChemSusChem **4**:1216–1240 (2011). doi:10.1002/cssc.201000447
10. Wang, W., Wang, S., Ma, X., Gong, J.: Recent advances in catalytic hydrogenation of carbon dioxide. Chem. Soc. Rev. **40**, 3703–3727 (2011)
11. Himeda, Y., Miyazawa, S., Hirose, T.: Interconversion between formic acid and H_2/CO_2 using rhodium and ruthenium catalysts for CO_2 fixation and H_2 storage. ChemSusChem **4**, 487–493 (2011). doi:10.1002/cssc.201000327
12. Matson, T.D., Barta, K., Iretskii, A.V., Ford, P.C.: One-pot catalytic conversion of cellulose and of woody biomass solids to liquid fuels. J. Am. Chem. Soc. **133**, 14090–14097 (2011). doi:10.1021/ja205436c
13. Wu, G.-P., Wei, S.-H., Ren, W.-M., Lu, X.-B., Xu, T.-Q., Darensbourg, D.J.: Perfectly alternating copolymerization of CO_2 and epichlorohydrin using cobalt(III)-based catalyst systems. J. Am. Chem. Soc. **133**, 15191–15199 (2011). doi:10.1021/ja206425j
14. Wu, J., Hazari, N., Incarvito, C.D.: Synthesis, properties, and reactivity with carbon dioxide of $(allyl)_2$ni(l) complexes. Organometallics **30**, 3142–3150 (2011). doi:10.1021/om2002238
15. Schaub, T., Paciello, R.A.: A process for the synthesis of formic acid by CO_2 hydrogenation: thermodynamic aspects and the role of CO. Angew. Chem. Int. Ed. **50**, 7278–7282 (2011). doi:10.1002/anie.201101292
16. Yu, K.M.K., Curcic, I., Gabriel, J., Tsang, S.C.E.: Recent advances in CO_2 capture and utilization. ChemSusChem **1**, 893–899. doi:10.1002/cssc.200800169
17. Cokoja, M., Bruckmeier, C., Rieger, B., Herrmann, W.A., Kühn, F.E.: Transformation of carbon dioxide with homogeneous transition-metal catalysts: a molecular solution to a global challenge? Angew. Chem. Int. Ed. **50**, 8510–8537 (2011). doi:10.1002/anie.201102010
18. Dibenedetto, A., Stufano, P., Nocito, F., Aresta, M.: Ru(II)-mediated hydrogen transfer from aqueous glycerol to CO_2: from waste to value-added products. ChemSusChem **4**, 1311–1315 (2011). doi:10.1002/cssc.201000434
19. Harder, S.: From limestone to catalysis: application of calcium compounds as homogeneous catalysts. Chem. Rev. **110**, 3852–3876 (2010). doi:10.1021/cr9003659
20. Jochmann, P., Dols, T.S., Spaniol, T.P., Perrin, L., Maron, L., Okuda, J.: Bis(allyl)calcium. Angew. Chem. Int. Ed. **48,** 5715–5719 (2009). doi:10.1002/anie.200901743
21. Jochmann, P., Dols, T.S., Spaniol, T.P., Perrin, L., Maron, L., Okuda, J.: Insertion of pyridine into the calcium allyl bond: regioselective 1,4-dihydropyridine formation and C–H bond activation. Angew. Chem. Int. Ed. **49**, 7795–7798. doi:10.1002/anie.201003704
22. Jochmann, P., Leich, V., Spaniol, T.P., Okuda, J.: Calcium-mediated dearomatization, C–H bond activation, and allylation of alkylated and benzannulated pyridine derivatives. Chem.-Eur. J. **17**, 12115–12122 (2011). doi:10.1002/chem.201101489
23. Shannon, R.: Revised effective ionic radii and systematic studies of interatomic distances in halides and chalcogenides. Acta. Crystallogr. Sect. A **32**, 751–767 (1976). doi:10.1107/S0567739476001551

24. Jochmann, P., Maslek, S., Spaniol, T.P., Okuda, J.: Allyl calcium compounds: synthesis and structure of bis(η^3-1-alkenyl)calcium. Organometallics **30**, 1991–1997 (2011). doi:10.1021/om200012k
25. Jochmann, P., Spaniol, T.P., Chmely, S.C., Hanusa, T.P., Okuda, J.: Preparation, structure, and ether cleavage of a mixed hapticity allyl compound of calcium. Organometallics **30**, 5291–5296 (2011). doi:10.1021/om200749f
26. Barta, N.S., Cook, G.R., Landis, M.S., Stille, J.R.: Studies of the regiospecific 3-aza-Cope rearrangement promoted by electrophilic reagents. J. Org. Chem. **57**, 7188–7194 (1992). doi:10.1021/jo00052a037
27. Conroy, H., Firestone, R.A.: The intermediate dienone in the para-Claisen rearrangement. J. Am. Chem. Soc. **78**, 2290–2297 (1956). doi:10.1021/ja01591a072
28. Blechert, S.: The hetero-Cope rearrangement in organic synthesis. Synth. 1989. **71**, 82 (1989). doi:10.1055/s-1989-27158
29. Onodera, G., Imajima, H., Yamanashi, M., Nishibayashi, Y., Hidai, M., Uemura, S.: Ruthenium-catalyzed allylation of aromatic compounds and allylic ether formation. Organometallics **23**, 5841–5848 (2004). doi:10.1021/om049358k
30. Dieter, J.W., Li, Z., Nicholas, K.M.: Iron-mediated aromatic allylation. Tetrahedron Lett. **28**, 5415–5418 (1987). doi:10.1016/s0040-4039(00)96742-x
31. Bechem, B., Patman, R.L., Hashmi, A.S.K., Krische, M.J.: Enantioselective carbonyl allylation, crotylation, and *tert*-prenylation of furan methanols and furfurals via iridium-catalyzed transfer hydrogenation. J. Org. Chem. **75**, 1795–1798 (2010). doi:10.1021/jo902697g
32. Ratios estimated by ^{1}H NMR and/or GC/MS analysis
33. Baird, M.S., Baxter, A.G.W., Hoorfar, A., Jefferies, I.: Ring-size and substituent effects in intramolecular reactions of alkylidenecarbenes (carbenoids). J. Chem. Soc. Perkin Trans. **1**, 2575–2581 (1991)
34. Su, L., Lei, C.-Y., Fan, W.-Y., Liu, L.-X.: $FeCl_3$-mediated reaction of alkynols with iodine: an efficient and convenient synthetic route to vinyl iodides. Synth. Commun. **41**, 1200–1207. doi:10.1080/00397911.2010.481739
35. Botteghi, C., Consiglio, G., Ceccarelli, G., Stefani, A.: Convenient synthetic approach to 3- and 4-alkyl-2,3-dihydrofurans. J. Org. Chem. **37**, 1835–1837 (1972). doi:10.1021/jo00976a040
36. Krompiec, S., Kuznik, N., Urbala, M., Rzepa, J.: Isomerization of alkyl allyl and allyl silyl ethers catalyzed by ruthenium complexes. J. Mol. Catal. A: Chem. **248**, 198–209 (2006). doi:10.1016/j.molcata.2005.12.022
37. Mazuela, J., Coll, M., Pàmies, O., Diéguez, M.: Rh-catalyzed asymmetric hydroformylation of heterocyclic olefins using chiral diphosphite ligands scope and limitations. J. Org. Chem. **74**, 5440–5445 (2009). doi:10.1021/jo900958k
38. Gual, A., Godard, C., Castillón, S., Claver, C.: Highly efficient rhodium catalysts for the asymmetric hydroformylation of vinyl and allyl ethers using C1-symmetrical diphosphite ligands. Adv. Synth. Catal. **352**, 463–477 (2010). doi:10.1002/adsc.200900608
39. Monnier, J.R., Moorehouse, C.S.: Vol. WO 962378 (A1) (Ed.: E. C. CO), USA (1996)
40. Junichi F.: Vol. WO 2009133950, C07B61/00; C07C41/32; C07C43/15; C07D307/28 ed. (Ed.: K. CO), Japan (2009)
41. Eliel, E.L., Nowak, B.E., Daignault, R.A., Badding, V.G.: Reductions with metal hydrides. XIV. Reduction of 2-Tetrahydropyranyl and 2-Tetrahydrofuranyl Ethers. J. Org. Chem. **30**, 2441–2447 (1965). doi:10.1021/jo01018a082
42. Hoveyda, A.H., Xu, Z., Morken, J.P., Houri, A.F.: Stereoselective zirconium-catalyzed ethylmagnesiation of homoallylic alcohols and ethers. The influence of internal Lewis bases on substrate reactivity. J. Am. Chem. Soc. **113**, 8950–8952. doi:10.1021/ja00023a055
43. Hoveyda, A.H., Morken, J.P., Houri, A.F., Xu, Z.: The mechanism of the zirconium-catalyzed carbomagnesiation reaction. Efficient and selective catalytic carbomagnesiation with higher alkyls of magnesium. J. Am. Chem. Soc. **114**, 6692–6697 (1992). doi:10.1021/ja00043a012
44. Hoveyda, AH., Xu, Z.: Stereoselective formation of carbon-carbon bonds through metal catalysis. The zirconium-catalyzed ethylmagnesiation reaction. J. Am. Chem. Soc. **113**, 5079–5080. doi:10.1021/ja00013a064

45. Morken, JP., Didiuk, MT., Hoveyda, AH: Zirconium-catalyzed asymmetric carbomagnesation. J. Am. Chem. Soc. **115**, 6997–6998 (1993). doi:10.1021/ja00068a077
46. Benn, R., Hoffmann, E.G.: ^{1}H NMR study of the fluxional behaviour of tetraallylhafnium and cyclooctatetraenediallylzirconium. J. Org. Chem. **193**, C33–C36 (1980). doi:10.1016/s0022-328x(00)85624-7
47. Hoffmann, E.G., Kallweit, R., Schroth, G., Seevogel, K., Stempfle, W., Wilke, G.: IR- und ^{1}H-NMR-spektroskopische untersuchungen an zirkon und hafniumallylen: IÜbergangsmetallallyle. J. Organomet. Chem. **97**, 183–202 (1975). doi:10.1016/s0022-328x(00)89465-6
48. Kablitz, H.-J., Wilke, G.: Übergangsmetallkomplexe: III. Über zirkonorganische komplexe mit cyclooctatetraen als ligand. J Organomet Chem **51**, 241–271 (1973). doi:10.1016/s0022-328x(00)93522-8
49. Mashima, K., Yasuda, H., Asami, K., Nakamura, A.: Structures of mono- and bis(2-butenyl) zirconium complexes in solution and threo selective insertion reaction of aliphatic aldehydes. Chem. Lett. **12**, 219–222 (1983). doi:10.1246/cl.1983.219
50. Osakada, K.: 1,4-hydrosilylation of pyridine by ruthenium catalyst: a new reaction and mechanism. Angew. Chem. Int. Ed. **50**, 3845–3846 (2011). doi:10.1002/anie.201008199
51. Gutsulyak, DV., van der Est, A., Nikonov, G.I.: Facile catalytic hydrosilylation of pyridines. Angew. Chem. Int. Ed. **50**, 1384–1387 (2011). doi:10.1002/anie.201006135

Co-expression of Cellulases in the Chloroplasts of *Nicotiana tabacum*

Johannes Klinger, Ulrich Commandeur and Rainer Fischer

Abstract Due to the increasing demand for alternative energy carriers, biomass is in the focus of research and industry as raw material for fuels and base chemicals. This has led to an increased use of feedstock for biofuel production and to conflict between using this feedstock for either food or for fuel generation. To avoid this problem, lignocellulose is seen as a promising raw material, as it is not used for food or feed production and harbors high amounts of sugars in form of cellulose and hemicellulose. Currently, the enzymes used for cellulose degradation are mainly produced directly in their natural hosts (e.g. *Trichoderma reesei*) or in genetically altered microorganisms. The high costs for the production of these enzymes make them a major obstacle for the economically feasible use of lignocellulose as a renewable raw material. As an alternative production platform, plants could be used to express these enzymes cheaply and directly in the raw material to be used for conversion. To alleviate the skepticism towards genetically altered plants, especially prevalent in Germany and Europe, chloroplast transformation offers the opportunity to combine efficient production of a set of cellulases within a single plant, while reducing the risk of releasing altered genetic information into the environment. This approach is used in this project to express seven cellulolytic enzymes, derived from the bacterium *Thermobifida fusca*, in the chloroplasts of *Nicotiana tabacum* and to analyze their activity on cellulosic substrates as well as their influence on plant growth.

1 Introduction

The rising prices of raw materials, especially for fossil energy carriers, have evolved into a serious economic problem over the last decades. This was observed, for example, when the oil price reached almost US$ 150 per barrel in 2008 [25].

J. Klinger · U. Commandeur (✉) · R. Fischer
Institute for Molecular Biotechnology, RWTH Aachen University, Worringer Weg 1, 52074 Aachen, Germany
e-mail: commandeur@molbiotech.rwth-aachen.de

M. Klaas et al. (eds.), *Fuels From Biomass: An Interdisciplinary Approach*,
Notes on Numerical Fluid Mechanics and Multidisciplinary Design 129,
DOI 10.1007/978-3-662-45425-1_7

In response to these developments, and due to the fact that fossil energy resources are limited, the search for alternative fuels is increasingly a research focus. In particular, the use of renewable raw materials, such as biomass, are of special interest and have led to ambitious government projects like the US "Energy Independence and Security Act" (EISA) of 2007. This act proclaimed the goal of raising the biofuel production in the US from ca. 15 billion liters per year to ca. 140 billion liters in 2022 [10]. In Europe, there is a similar directive from the European Commission (EC), setting goals for the amount of bioenergy to be consumed in each country by 2020 [15]. In case of Germany, the Federal Ministry for Education and Research is supporting initiatives like the excellence cluster "Tailor Made Fuels from Biomass" (TMFB, http://www.fuelcenter.rwth-aachen.de/) and has developed programs such as the "National Research Strategy Bioeconomy 2030" [4]. According to Yuan et al. [71], world bioethanol production was 38 billion liters in 2007, mainly derived from sugar (sugarcane, sugar beet) or corn starch. In another report, an estimated world bioethanol production of roughly 45 billion liters from sugars was calculated for 2006, mainly produced in Brazil and the US with a predicted bioethanol consumption of 113.6 billion liters from these sources for 2022 in Brazil alone, assuming cellulosic materials are not used to an industrially relevant scale [18].

Using first generation biofuels, i.e. these derived from starch or sugar, leads to a competition in resource utilization, as those materials can also be used to produce food or feed to nourish the population [48]. Additionally, there are land use effects to be considered which could potentially lead to increased greenhouse gas emissions if large areas of arable land are used to grow biofuel crops, rather than food or feed crops [51]. To avoid these conflicts, the second generation of biofuels, based on lignocellulose derived from agricultural residues or waste materials, offers a promising solution. Several pilot plants using lignocellulosic materials have been established in the US and Europe [5, 49]. Unfortunately, the more complex structure of lignocellulose compared to starch leads to higher processing costs for these materials. The enzymes used for cellulose degradation currently constitute about 40 % of these costs, making them a key obstacle in the development of an economically feasible second generation bioethanol production [49]. This chapter aims to give a short introduction to plant biomass, and its use in lignocellulosic biofuel production, as well as an overview of cellulases and their production in plants.

2 The Plant Cell

The basic structure of plant cells is of fundamental importance when considering plant biomass degradation (Fig. 1). Each plant cell is encapsuled by a primary and a secondary cell wall (the combination is hereafter referred to as the cell wall) consisting mainly of cellulose, lignin and hemicelluloses [34]. This cell wall and the cell membrane separate the cell from its environment and maintain its physical structure. Inside the cell there are different compartments and organelles, all surrounded by the cytosol. The nucleus contains the major part of the genetic information of the cell

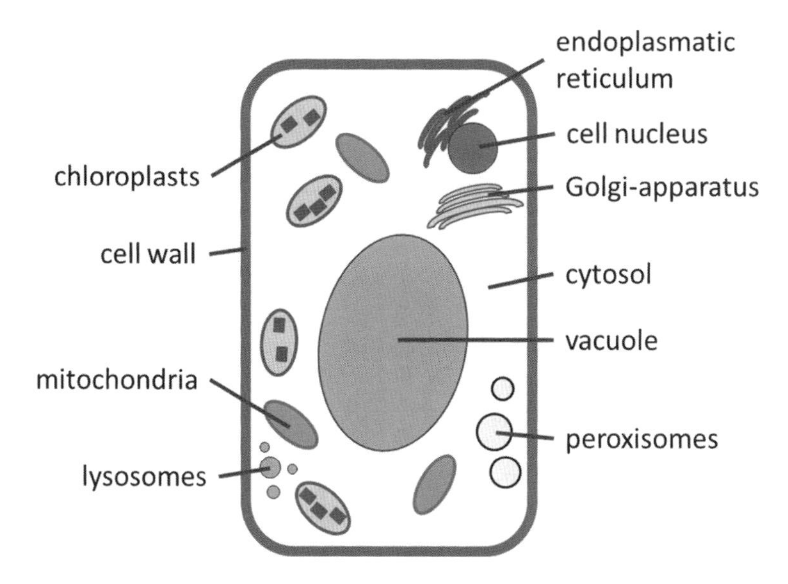

Fig. 1 Schematic picture of a plant cell

and is the center for most gene expression. Directly connected to the nucleus is the endoplasmic reticulum (ER), a highly branched compartment. It is separated from the cytosol by a membrane and functions as the location where post-translational protein modifications are performed (e.g. glycosylation) and as part of the system for transporting proteins to the outside of the cell. Additionally, the endoplasmic reticulum takes part in the production of cellular membrane components. The Golgi apparatus is another place of post-translational protein modification and proteins on their way to the outside of the cell are further modified in this compartment before they are transported to the cell wall. Additionally, the Golgi apparatus, like the ER, is involved in cellular membrane synthesis. The peroxisomes are vesicles containing various oxidases, enzymes which are used for fatty acid degradation. They are also the place where cellular H_2O_2 is removed by chemical reactions, thus detoxifying the cell [1].

Mitochondria are organelles producing energy for the cell, mainly in form of ATP, a co-substrate for many cellular reactions. They also contain a small genome coding for most enzymes of the respiratory chain of the mitochondria. Chloroplasts are the organelles where photosynthesis and cellular carbon fixation occur. They too contain a small genome which exists in several copies per chloroplast. In tobacco leaf cells, an average of 100 chloroplasts harboring about 100,000 genomes is reported [53]. Separated from the rest of the cell by two membranes, the chloroplast interior resembles more closely a prokaryotic than an eukaryotic, leading to the endosymbiont thesis about their origin, which states that photosynthetic cyanobacteria were taken up by plant cells and subsequently developed into cellular organelles [6, 60]. Finally, the vacuole contains a reservoir for cell products and maintains the fluid balance of the plant cell. Thus, the plant cell is a highly compartmentalized and complex network of protein interactions and chemical reactions, providing different subcellular compartments for genetic manipulation and protein localization.

3 Lignocellulose

The conflict of using plant biomass resources, like crops, for either food or fuel production led to a search for alternative raw materials with the potential to be degraded into sugars. Lignocellulose is seen as a very promising candidate for this challenge. This complex heteropolymer is comprised on average of ca. 25 % lignin, ca. 30 % hemicellulose and ca. 45 % cellulose [34, 72] and is part of every plant cell wall, making it a highly abundant raw material which can be derived from agricultural residues as well as forestry or biological waste materials. If the lignocellulosic parts of cereals are considered as raw material, the usable biomass derived from these plants could be doubled [27]. Currently, perennial grasses such as switch grass (*Panicum virgatum*) or miscanthus (*Miscanthus* spp.) are considered as an alternative to conventional crops for biofuel production because of their high biomass yield and broad climate adaption. Furthermore, as these grasses are perennial, the risk of erosion would be considerately lowered, reducing the environmental impact of growing crops for fuel. Also forestry resources, such as willow and poplar, are also in the scientific focus for this purpose [9, 71].

Cellulose is the most abundant biopolymer on earth comprising about half of the global biomass [19, 41]. It consists of long chains containing on average 500–14,000 glucose units, connected by β-1,4 glycosidic bonds. These chains agglomerate to form approximately 4 nm wide, mainly crystalline micro-fibrils which are stabilized by hydrogen bonds and van der Waals interactions. This crystalline structure makes the cellulose more recalcitrant to degradation (reviewed in Jordan et al. [34]; Somerville [54]). The micro-fibrils are surrounded by hemicellulose which connects the cellulose with the lignin (Fig. 2). In contrast to cellulose, hemicellulose is a more heterogenic polymer consisting of C5 sugars (e.g. xylose, arabinose), C6 sugars (e.g. glucose, mannose, galactose) and sugar related acids connected by various glycosidic bonds [72]. It coats the cellulose fibers and prevents further agglomeration of the microfibrils, thus enhancing their flexibility [34].

The third lignocellulose component, the lignin, is a highly complex polymer comprised of phenolic compounds instead of C5 or C6 sugars. The main components are *p*-coumaryl alcohol, sinapyl alcohol and coniferyl alcohol, all of which are connected via ether bonds. Lignin surrounds the polysaccharides and holds the fibers together, working as natural glue via various types of covalent chemical bonds with hemicellulose residues and thus linking the two polymers [35]. This is a major factor in lignocellulose recalcitrance or resistance to degradation, and makes a physicochemical pretreatment of lignocellulose necessary before the degradation into sugars is feasible [26, 62].

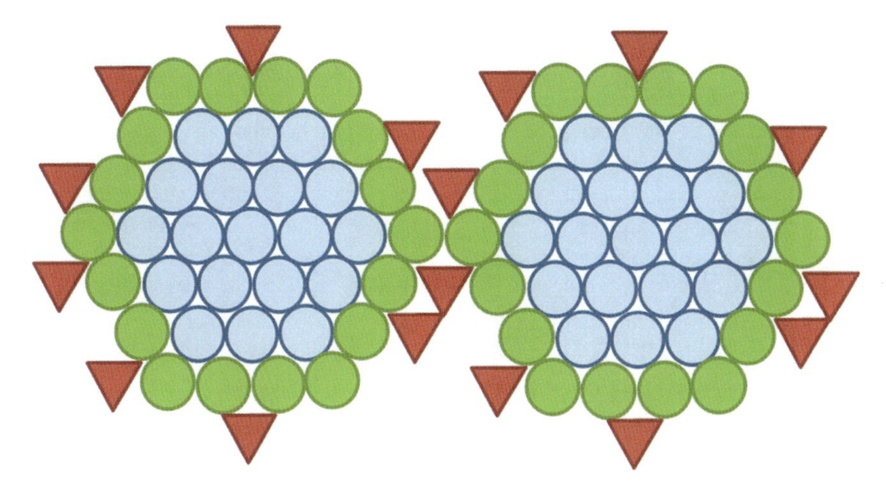

Fig. 2 Lignocellulose structure; cellulose in *blue*, hemicellulose in *green*, lignin in *red*

4 Enzymatic Hydrolysis of Lignocellulose

The valorization process for lignocellulose can roughly be divided into four steps. The first step is the harvest of the lignocellulosic biomass and its transportation to the processing site. There, a physicochemical pretreatment step is performed to separate the biomass components. The cellulose is then hydrolyzed enzymatically to generate sugars, which is followed by a fermentation step to produce the desired biofuels or biofuel precursors (reviewed in Jordon et al. [34]; Sainz [49]).

The highly complex structure of lignocellulose, which was discussed before, is the main obstacle for its use as renewable resource [62]. To degrade cellulose into its monomeric glucose units, lignin and hemicellulose have to be removed. In particular, lignin is a problem as it protects the cellulose and reduces its accessibility. Additionally, cellulolytic proteins can bind to lignin in an irreversible manner and are thus inactivated [49, 69]. Another beneficial effect of many pretreatment methods is the reduction in cellulose crystallinity. This leads to more amorphous regions, where cellulolytic enzymes can access their substrate and begin the hydrolysis.

To pretreat the cellulose, several methods have been established using high temperature, high pressure and extreme pH-values (reviewed in Yang and Wyman [69]). This treatment removes the lignin and dissolves the hemicellulose, leaving cellulose for further processing. After these harsh conditions, washing steps have to be performed to separate the cellulose from the other potentially toxic or enzyme inhibiting components and to remove inhibiting degradation products before enzymatic hydrolysis can be performed [68].

The enzymatic cellulose hydrolysis is the next step of the lignocellulose valorization process. Here, several classes of cellulolytic enzymes are utilized to achieve efficient cellulose degradation. Endoglucanases or endocellulases (EC3.2.1.4) attack the amorphous regions of the cellulose fibers and cleave the β-1,4-glycosidic bonds between the glucose subunits by integration of a water molecule [39]. Due to the physical structure of their catalytic domain which forms a catalytic cleft, most endoglucanases do not work processively but bind to the amorphous regions of the cellulose repeatedly and cleave single bonds along the chain [28, 49]. Exoglucanases, also called cellobiohydrolases or exocellulases, instead act in a processive manner, i.e. by binding once and then proceeding along a cellulose chain, and degrade single cellulose strands starting from their ends, generally releasing short glucose oligomers like cellobiose or cellotetraose [50]. Exoglucanases consist of a carbohydrate-binding module (CBM) connected via a flexible linker region to a tunnel shaped catalytic domain. While the CBM slides down the chain, it leads the cellulose through the tunnel in the catalytic domain where the β-1,4-glycosidic bonds are cleaved and the resulting sugars are released. Depending on the end of the cellulose chain attacked, exocellulases are divided into class I (attacking the reducing end) and class II (attacking the non reducing end) [3, 28, 49]. Additionally, there are enzymes called processive endoglucanases, which initially cleave a hydrogen bond within a chain in an amorphous cellulose region and then continue to degrade further bonds while sliding down the chain like an exoglucanase [50]. A third class of enzymes supports the cellulases. These enzymes are called β-glucosidases (EC3.2.1.21) and cleave the β-1,4-glycosidic bonds in sugar oligomers such as cellobiose to release glucose monomers. This enhances the efficiency of endo- and exocellulases, which are inhibited by the sugar oligomers they release [23, 29].

To achieve an economically relevant level of cellulose hydrolysis, a combination of at least the three previously named enzyme classes is required (Fig. 3). In general, even in such combinations, the activity of cellulases is significantly lower than the activity of the amylases used for starch hydrolysis in first generation biofuel processes [24]. Thus, enzyme activity is a field where researchers expect significant improvements to be possible, in order to strengthen the effectiveness of

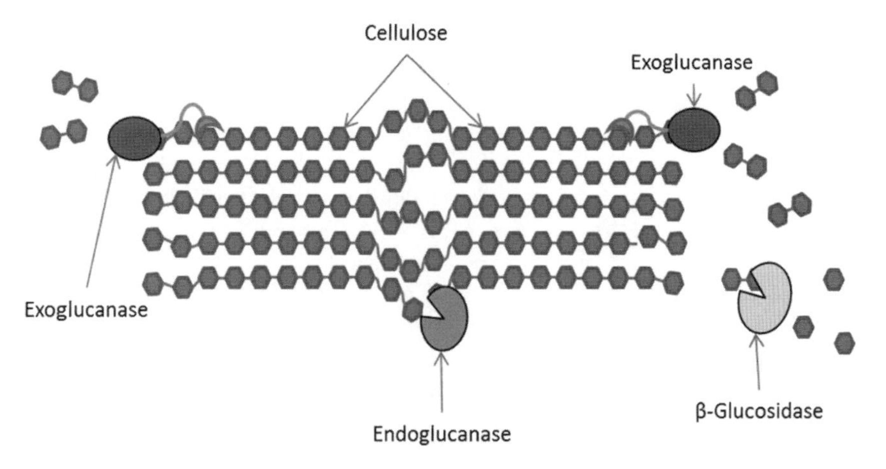

Fig. 3 Schematic picture of enzymatic hydrolysis of cellulose (adapted from Lynd et al. [42])

lignocellulose valorization [59]. Current research focuses on improvements of the stability of the cellulases, to further reduce process costs and increase economic performance ([2]; reviewed in Wen et al. [66]). An alternative approach towards improving enzymatic lignocellulose hydrolysis is the use of accessory enzymes, like expansins or swollenin. These enzymes bind to cellulose and disrupt the hydrogen bonds between the glucose chains, leading to the generation of more amorphous regions which can be utilized as starting points for improved enzymatic cellulose hydrolysis [33].

5 Plant Based Cellulase Production

Currently, cellulase production is usually performed using fermentative microorganisms, including the natural hosts, such as *Trichoderma reesei*, or genetically modified organisms. This relatively laborious method for cellulase production leads to high costs for the enzymes and is a major obstacle for an economically feasible lignocellulose valorization process, with enzyme production contributing up to 40 % of the overall process costs, as mentioned previously. Reduction of these expenses is thus a key aspect for making lignocellulose a competitive, renewable raw material, with each of the currently used production systems having challenges to be overcome (reviewed in Garvey et al. [17]). To achieve lower production costs for the cellulolytic enzymes, plants could be used for their expression, instead of the current microbial fermentation methods. This way, no microorganism growth media would be required and the technical effort for expression would be greatly reduced. The production could be easily adjusted to economic needs, simply by increasing or decreasing the size of the cultivated area for these plants [13, 61]. Further, the production of cellulases within plants potentially used for lignocellulose valorization could help to reduce processing costs, even when the expression levels are low, if enzyme activity acts to pretreat the plants during growth and thus reduces the amount of expensive cellulases needed for efficient hydrolysis.

To transform plants into natural bioreactors for heterologous protein production, two strategies are commonly used. First, the genetic information contained in the cell nucleus can be altered to express the target genes in a eukaryotic environment. This might be preferable for genes originating from eukaryotic sources where the expressed proteins might require post-translational modifications, such as glycosylation, to obtain full activity. Alternatively, the genome contained in the chloroplasts can be modified to produce enzymes in a more prokaryotic environment, without major post-translational modification of the proteins [52].

The transformation of the plant cell nucleus is usually achieved using *Agrobacterium tumefaciens* based techniques. This soil bacterium possesses a natural mechanism for nuclear transformation of plant cells, which it uses to modify their genetic information and generate living space for itself within the plant. As described by Caplan et al. [8], genetic manipulation of this bacterium can lead to a very efficient integration of a gene of interest into the plant cell nuclear genome.

With this technique, many genetically altered plants were generated for various purposes, including the production of cellulases (reviewed in Sharma and Sharma [52]; Yuan et al. [71]).

Nuclear transformed plants have the advantage that they are well suited for the expression of enzymes derived from eukaryotic hosts such as *T. reesei*, which is currently used for commercial cellulase production. Compared to bacteria, the already mentioned ability of plants to perform post-translational modifications, such as glycosylation of proteins or disulfide bond formation, is a major advantage of this system for the expression of eukaryotic genes. An additional benefit of plants as bioreactors is their lack of human pathogens, making them highly interesting for pharmaceutical production [52].

However, nuclear transformed plants also show some disadvantages. The transformation and selection process to obtain a stable transformed plant takes months or even years [30, 32], which is significantly longer than the time needed to generate genetically altered bacteria. Additional drawbacks are so called position effects, which occur during the random integration of the new genetic information into the plant nuclear genome. This can severely influence the expression of target genes [1, 12]. Furthermore, "silencing" can be observed for heterologously expressed genes. In this case, the expression of the gene is suppressed on the transcriptional or post-transcriptional level leading to greatly diminished protein production. Thus, it is possible to create a genetically modified plant which shows little or no expression of the gene of interest, despite the stable integration of the transgene [12]. The containment of the genetic information of genetically modified plants is an additional potential problem, as these plants can release their altered genome via pollen and thus transport the genetic information in an uncontrolled manner to undesired recipients. This way, the altered genetic information could potentially enter the human food chain, which is in Europe and other countries a topic of great public concern [14], leading to a politically difficult situation when looking to cultivate genetically altered plants, an aspect that should not be underestimated.

Despite these possible drawbacks, there are many examples for the successful generation of nuclear altered plants producing cellulases in various species, including tobacco, alfalfa, rice, corn or potatoes [16, 76]. A major focus lies on the expression of the endoglucanase E1 derived from the bacterium *Acidothermus cellulolyticus*. Here, expression of E1 constituted an average of 2 % of total soluble protein (TSP) when expressed in tobacco. In these experiments, the cellulase was transported to subcellular compartments like the vacuole or the apoplast, to prevent damage to the cells and to investigate the effects of different protein localizations [59, 75].

An alternative method to prevent damage to the plant host is the use of cellulolytic enzymes derived from thermophilic hosts, such as *Thermobifida fusca* or *Sulfolobus solfataricus*, which show minimal or no activity when expressed at the mesophilic temperatures at which the transformed plants grow, with the cellulases activated by a rise in temperature [37, 59]. Another possible method is to use inducible promoters for target gene expression. This way, the expression of the cellulase gene is induced at a desired time point and the plant can grow without any burden caused by the cellulase expression. This method was demonstrated to significantly reduce growth defects in the expression of *T. reeesei* endocellullase Cel5A in tobacco plants [36].

In addition to nuclear transformation, plant cells offer the possibility to alter the genetic information contained within the chloroplast genome. This method provides a high copy number, of up to 10,000 copies, of the target gene [44, 53] in something similar to a prokaryotic environment, which might be favorable for bacterial enzymes. To obtain chloroplast transformation, the standard method established is particle bombardment, developed by Boynton et al. [7] for the algae *Chlamydomonas reinhardtii* and later adapted for *Nicotiana tabacum* [57]. Here the genetic information to be integrated into the plant genome is coated onto gold or tungsten particles and introduced into the target material by particle acceleration. The DNA dissolves from the particles within the cells and integrates by homologous recombination at defined sites into the chloroplast genome (Fig. 4).

Due to this mechanism of integration, position effects like those observed for nuclear transformation by *A. tumefaciens* are avoided [43]. To accumulate the transformed plant cells, several rounds of selection are performed until all chloroplast genomes contain the altered genetic information [65]. Similarly to nuclear transformation strategies, chloroplast transformation has both benefits and drawbacks.

One drawback of chloroplast transformation is that, compared to nuclear transformation, the transformation efficiency is relatively low [7, 47, 57] and the selection process is significantly longer for this technique [30, 65]. In contrast, one of the benefits of chloroplast transformation is that it can yield significantly higher expression levels than nuclear transformation [56]. Moreover, this method leads to a much stronger containment of the altered genetic information, as the chloroplast genome is usually not transported in pollen, and thus features a greatly reduced risk of the generation of uncontrolled genetically modified plants [11, 12]. The missing protein glycosylation in chloroplasts can be an advantage for bacterial proteins, but may be a drawback for enzymes from higher organisms where glycosylation might be required to obtain fully active and stable proteins [11]. Other post-translational modifications, like lipid modifications or the formation of disulfide bonds, can occur in this organelle [63]. Furthermore, polycistronic expression of genes is possible in the chloroplast. This is a feature also found in bacteria, where different

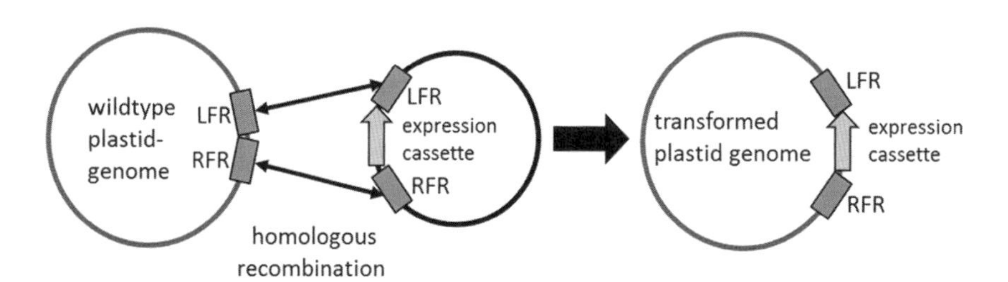

Fig. 4 Specific integration of the expression cassette into the chloroplast genome via homologous recombination (adapted from Maliga [43]); *LFR* left flanking region, *RFR* right flanking region, *arrow* expression cassette

genes are under the control a single promoter, leading to the generation of one long mRNA coding for all genes of the operon [38, 46]. For cellulase production, natural bacterial or archaeal operons (e.g. *T. fusca*, *S. solfataricus*) could be used, alternatively an artificial polycistronic cassette would allow the generation of synthetic operons, coding for endocellulases, exocellulases and β-glucosidases, in one plant. This would allow the production of a complete set of cellulolytic enzymes within a single plant.

Over the last decade, some results on the use of chloroplast transformation for cellulase production have been published. For example, Yu et al. [70] showed the expression of two *T. fusca* cellulases in the chloroplasts of *N. tabacum*. They observed the expression of active enzymes with a concentration of 2–4 % of TSP. An improvement in expression levels of 150 %, as compared to nuclear expression of the same enzymes in plants. More recently, Verma et al. [64] showed the expression of various cellulolytic enzymes and accessory proteins in tobacco chloroplasts. They demonstrated effective hydrolysis of several cellulosic substrates can be obtained by optimizing the protein mix used for degradation. Another example of chloroplast transformation was shown by Gray et al. [22], where downstream box fusions were used to express *T. fusca* endoglucanase Cel6A in tobacco chloroplasts. In these experiments, the expression levels reached up to 10 % of TSP for the full size protein without any visible protein degradation. In a later study, Gray et al. [21] achieved expression levels of up to 12 % of TSP for the β-glucosidase BglC from *T. fusca* using the same downstream box approach as before. Interestingly, the fusion resulting in the highest expression levels differed between Cel6A and BglC, indicating that such fusions have to be tested for each gene individually.

In another study, Petersen and Bock [45] expressed several cellulolytic enzymes, also derived from *T. fusca*, to achieve effective cellulose hydrolysis. Each gene was transformed separately to generate plant lines for each enzyme. The expression levels ranged from 5 to 40 % of TSP, leading to plants with expression levels of active enzymes too high to be able to grow on normal soil, instead they required synthetic media containing high amounts of sucrose for growth. These experiments show that cellulase expression at high levels can be achieved in tobacco chloroplasts and thus might be a promising alternative to nuclear transformed plants on the way towards an economically feasible lignocellulose valorization industry.

6 Project Description

Since the first successful chloroplast transformation in 1988, tobacco has developed into the model plant for this technique [7, 58]. This is attributed to the relatively simple and efficient transformation procedure, compared to other species which have been investigated [40, 47] and to the amount of information already present about tobacco plants. In this project, chloroplast transformation will be used to generate transgenic *N. tabacum* cv *petit havana* plants expressing the desired cellulases. In contrast to the experiments described above, this project focuses on the expression of multiple cellulolytic enzymes within a single plant to achieve an inexpensive production of efficient enzyme combinations, rather than the production of single enzymes. The expression performance of transplastomic plants will be assessed and the cellulose hydrolysis efficiency of the enzymes, both when expressed singly and in combination shall be investigated on various test substrates, to compare the results with other currently achievable enzyme production levels and activities.

The vector used for the experiments is a modified version of the pRB95 plasmid, which has been successfully used for chloroplast transformation in tobacco previously. Due to the designated flanking regions of the expression cassette, the foreign genes will be integrated in the chloroplast genome between the genes trnfM and trnG in the large single copy (LSC) region [47], which has been successfully used for stable chloroplast transformation before [45, 47, 73].

The genes which will be integrated into the tobacco chloroplast genome are derived from the thermophilic bacterium *T. fusca*, which is a thoroughly studied resource for cellulolytic genes (reviewed in Gomez del Pulgar and Saadeddin [20]; Wilson [67]). For this project, six cellulases secreted by *T. fusca* and one β-glucosidase have been chosen for expression. If successfully expressed, this cellulase mixture would contain three endoglucanases, two exoglucanses (one attacking the reducing end of the cellulose chain and one acting on the non-reducing end) and one processive endoglucanase. The combination of these enzymes has been shown to act synergistically on test substrates, such as filter paper [31]. Thus, a combined expression of these enzymes should result in a cellulase cocktail with high efficiency. To further enhance the cellulase activity a β-glucosidase is also expressed to prevent product inhibition of the enzymes by accumulation of cellobiose or other oligosaccharides [50, 55].

To obtain an efficient expression of the genes in the synthetic operon, an intercistronic expression element (IEE) derived from the tobacco chloroplast *psbB* operon will be inserted together with an additional ribosome binding site between the individual genes (Fig. 5). This element has been shown to significantly increase the expression of genes in synthetic operons [74]. As a preliminary experiment, all chosen cellulolytic enzymes will be expressed transiently in the ER of *N. tabacum* using the *A. tumefaciens* transformation method. This investigation will assess the production and activity of these enzymes when expressed in a eukaryotic

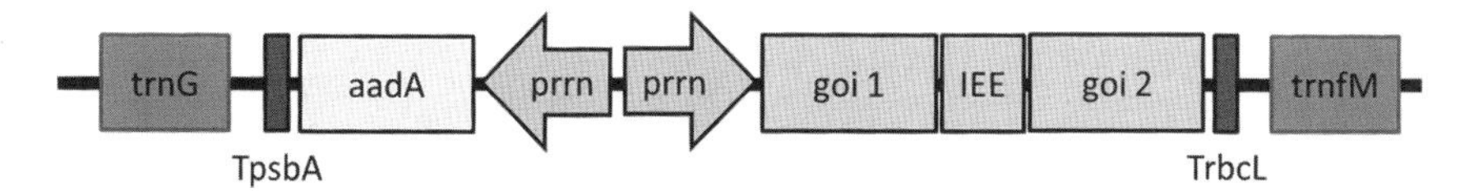

Fig. 5 Schematic drawing of an chloroplast expression cassette; *trnG* tRNA(Gly), *TpsbA* 3′ UTR of *psbA*, *aadA* aminoglycoside 3″-adenylyltransferase, *Prrn* ribosomal RNA promoter, *goi* gene of interest, *IEE* intercistronic expression element, *Trbcl* 3′ UTR of *rbcL*, *trnfM* tRNA(Met)

environment, and enable comparison of this transient method of production with the chloroplast expression. Additionally, the enzymes will be expressed in *Escherichia coli* to generate a similar set of data, for later comparison of bacterial expression and chloroplast expression. The chloroplast transformation will be performed for each enzyme individually, to analyze the expression and activity when expressed alone, and in combinations of enzymes designed to obtain an effective cellulolytic cocktail. The expression results will be analyzed and compared to the data generated for transient nuclear expression in tobacco and to the information obtained from the *E. coli* experiments. Additionally the growth behavior of the transformed plants will be analyzed to gain information about how the additional metabolic burden affects the plants.

Overall, this series of experiments should give valuable information about the expression of multiple cellulase genes within the chloroplasts of tobacco and compare it to currently used systems. Thus, this project may contribute to solve the current situation concerning first generation and second generation biofuels and might help to decrease the cost of lignocellulose valorization to make it a more economical and sustainable process.

References

1. Alberts, B., Johnson, A., Lewis, J., Raff, M., Roberts, K., Walter, A.P.: Molecular biology of the cell, 4th edn. Garland Science, NY (2002). http://www.ncbi.nlm.nih.gov/books/NBK21054/
2. Anbar, M., Lamed, R., Bayer, E.A.: Thermostability enhancement of *Clostridium thermocellum* cellulosomal endoglucanase Cel8A by a single glycine substitution. ChemCatChem **2**(8), 997–1003 (2010)
3. Barr, B.K., Hsieh, Y.-L., Ganem, B., Wilson, D.B.: Identification of two functionally different classes of exocellulases. Biochemistry **35**(2), 586–592 (1996)
4. Bundesministerium für Bildung und Forschung, BMBF. National research strategy bioeconomy 2030 (2011) http://www.bmbf.de/pub/bioeconomy_2030.pdf
5. Bundesministerium für Bildung und Forschung, BMBF. Funding success: Biofuel made from straw (2012). http://www.bmbf.de/en/17786.php?hilite=bioenergy
6. Bock, R., Timmis, J.N.: Reconstructing evolution: Gene transfer from plastids to the nucleus. BioEssays **30**(6), 556–566 (2008)
7. Boynton, J.E., Gillham, N.W., Harris, E.H., Hosler, J.P., Johnson, A.M., Jones, A.R., Randolph-Anderson, B.L., Robertson, D., Klein, T.M., Shark, K.B., Sanford, J.C.: Chloroplast transformation in *Chlamydomonas* with high velocity microprojectiles. Science **240**(4858), 1534–1538 (1988)
8. Caplan, A., Herrera-Estrella, L., Inzé, D., Van Haute, E., Van Montagu, M., Schell, J., Zambryski, P.: Introduction of genetic material into plant cells. Science **222**(4625), 815–821 (1983)
9. Carroll, A., Somerville, C.: Cellulosic biofuels. Annu. Rev. Plant Biol. **60**(1), 165–182 (2009)
10. Congress of the United States of America (2207) Energy independence and security act of 2007. Pub. L. No. 110–140, 121 Stat. **1492**, 1783–1784 (2007, December 19) (*codified at* 42 U.S.C. §17381)
11. Daniell, H.: Transgene containment by maternal inheritance: Effective or elusive? Proc Natl Acad Sci USA **104**(17), 6879–6880 (2007)

12. Daniell, H., Khan, M.S., Allison, L.: Milestones in chloroplast genetic engineering: an environmentally friendly era in biotechnology. Trends Plant Sci. **7**(2), 84–91 (2002)
13. Desai, P.N., Shrivastava, N., Padh, H.: Production of heterologous proteins in plants: strategies for optimal expression. Biotech. Adv. **28**(4), 427–435 (2010)
14. Domingo, J.L.: Health risks of GM foods: many opinions but few data. Science **288**(5472), 1748–1749 (2000)
15. European Comission (2009) Directive 2009/28/EC of the European Parliament and of the Council of 23 April 2009 on the promotion of the use of energy from renewable sources and amending and subsequently repealing directives 2001/77/EC and 2003/30/EC
16. Egelkrout, E., McGaughey, K., Keener, T., Ferleman, A., Woodard, S., Devaiah, S., Nikolov, Z., Hood, E., Howard, J.: Enhanced expression levels of cellulase enzymes using multiple transcription units. Bioenerg. Res. **6**(2), 1–12 (2012)
17. Garvey, M., Klose, H., Fischer, R., Lambertz, C., Commandeur, U.: Cellulases for biomass degradation: comparing recombinant cellulase expression platforms. Trends Biotechnol. **31** (10), 581–593 (2013)
18. Goldemberg, J., Guardabassi, P.: Are biofuels a feasible option? Energy Policy **37**(1), 10–14 (2009)
19. Goldstein, I.S.: Organic Chemicals from Biomass. CRC Press, Inc. FL (1981) 310 pp
20. Gomez del Pulgar, E.M., Saadeddin, A.: The cellulolytic system of *Thermobifida fusca*. Crit. Rev. Microbiol. **40**(3), 236–247 (2014)
21. Gray, B., Yang, H., Ahner, B., Hanson, M.: An efficient downstream box fusion allows high-level accumulation of active bacterial beta-glucosidase in tobacco chloroplasts. Plant Mol. Biol. **76**(3–5), 1–11 (2011)
22. Gray, B., Ahner, B.A., Hanson, M.: High-level bacterial cellulase accumulation in chloroplast-transformed tobacco mediated by downstream box fusions. Biotechnol. Bioeng. **102**(4), 1045–1054 (2009)
23. Gusakov, A.V., Sinitsyn, A.P.: A theoretical analysis of cellulase product inhibition: Effect of cellulase binding constant, enzyme/substrate ratio, and β-glucosidase activity on the inhibition pattern. Biotech. Bioeng. **40**(6), 663–671 (1992)
24. Gusakov, A.V.: Alternatives to *Trichoderma reesei* in biofuel production. Trends Biotechnol. **29**(9), 419–425 (2011)
25. Hamilton, J. D.: Causes and consequences of the oil shock of 2007-08. Natl. Bur. Econ. Res. Working Pap. Ser. **40**, 215–283 (2009)
26. Hendriks, A.T.W.M., Zeeman, G.: Pretreatments to enhance the digestibility of lignocellulosic biomass. Bioresour. Technol. **100**(1), 10–18 (2009)
27. Henry, R.J.: Evaluation of plant biomass resources available for replacement of fossil oil. Plant Biotechnol. J. **8**(3), 288–293 (2010)
28. Himmel, M.E., Ding, S.-Y., Johnson, D.K., Adney, W.S., Nimlos, M.R., Brady, J.W., Foust, T.D.: Biomass recalcitrance: engineering plants and enzymes for biofuels production. Science **315**(5813), 804–807 (2007)
29. Holtzapple, M.T., Caram, H.S., Humphrey, A.E.: Determining the inhibition constants in the HCH-1 model of cellulose hydrolysis. Biotech. Bioeng. **26**(7), 753–757 (1984)
30. Horsch, R., Fry, J., Hoffmann, N., Eichholtz, D., Rogers, S.A., Fraley, R.: A simple and general method for transferring genes into plants. Science **227**, 1229–1231 (1985)
31. Irwin, D.C., Zhang, S., Wilson, D.B.: Cloning, expression and characterization of a family 48 exocellulase, Cel48A, from *Thermobifida fusca*. Eur. J. Biochem. **267**(16), 4988–4997 (2000)
32. Ishida, Y., Hiei, Y., Komari, T.: *Agrobacterium*-mediated transformation of maize. Natl. Protoc. **2**(7), 1614–1621 (2007)
33. Jaeger, G., Girfoglio, M., Dollo, F., Rinaldi, R., Bongard, H., Commandeur, U., Fischer, R., Spiess, A.C., Buechs, J.: How recombinant swollenin from *Kluyveromyces lactis* affects cellulosic substrates and accelerates their hydrolysis. Biotechnol. Biofuels **4**(1), 33 (2011)
34. Jordan, D.B., Bowman, M.J., Braker, J.D., Dien, B.S., Hector, R.E., Lee, C.C., Mertens, J.A., Wagschal, K.: Plant cell walls to ethanol. Biochem. J. **442**, 241–252 (2012)

35. Jørgensen, H., Kristensen, J.B., Felby, C.: Enzymatic conversion of lignocellulose into fermentable sugars: challenges and opportunities. Biofuels Bioprod. Bioref. **1**(2), 119–134 (2007)
36. Klose, H., Gunl, M., Usadel, B., Fischer, R., Commandeur, U.: Ethanol inducible expression of a mesophilic cellulase avoids adverse effects on plant development. Biotechnol. Biofuels **6** (1), 53 (2013)
37. Klose, H., Röder, J., Girfoglio, M., Fischer, R., Commandeur, U.: Hyperthermophilic endoglucanase for *in planta* lignocellulose conversion. Biotechnol. Biofuels **5**, 63 (2012)
38. Krichevsky, A., Meyers, B., Vainstein, A., Maliga, P., Citovsky, V.: Autoluminescent plants. PLoS ONE **5**(11), e15461 (2010)
39. Lombard, V., Golaconda, R.H., Drula, E., Coutinho, P.M., Henrissat, B.: The carbohydrate-active enzymes database (CAZy) in 2013. Nucleic Acids Res. **42**, D490–D495 (2014)
40. Lutz, K., Azhagiri, A., Maliga, P.: "Transplastomics in Arabidopsis: Progress toward developing an efficient method." Chloroplast Research in Arabidopsis. R. P. Jarvis. Humana Press. NY **774**, 133–147 (2011)
41. Lutzen, N.W., Nielsen, M.H., Oxenboell, K.M., Schulein, M., Stentebjerg-Olesen, B.: Cellulases and their application in the conversion of lignocellulose to fermentable sugars. Philos. Trans. R. Soc. Lond. B Biol. Sci. **300**(1100), 283–291 (1983)
42. Lynd, L.R., Weimer, P.J., van Zyl, W.H., Pretorius, I.S.: Microbial cellulose utilization: fundamentals and biotechnology. Microbiol Mol Biol R **66**(3), 506–577 (2002)
43. Maliga, P.: Engineering the plastid genome of higher plants. Curr. Opin. Plant Biol. **5**(2), 164–172 (2002)
44. Maliga, P.: Plastid transformation in higher plants. Annu. Rev. Plant Biol. **55**(1), 289–313 (2004)
45. Petersen, K., Bock, R.: High-level expression of a suite of thermostable cell wall-degrading enzymes from the chloroplast genome. Plant Mol. Biol. **76**(3), 311–321 (2011)
46. Quesada-Vargas, T., Ruiz, O.N., Daniell, H.: Characterization of heterologous multigene operons in transgenic chloroplasts. Transcription, processing, and translation. Plant Physiol. **138**(3), 1746–1762 (2005)
47. Ruf, S., Hermann, M., Berger, I.J., Carrer, H., Bock, R.: Stable genetic transformation of tomato plastids and expression of a foreign protein in fruit. Natl. Biotech. **19**(9), 870–875 (2001)
48. Runge, C.F., Senauer, B.: How biofuels could starve the poor. Foreign affairs **86**(3), 41–53 (2007)
49. Sainz, M.: Commercial cellulosic ethanol: The role of plant-expressed enzymes. In Vitro Cell Dev Biol: Plant **45**(3), 314–329 (2009)
50. Sakon, J., Irwin, D., Wilson, D.B., Karplus, P.A.: Structure and mechanism of endo/exocellulase E4 from *Thermomonospora fusca*. Natl. Struct. Mol. Biol. **4**(10), 810–818 (1997)
51. Searchinger, T., Heimlich, R., Houghton, R.A., Dong, F., Elobeid, A., Fabiosa, J., Tokgoz, S., Hayes, D., Yu, T.-H.: Use of U.S. croplands for biofuels increases greenhouse gases through emissions from land-use change. Science **319**(5867), 1238–1240 (2008)
52. Sharma, A.K., Sharma, M.K.: Plants as bioreactors: Recent developments and emerging opportunities. Biotechnol. Adv. **27**(6), 811–832 (2009)
53. Shaver, J., Oldenburg, D., Bendich, A.: Changes in chloroplast DNA during development in tobacco, Medicago truncatula, pea, and maize. Planta **224**(1), 72–82 (2006)
54. Somerville, C.: Cellulose synthesis in higher plants. Annu. Rev. Cell Dev. Biol. **22**(1), 53–78 (2006)
55. Spiridonov, N.A., Wilson, D.B.: Cloning and biochemical characterization of BglC, a β-glucosidase from the cellulolytic actinomycete *Thermobifida fusca*. Curr. Microbiol. **42** (4), 295–301 (2001)
56. Sun, Y., Cheng, J.J., Himmel, M.E., Skory, C.D., Adney, W.S., Thomas, S.R., Tisserat, B., Nishimura, Y., Yamamoto, Y.T.: Expression and characterization of *Acidothermus cellulolyticus* E1 endoglucanase in transgenic duckweed Lemna minor 8627. Bioresour. Technol. **98**(15), 2866–2872 (2007)

57. Svab, Z., Hajdukiewicz, P., Maliga, P.: Stable transformation of plastids in higher plants. Proc. Natl. Acad. Sci. USA **87**(21), 8526–8530 (1990)
58. Svab, Z., Maliga, P.: High-frequency plastid transformation in tobacco by selection for a chimeric aadA gene. Proc. Natl. Acad. Sci. USA **90**(3), 913–917 (1993)
59. Taylor II, L.E., Dai, Z., Decker, S.R., Brunecky, R., Adney, W.S., Ding, S.-Y., Himmel, M.E.: Heterologous expression of glycosyl hydrolases in planta: a new departure for biofuels. Trends Biotechnol. **26**(8), 413–424 (2008)
60. Timmis, J.N., Ayliffe, M.A., Huang, C.Y., Martin, W.: Endosymbiotic gene transfer: organelle genomes forge eukaryotic chromosomes. Natl. Rev. Genet. **5**(2), 123–135 (2004)
61. Twyman, R.M., Stoger, E., Schillberg, S., Christou, P., Fischer, R.: Molecular farming in plants: host systems and expression technology. Trends Biotechnol. **21**(12), 570–578 (2003)
62. Vega-Sánchez, M.E., Ronald, P.C.: Genetic and biotechnological approaches for biofuel crop improvement. Curr. Opin. Biotechnol. **21**(2), 218–224 (2010)
63. Verma, D., Daniell, H.: Chloroplast vector systems for biotechnology applications. Plant Physiol. **145**(4), 1129–1143 (2007)
64. Verma, D., Kanagaraj, A., Jin, S., Singh, N.D., Kolattukudy, P.E., Daniell, H.: Chloroplast-derived enzyme cocktails hydrolyse lignocellulosic biomass and release fermentable sugars. Plant Biotechnol. J. **8**(3), 332–350 (2010)
65. Verma, D., Samson, N.P., Koya, V., Daniell, H.: A protocol for expression of foreign genes in chloroplasts. Natl. Protoc. **3**(4), 739–758 (2008)
66. Wen, F., Nair, N.U., Zhao, H.: Protein engineering in designing tailored enzymes and microorganisms for biofuels production. Curr. Opin. Biotechnol. **20**(4), 412–419 (2009)
67. Wilson, D.B.: Studies of Thermobifida fusca plant cell wall degrading enzymes. Chem. Rec. **4** (2), 72–82 (2004)
68. Ximenes, E., Kim, Y., Mosier, N., Dien, B., Ladisch, M.: Deactivation of cellulases by phenols. Enzyme Microb. Technol. **48**(1), 54–60 (2011)
69. Yang, B., Wyman, C.E.: Pretreatment: The key to unlocking low-cost cellulosic ethanol. Biofuels Bioprod. Bioref. **2**(1), 26–40 (2008)
70. Yu, L.-X., Gray, B.N., Rutzke, C.J., Walker, L.P., Wilson, D.B., Hanson, M.R.: Expression of thermostable microbial cellulases in the chloroplasts of nicotine-free tobacco. J. Biotechnol. **131**(3), 362–369 (2007)
71. Yuan, J.S., Tiller, K.H., Al-Ahmad, H., Stewart, N.R., Stewart Jr, C.N.: Plants to power: bioenergy to fuel the future. Trends Plant Sci. **13**(8), 421–429 (2008)
72. Zaldivar, J., Nielsen, J., Olsson, L.: Fuel ethanol production from lignocellulose: a challenge for metabolic engineering and process integration. Appl. Microbiol. Biotechnol. **56**(1), 17–34 (2001)
73. Zhou, F., Badillo-Corona, J.A., Karcher, D., Gonzalez-Rabade, N., Piepenburg, K., Borchers, A.M.I., Maloney, A.P., Kavanagh, T.A., Gray, J.C., Bock, R.: High-level expression of human immunodeficiency virus antigens from the tobacco and tomato plastid genomes. Plant Biotechnol. J. **6**(9), 897–913 (2008)
74. Zhou, F., Karcher, D., Bock, R.: Identification of a plastid intercistronic expression element (IEE) facilitating the expression of stable translatable monocistronic mRNAs from operons. Plant J **52**(5), 961–972 (2007)
75. Ziegelhoffer, T., Raasch, J.A., Austin-Phillips, S.: Dramatic effects of truncation and sub-cellular targeting on the accumulation of recombinant microbial cellulase in tobacco. Mol. Breeding **8**(2), 147–158 (2001)
76. Ziegelhoffer, T., Will, J., Austin-Phillips, S.: Expression of bacterial cellulase genes in transgenic alfalfa (Medicago sativa L.), potato (Solanum tuberosum L.) and tobacco (Nicotiana tabacum L.). Mol. Breeding **5**(4), 309–318 (1999)

Cleavage and Diastereoselective Synthesis of Mono- and Dilignol *β-O*-4 Model Compounds

Jakob Mottweiler, Julien Buendia, Erik Zuidema and Carsten Bolm

Abstract A short and convenient synthetic pathway affording diastereomerically pure 1,3-dilignols in both their *erythro* and *threo* form has been developed. Additionally, $H_2Ru(CO)(PPh_3)_3$ has been identified as a promising catalyst for the cleavage of lignin model compounds. The greater accessibility of 1,3-dilignols will facilitate future lignin cleavage studies of ruthenium catalysts and other transition metal systems, employing model compounds that closely resemble the *β-O*-4 linkage within lignin.

1 Introduction

The use of biomass as a source for fuels, chemicals and energy has become increasingly relevant considering the ever decreasing world petroleum reserves [1, 2]. This consequently led to extensive research for the valorization of lignocellulosic biomass within the last decade, and major breakthroughs have been achieved for the conversion of cellulose and hemicellulose into fuels [3–7]. Despite the great natural availability of lignin (in the year 2004 50 million tons of lignin were extracted by the

J. Mottweiler · J. Buendia · E. Zuidema · C. Bolm (✉)
Institut für Organische Chemie, RWTH Aachen University, Landoltweg 1,
52056 Aachen, Germany
e-mail: carsten.bolm@oc.rwth-aachen.de

J. Mottweiler
e-mail: jakob.mottweiler@oc.rwth-aachen.de

J. Buendia
CNRS Centre de Recherche de Gif-sur-Yvette, Institut de Chimie des Substances Naturelles,
UPR 2301, Avenue de la Terrasse, Bât. 27, 91198 Gif-sur-Yvette Cedex, France
e-mail: julien.buendia@oc.rwth-aachen.de; julien.buendia@cnrs.fr

E. Zuidema
SABIC Technology and Innovation, Urmonderbaan 22, 6167RD Geleen, The Netherlands
e-mail: erik.zuidema@oc.rwth-aachen.de; Erik.zuidema@sabic-europe.com

M. Klaas et al. (eds.), *Fuels From Biomass: An Interdisciplinary Approach*,
Notes on Numerical Fluid Mechanics and Multidisciplinary Design 129,
DOI 10.1007/978-3-662-45425-1_8

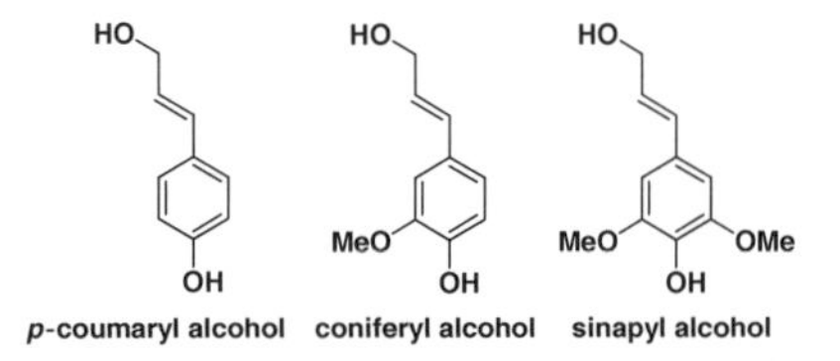

Fig. 1 Monomeric building blocks of lignin

paper industry), its valorization still remains at an initial stage [8, 9]. Lignin is an amorphous three-dimensional polymer consisting of methoxylated phenylpropane units which represent 15–30 wt% of the lignocellulosic biomass. The biosynthesis of lignin is believed to proceed through a radical mechanism in which, depending on the plant type, *p*-coumaryl, coniferyl and sinapyl alcohols are incorporated as monomeric building units in different ratios (Fig. 1) [9–15].

This leads to a great structural diversity of lignin and renders analytics even more challenging [10–15]. The *β-O*-4 unit is the most predominant linkage within lignin. Depending on the wood type, it accounts for 45–60 % of the interconnecting bonds [9]. Therefore, *β-O*-4 model compounds were used for initial lignin cleavage studies with transition metal catalysts.

2 Discussion

Previous investigations performed in the group on lignin cleavage showed that among the various screened catalysts the ruthenium complex $H_2Ru(CO)(PPh_3)_3$ was the most efficient one (Fig. 2) [16, 17]. To gain a deeper understanding of the reaction behavior, theoretical studies were conducted on the dehydration of alcohols using this ruthenium catalyst [18].

For the optimization of the reaction conditions, monolignol model compound **1** was chosen. This substrate is accessible by a three step synthesis (Fig. 3) [19, 20].

Previous studies showed that after 72 h, the reaction afforded the cleavage products 3,4-dimethoxyacetophenone and 2-methoxyphenol in good yields (75 % in toluene and 77 % in dioxane) [16, 17]. The oxidation product **2a** was formed as a side product in 7 and 8 % yield, respectively. However, fluctuations in the conversion were observed, and the yields proved difficult to reproduce. Therefore, slight alterations of the reaction conditions were made (Fig. 4).

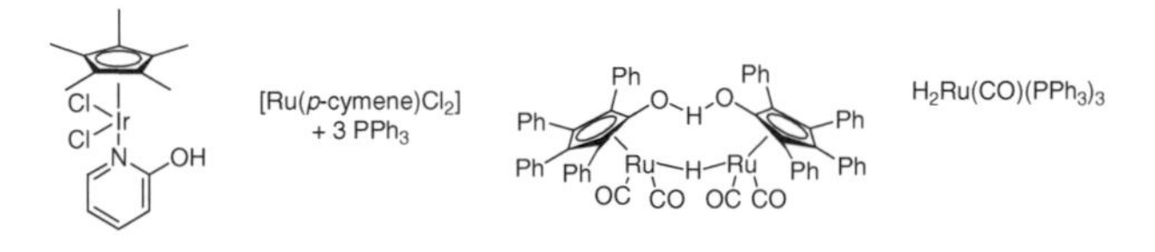

Fig. 2 Transition metal catalysts tested in the Bolm group

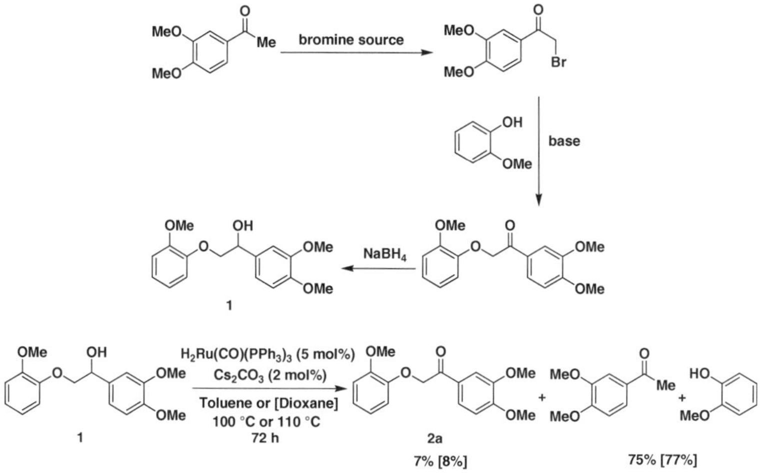

Fig. 3 Synthesis of lignin model compound **1** and the initial reaction conditions

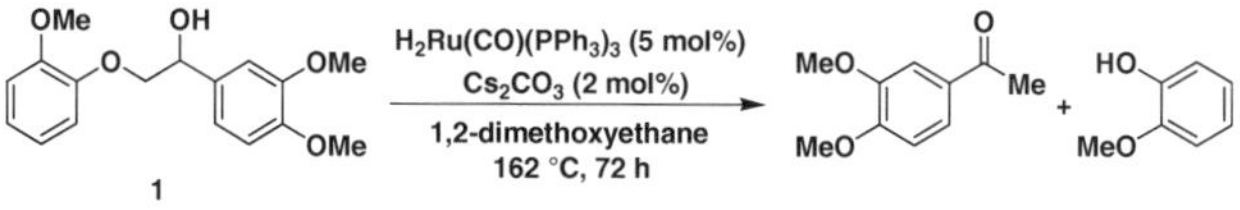

Fig. 4 Cleavage reaction of monolignol **1** under optimized conditions

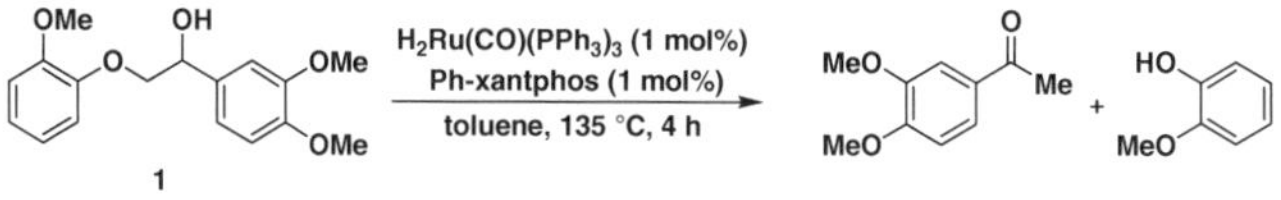

Fig. 5 Reaction conditions used by Bergman and Ellman [21]

This led to complete conversion of the starting material in which the cleavage products 3,4-dimethoxyacetophenone and 2-methoxyphenole were exclusively formed. However, Bergman and Ellman simultaneously described the cleavage of this lignin model compound with the same catalyst, using improved reaction conditions (Fig. 5) [21].

This led us to shift the focus of our research to more functionalized *β*-*O*-4 dilignol model compounds, which displayed a greater structural resemblance with lignin (Fig. 6).

Various strategies are known for the synthesis of the corresponding *erythro* and *threo* *β*-*O*-4 dilignol model compounds [19, 22–33, 34–38].

Adler and co-workers were the first to describe the synthesis of such diols [27, 29–33]. In the first synthetic step an acetophenone derivative, benzyl acetovanillone, was brominated and subsequently set to react with 2-methoxyphenol under

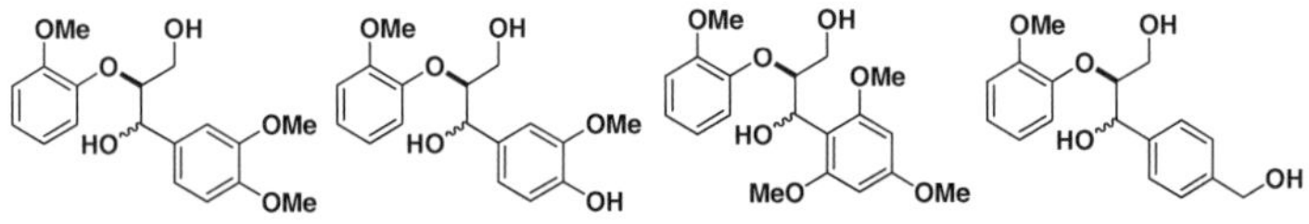

Fig. 6 *β-O*-4 dilignol model compounds

Fig. 7 Reaction conditions by Adler and co-workers to form *β*-hydroxy ketone **3**

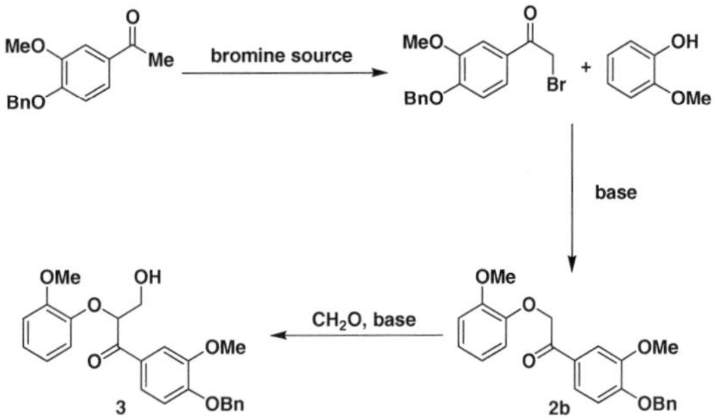

basic reaction conditions to afford keto aryl ether **2b** (Fig. 7). This was followed by an aldol addition yielding *β*-hydroxy ketone **3**.

The reduction of **3** with $NaBH_4$ led to diol **4a** as a mixture of the *erythro* and *threo* isomers. However, the separation of the corresponding diastereomers by column chromatography or crystallization proved to be very challenging. Therefore diol **4a** was converted to the acetonides **5aT** and **5aE** (Fig. 8) which were separated by column chromatography [22, 30]. The desired diols were then obtained after hydrolysis of the corresponding acetonides.

To circumvent the need of derivatizing the diol **4a** Helm and co-workers described a diastereoselective reduction of the protected *β*-hydroxy ketone **6** (Fig. 9) [39, 40]. In the presence of diisopinocampheyl chloroborane (Dip-Cl^{TM})

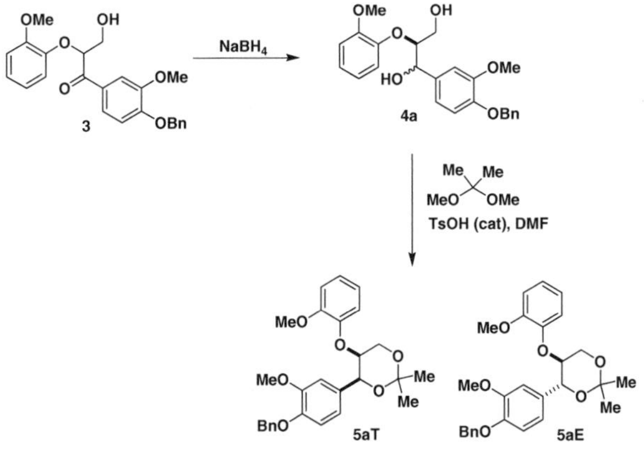

Fig. 8 Reduction of *β*-hydroxy ketone **3** and subsequent conversion to acetonides **5aT** and **5aE**

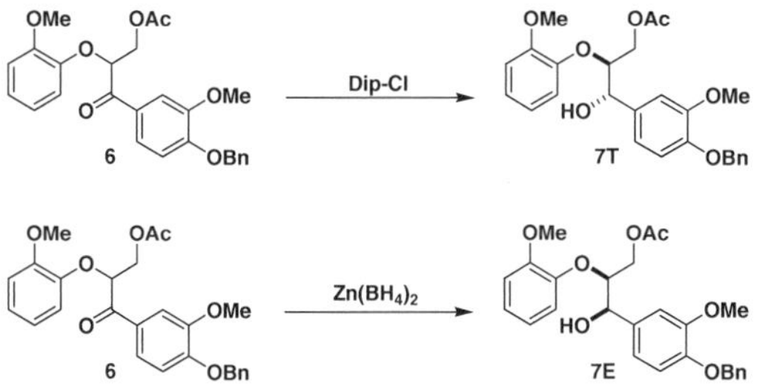

Fig. 9 Diastereoselective reduction by Helm and co-workers

they reported a yield of 80 % for the *threo* dilignol **7T** and a diastereoselectivity of >98 %. When changing the reducing agent to $Zn(BH_4)_2$ the *erythro* dilignol **7E** was obtained in 86 % yield with a diastereoselectivity of 95 % [39, 40].

An *erythro* selective synthesis of diol **4bE** has been described by Lundquist and co-workers (Fig. 10) [28, 34–38]. The starting reagent for this synthesis is the aryloxy ester **8** which reacts with veratraldehyde in the presence of sodium hydride to form the acid **9** in 41 % yield. After hydroboration and consecutive oxidation of the resulting borane, **4bE** is obtained in 68 % and total diastereoselectivity [28, 34–38].

In the same year Nakatsubo et al. published a three step synthesis for dilignol model compounds starting from 2-methoxyphenol and ethyl chloroacetate which reacted under basic conditions to give the nucleophilic substitution product aryloxy ester **10a** (Fig. 11) [26]. The corresponding lithium enolate was formed with LDA and upon addition of benzyl vanillin a mixture of *erythro* and *threo* β-hydroxy ester **11a** was obtained. The formation of *erythro* β-hydroxy ester **11aE** was hereby favored. The reported yields of *threo* β-hydroxy ester **11aT** and other similar *threo* β-hydroxy esters after column chromatography or crystallization were low [19, 23].

When comparing the different reaction pathways developed by Adler, Lundquist and Nakatsubo it became apparent that the routes of Adler and Lundquist required more synthetic steps and had notably lower overall yields than the one of Nakatsubo. On the other hand, with the synthesis of Nakatsubo and co-workers diastereomerically

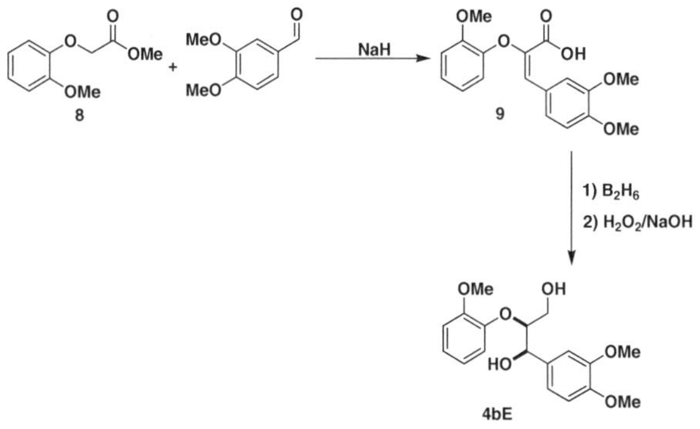

Fig. 10 Reaction conditions by Lundquist and co-workers

Fig. 11 Reaction condition by Nakatsubo

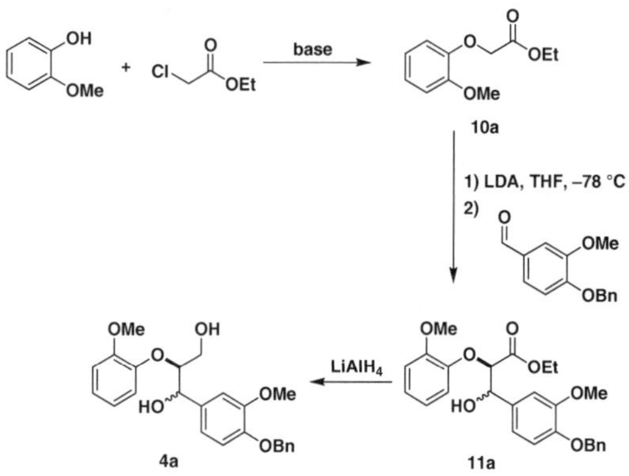

pure *threo* dilignols were only attainable in low yields due to the unfavored formation and difficult purification of the corresponding *threo* β-hydroxy esters in the preceding step.

The protocol of Nakatsubo served as starting point for the development of an improved synthesis of β-*O*-4 dilignol model compounds which made both *erythro* and *threo* dilignols easily accessible in almost equal yields [41].

Initially, we tried to optimize the previous reaction conditions [19, 22–26]. Nakatsubo reported the formation of ester **10a** in 90 % yield after 170 h at room temperature starting from commercially available ethyl chloroacetate. After a screening of the reaction conditions the reaction time could be significantly reduced while maintaining high yields with 90 %. The reaction was performed starting from ethyl bromoacetate and 2-methoxyphenol in the presence of K_2CO_3 for 8 h under reflux (Fig. 12). The yields remained constant both on a millimol and multi gram scale [41].

Ester **10a** was then transformed into the corresponding lithium enolate using LDA at −78 °C and various functionalized aromatic aldehydes were tested for their reaction behavior. Without exception they formed *erythro*-rich diastereomeric mixtures with the *erythro*:*threo* ratios of β-hydroxy esters **11** ranging from 3:1 to 5:1 which is in accordance with the literature [19, 22–26]. β-Hydroxy ester **11b** was obtained as a solid that could be recrystallized in ethyl acetate to afford *erythro* ester **11bE** in 52 % yield and an *erythro*/*threo* diastereomer ratio of >98:2. In this way we were able to achieve higher yields of diastereomerically pure ester **11bE** than previously described in literature in a more convenient manner while being applicable for multi gram syntheses. Likewise, β-hydroxy ester **11a** was isolated in its pure *erythro* diastereomeric form (*erythro*/*threo* > 98:2) after crystallization in 30 % yield. However, after column chromatography esters **11c–e** were only attainable as pure *erythro* diastereomers in low yields, and the corresponding pure *threo* diastereomers remained unavailable.

The reduction of diastereomerically pure β-hydroxy ester **11bE** was tested with $LiAlH_4$ and $NaBH_4$ (Fig. 13) [42]. Both reactions yielded dilignol **4bE** in 90 % without change in the stereochemistry.

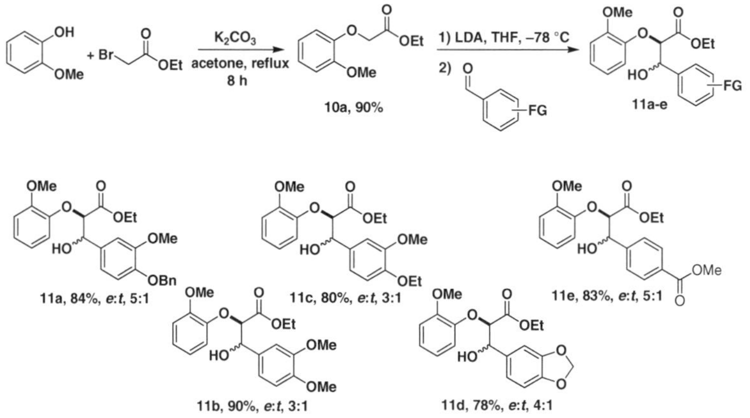

Fig. 12 Synthesis of *β*-hydroxy esters **11a–e**

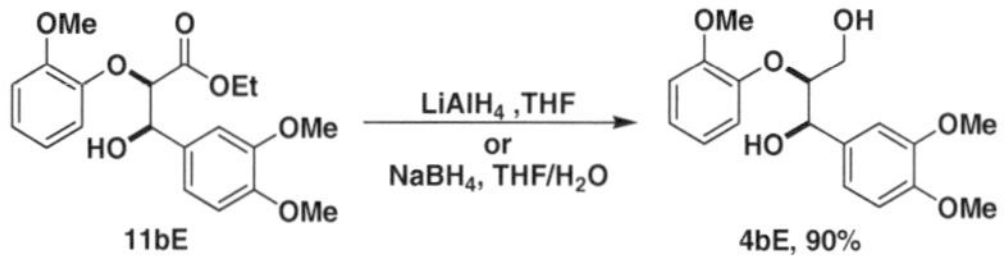

Fig. 13 Reduction of *erythro* *β*-hydroxy ester **4bE**

With regard to the fact that all *β*-hydroxy esters stemming from ethyl aryloxy ester **10a** afforded *erythro* rich diastereomeric mixtures and bearing in mind that for the corresponding methyl aryloxy ester derivatives the same selectivities are described in literature the focus was shifted to sterically more hindered aryloxy esters to potentially enhance the formation of *threo* *β*-hydroxy esters. Starting from ethyl aryloxy ester **10a**, acid **12** was obtained in 99 % after saponification. Acid **12** served as starting material for both the synthesis of 1-adamantyl and *tert*-amyl aryloxy esters **10b** and **10c** (Fig. 14).

Subsequent 1,2-addition of the lithium enolate of ester **10b** with veratraldehyde yielded a 1:1 mixture of the *erythro* and *threo* diastereomers **13bE** and **13bT** in 80 % overall yield (Fig. 15). To our delight both diastereomers could be completely separated from each other by column chromatography using a gradient of elution. The corresponding *tert*-amyl aryloxy ester **10c** also led to the formation of both

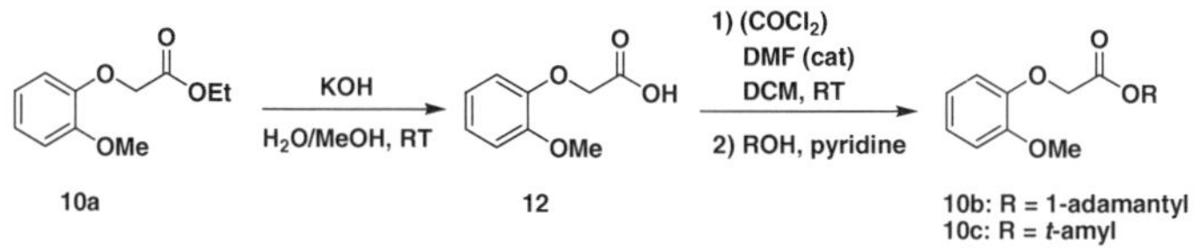

Fig. 14 Synthesis of aryloxy esters **10b** and **10c**

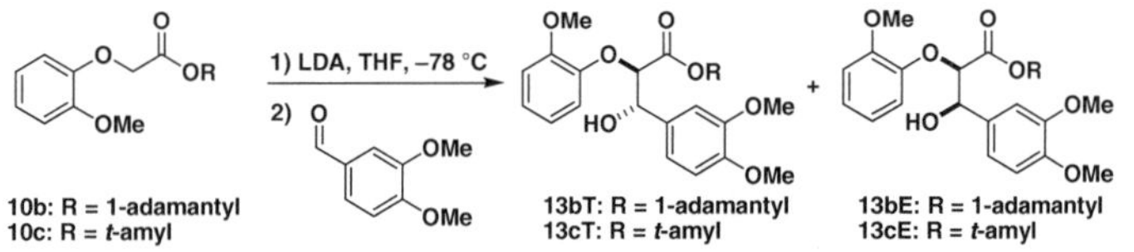

Fig. 15 Synthesis of *erythro* and *threo* β-hydroxy esters **13b** and **13c**

diastereomers **13cE** and **13cT** in an equimolar ratio and with an overall yield of 75 %. Once again both diastereomers were completely separated by column chromatography [41].

We then decided to test the reactivity of the more easily accessible *tert*-butyl-aryloxy ester **10d** (Fig. 16). It was synthesized in one step from commercially available *tert*-butyl bromoacetate and 2-methoxyphenol in 93 % yield. The subsequent addition with vertraldehyde afforded a 1:1 mixture of *erythro* and *threo* *tert*-butyl β-hydroxy esters **14bE** and **14bT** in 90 % overall yield. Moreover, both diastereomers were completely separated by column chromatography.

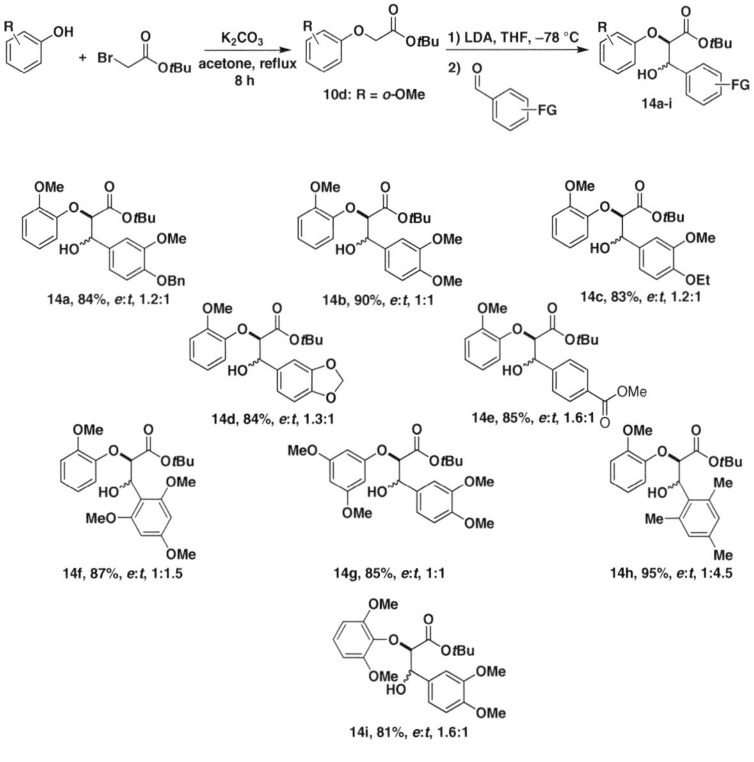

Fig. 16 Synthesis of β-hydroxy esters **14a–i**

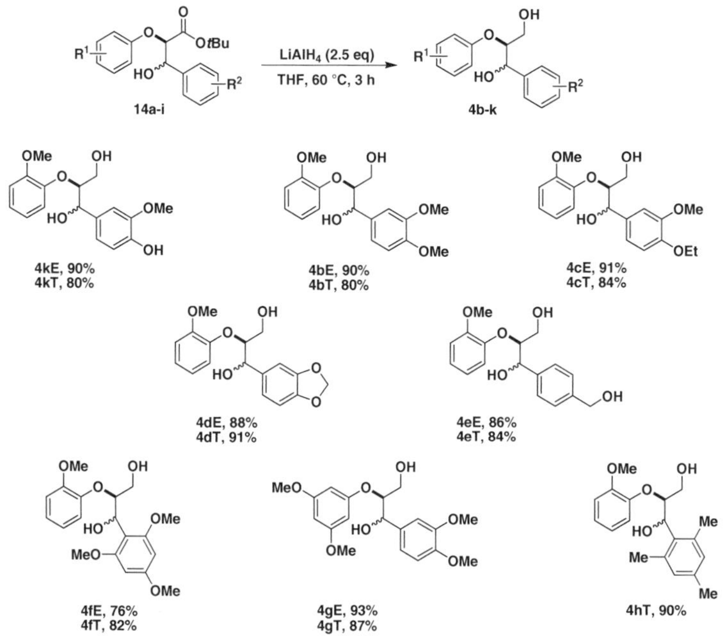

Fig. 17 Reduction of the *tert* butyl β-hydroxy esters **14**

With the optimized reaction conditions in hand the scope of the reaction was expanded using *tert*-butyl ester **10d** and various differently functionalized benzaldehyde derivatives. *tert*-butyl-β-hydroxy esters **14a–i** were obtained in good to excellent yields. In most cases the *erythro* diastereomers were formed in slight excess except for esters **14f** and **14h**, which results from sterically hindered aldehydes favoring the *threo* transition state in the 1,2-addition. Ester **14i** was the only example in which the two diastereomers could not be separated from each other by column chromatography or crystallization.

In the last step the diastereomerically pure *tert*-butyl β-hydroxy esters **14** were reduced to the desired dilignols **4** with $LiAlH_4$ in very good yields and with total retention of the stereochemistry (76–93 %, Fig. 17) [41].

3 Summary

During the course of this study we were able to establish an improved synthesis for *erythro* and *threo* β-*O*-4 dilignol model compounds, the results of which were accepted for publication in the journal “Chemistry-A European Journal” [41]. These model compounds are currently successfully applied in transition metal catalyzed lignin cleavage studies.

Acknowledgments The authors are grateful to the Fonds der Chemischen Industrie and the Cluster of Excellence (Tailor-Made Fuels from Biomass) funded by the Excellence Initiative of the German federal and state governments for their financial support. J.M. thanks the NRW Graduate School BrenaRo for a predoctoral stipend.

References

1. Perlack, R.D., Wright, L.L., Turhollow, A.F., Graham, R.L., Stokes, B.J., Erbach, D.C.: Biomass as feedstock for a bioenergy and bioproducts industy: the technical feasibility of a billion-ton annual supply (2005). doi:10.2172/885984
2. Hicks, J.C.: Advances in C-O bond transformations in lignin-derived compounds for biofuels production. J. Phys. Chem. Lett. **2**, 2280–2287 (2011). doi:10.1021/jz2007885
3. Chheda, J.N., Huber, G.W., Dumesic, J.A.: Liquid-phase catalytic processing of biomass-derived oxygenated hydrocarbons to fuels and chemicals. Angew. Chem. Int. Ed. **46**, 7164–7183 (2007). doi:10.1002/anie.200604274
4. Corma, A., Iborra, S., Velty, A.: Chemical routes for the transformation of biomass into chemicals. Chem. Rev. **107**, 2411–2502 (2007). doi:10.1021/cr050989d
5. Huber, G.W., Iborra, S., Corma, A.: Synthesis of transportation fuels from biomass: chemistry. Catal. Eng. Chem. Rev. **106**, 4044–4098 (2006). doi:10.1021/cr068360d
6. Huber, G.W., Corma, A.: Synergies between bio- and oil refineries for the production of fuels from biomass. Angew. Chem. Int. Ed. **46**, 7184–7201 (2007). doi:10.1002/anie.200604504
7. Mäki-Arvela, P., Holmbom, P., Salmi, T., Murzin, D.Y.: Recent progress in synthesis of fine and specialty chemicals from wood and other biomass by heterogeneous catalytic processes. Catal. Rev. Sci. **49**, 197–340 (2007). doi:10.1080/01614940701313127
8. Amen-Chen, C., Pakdel, H., Roy, C.: Production of monomeric phenols by thermochemical conversion of biomass: a review. Bioresour. Technol. **79**, 277–299 (2001). doi:10.1016/S0960-8524(00)00180-2
9. Zakzeski, J., Bruijnincx, P.C.A., Jongerius, A.L., Weckhuysen, B.M.: The catalytic valorization of lignin for the production of renewable chemicals. Chem. Rev. **110**, 3552–3599 (2010). doi:10.1021/cr900354u
10. Capanema, E.A., Balakshin, M.Y., Kadla, J.F.: A comprehensive approach for quantitative lignin characterization by NMR spectroscopy. J. Agric. Food Chem. **52**, 1850–1860 (2004). doi:10.1021/jf035282b
11. Capanema, E.A., Balakshin, M.Y., Kadla, J.F.: Quantitative characterization of a hardwood milled wood lignin by nuclear magnetic resonance spectroscopy. J. Agric. Food Chem. **53**, 9639–9649 (2005). doi:10.1021/jf0515330
12. Chakar, F.S., Ragauskas, A.J.: Review of current and future softwood kraft liginin process chemistry. Ind. Crops Prod. **20**, 131–141 (2004). doi:10.1016/j.indcrop.2004.04.016
13. Holmgren, A., Brunow, G., Henriksson, G., Zhang, L., Ralph, J.: Non-enzymatic reduction of quinone methides during oxidative coupling of monolignols: implications for the origin of benzyl structures in lignins. J. Org. Biomol. Chem. **4**, 3456–3461 (2006). doi:10.1039/b606369a
14. Rencoret, J., Gisela, M., Gutiérrez, A., Nieto, L., Jiménez-Barbero, J., Martínez, Á.T., del Río, J.C.: Isolation and structural characterization of the milled-wood lignin from Paulownia fortunei wood. Ind. Crops Prod. **30**, 137–143 (2009). doi:10.1016/j.indcrop.2009.03.004
15. Sakakibara, A.: A structural model of softwood lignin. Wood Sci. Technol. **14**, 89–100 (1980). doi:10.1007/BF00584038
16. Zuidema, E., Bolm, C.: A Catalytic Approach to the Depolymerization of Lignin. Poster at The Netherlands conference on chemistry and catalysis, 264 Noordwijkerhout, Netherlands, March 2 (2009)

17. Zuidema, E., Bolm, C.: A Catalytic Approach to the Depolymerization of Lignin. CaRLa Winterschool, Heidelberg, Germany, March 14–20 (2009)
18. Johansson, A.J., Zuidema, E., Bolm, C.: On the mechanism of ruthenium-catalyzed formation of hydrogen from alcohols: a DFT study. Chem. Eur. J. 16:13487–13499. doi:10.1002/chem.201000593
19. Pardini, V.L., Smith, C.Z., Utley, J.H.P., Vargas, R.R., Viertler, H.: Electroorganic reactions. 38. Mechanism of electrooxidative cleavage of lignin model dimers. J. Org. Chem. **56**, 7305–7313 (1991). doi:10.1021/jo00026a022
20. Lee, J.C., Bae, Y.H., Chang, S.-K.: Efficient β-halogenation of carbonyl compounds by N-bromosuccinimide and N-chlorosuccinimde. Bull. Korean Chem. Soc. **24**, 407–408 (2003). doi:10.5012/bkcs.2003.24.4.407
21. Nichols, J.M., Bishop, L.M., Bergman, R.G., Ellman, J.A.: Catalytic C-O bond cleavage of 2-Aryloxy-1-arylethanols and its application to the depolymerization of lignin-related polymers. J. Am. Chem. Soc. **132**, 12554–12555 (2010). doi:10.1021/ja106101f
22. Baciocchi, E., Fabbri, C., Lanzalunga, O.: Lignin peroxidase-catalyzed oxidation of nonphenolic trimeric lignin model compounds: fragmentation reactions in the intermediate radical cations. J. Org. Chem. **68**, 9061–9069 (2003). doi:10.1021/jo035052w
23. Cho, D.W., Parthasarathi, R., Pimentel, A.S., Maestas, G.D., Park, H.J., Yoon, U.C., Dunaway-Mariano, D., Gnanakaran, S., Langan, P., Mariano, P.S.: Nature and kinetic analysis of carbon-carbon bond fragmentation reactions of cation radicals derived from SET-oxidation of lignin model compounds. J. Org. Chem. **75**, 6549–6562 (2010). doi:10.1021/jo1012509
24. Cho, D.W., Latham, J.A., Park, H.J., Yoon, U.C., Langan, P., Dunaway-Mariano, D., Mariano, P.S.: Regioselectivity of enzymatic and photochemical single electron transfer promoted carbon-carbon bond fragmentation reactions of tetrameric lignin model compounds. J. Org. Chem. **76**, 2840–2852 (2011). doi:10.1021/jo200253v
25. Ciofi-Baffoni, S., Banci, L., Brandi, A.: Synthesis of oligomeric mimics of lignin. J. Chem. Soc. Perkin Trans. **19**, 3207–3217 (1998). doi:10.1039/A805027I
26. Nakatsubo, F., Sato, K., Higuchi, T.: Synthesis of guaiacylglycerol-β-guaiacyl ether. Holzforschung **29**, 165–168 (1975). doi:10.1515/hfsg.1975.29.5.165
27. Adler, E., Lindgren, B.O., Saedlén, U.: The beta-guaiacyl ether of alpha-veratrylglycerol as a lignin model. Svensk Papperstidning **55**, 245–254 (1952)
28. Ahvonen, T., Brouno, G., Kristersson, P., Lundquist, K.: Stereoselective syntheses of lignin model compounds of the β-*O*-4 and β-1 types. Acta Chem. Scand. **B37**, 845–849 (1983)
29. Adler, E., Eriksoo, E.: Guaiacylglycerol and its β-guaiacyl ether. Acta Chem. Scand. **9**, 341–342 (1955)
30. Kawai, S., Okita, K., Sugishita, K., Tanaka, A., Ohashi, H.: Simple method for synthesizing phenolic β-*O*-4 dilignols. J. Wood Sci. **45**, 440–443 (1999). doi:10.1007/BF01177919
31. Kishimoto, T., Uraki, Y., Ubukata, M.: Easy synthesis of β-*O*-4 type lignin related polymers. Org. Biomol. Chem. **3**, 1067–1073 (2005). doi:10.1039/B416699J
32. Kratzl, K., Kisser, W., Gratzl, J., Silbernagel, H.: Der β-Guajacyläther des Guajacylglycerins, seine Umwandlung in Coniferylaldehyd und verschiedene andere Arylpropanderivate. Monatsh. Chem. **90**, 771–782 (1959)
33. Yokoyama, T., Matsumoto, Y.: Revisiting the mechanism of β-*O*-4 bond cleavage during acidolysis of lignin. Part 1: kinetics of the formation of enol ether from non-phenolic C_6–C_2 type model compounds. Holzforschung **62**, 164–168 (2008). doi:10.1515/HF.2008.037
34. Berndtsson, I., Lundquist, K.: On the synthesis of lignin model compounds of the arylglycerol-β-aryl ether type. Acta Chem. Scand. **B31**, 725–726 (1977)
35. Chen, X., Ren, X., Peng, K., Pan, X., Chan, A.S.C., Yang, T.K.: A facile enantioselective approach to neolignans. Tetrahedron Asymmetry **14**, 701–704 (2003). doi:10.1016/S0957-4166(03)00085-5
36. Gratzl, J., Fried-Matzka, M., Miksche, G.E.: Two diastereomeric forms of guaiacylglycerol β-(2-methoxyphenyl) ether and of guaiacylglycerol. Acta Chem. Scand. **20**, 1038–1043 (1966)
37. Lundquist, K., Remmerth, S.: New synthetic routes to lignin model compounds of the arylglycerol-β-aryl ether type. Acta Chem. Scand. **B29**, 276–278 (1975)

38. Pearl, I.A., Gratzl, J.: Lignin and related products. XVI. Synthesis of lignin model compounds of the phenylglycerol *β*-ether and related series. J. Org. Chem. **27**, 2111–2114 (1962). doi:10.1021/jo01053a051
39. Helm, R.F., Ralph, J.: Stereospecificity for the zinc borohydride reduction of α-aryloxy-*β*-hydroxy ketones. J. Wood Chem. Technol. **13**, 593–601 (1993). doi:10.1080/02773819308020536
40. Helm, R.F., Li, K.: Complete threo-stereoselectivity for the preparation of *β*-O-4 lignin model dimers. Holzforschung **49**, 533–536 (1995)
41. Buendia, J., Mottweiler, J., Bolm, C.: Preparation of diastereomerically pure dilignol model compounds. Chem. Eur. J. **17**, 13877–13882 (2011). doi:10.1002/chem.201101579
42. Son, S., Toste, F.D.: Non-oxidative vanadium-catalyzed C–O bond cleavage: application to degradation of lignin model compounds. Angew. Chem. Int. Ed. **49**, 3791–3794 (2010). doi:10.1002/anie.201001293

Feasibility Study of Auto Thermal Reforming of Biogas for HT PEM Fuel Cell Applications

Nan Kishore Nalluraya, Heinrich Köhne, Stephan Köhne and Martin Konrad

Abstract Biogas is easily transportable, storable and CO_2 neutral. The scope of this work is to show the feasibility of H_2 production through Auto Thermal Reforming (ATR) of biogas and its utilisation in High Temperature Proton Exchange Membrane Fuel Cell (HT-PEM FC) application. In this study, a fuel cell system was modelled and simulated using MATLAB/Simulink to find the feasibility of biogas reforming for High Temperature Proton Exchange Membrane (HT PEM) Fuel Cell Application. The main criteria considered are high yield of H_2 and lowest possible CO. The optimum temperature, Steam to Carbon Ratio (SCR), Air-Fuel Ratio (AFR), and reforming temperature were found out with the help of simulation. A test reference 5 kWth Auto Thermal Reformer (ATR) with Water Gas Shift (WGS) reactor was built according to the simulation guidelines in order to produce H_2 to feed a 1 kW_{el} HT-PEM Fuel Cell with anode gas. The reliability and the durability of the system were tested with a start-and-stop strategy and a continuous mode respectively. The electrical efficiency of the whole Fuel Cell system was simulated to around 30 %. The experimental work validated the simulation results within acceptable margins. The experimental study shows that it is not only feasible to produce on-board H_2 with biogas but also that the start-and-stop mode of operation does not damage the fuel cell which makes it even suitable for automotive application.

N.K. Nalluraya
Fakultaet fuer Georessourcen und Materialtechnik, RWTH Aachen University, Intzestrasse 1, 52072 Aachen, Germany
e-mail: nan.kishore.nalluraya@rwth-aachen.de

H. Köhne (✉)
Energie- und Stofftransport, RWTH Aachen University, Kopernikusstrasse 10, 52074 Aachen, Germany
e-mail: hkoehne@gmx.de

S. Köhne · M. Konrad
FCPower Fuel Cell Power GmbH, Schumanstrasse 18d, 52146 Wuerselen, Germany

M. Klaas et al. (eds.), *Fuels From Biomass: An Interdisciplinary Approach*,
Notes on Numerical Fluid Mechanics and Multidisciplinary Design 129,
DOI 10.1007/978-3-662-45425-1_9

1 Introduction

With the rising awareness of global warming and its effect on the environment, the scientific community has started to take steps to mitigate or to prolong the adversity. One of the various ways to do so is to increase the efficiency of the energy converters such as motors and generators and another way is to use CO_2 neutral fuels. The fuel cell, first demonstrated by Sir William Grove in 1839, promises clean and efficient power generation. Although a typical fuel cell runs on hydrogen and oxygen, the practicality of the hydrogen network limits the size of the fuel cell. Therefore there is a need for a fuel cell system that combines fuel processing to produce hydrogen on site from easily available fuels which can be any of the hydrocarbons with high H/C ratio. Biogas which mainly contains 50–75 % methane is CO_2 neutral. By driving a fuel cell with Biogas, one not only has an efficient but also environmental-friendly power generation.

The aim of this research is to find the feasibility of auto thermal reforming biogas to drive a 1 kW High Temperature Proton Exchange Membrane (HT PEM) fuel cell.

2 Theory

Brief theoretical descriptions of the fuel and the components of the fuel cell system are given below.

2.1 Biogas and Its Processing

The term Biogas generically refers to any gas derived from anaerobic degradation of biomass. Biomass is any microbiologically degradable substrate such as manure (cattle and pig dung and solid manure, poultry manure) and organic waste from agro-industry (fruit and vegetable waste, grass cuttings, grains, pulp, mash), and food industry (animal products, oils and fats) or energy crops (renewable resources).

Anaerobic biogas synthesis takes place in four steps, namely hydrolysis, acidogenesis, acetogenesis and methanogenesis. Hydrolysis is characterised by the breaking the long chains of the polymeric proteins, fats and carbohydrates down into simpler monomer units such as amino acids, fatty acids and sugars with the help of enzymes released by the specific bacteria. These monomers are further broken down to lower fatty acids (Acetic acid, Propanioc acid, Butyric acid) carbon dioxide and hydrogen along with traces of lactic acid and hydrogen in next step, Acidogenesis. In the third step, Acetogenesis, enzymes produced by specific bacteria convert the product from acidogenesis to forerunners of biogas namely acetic acid, acetates, hydrogen and carbon dioxide. In the last process, Methanogenesis,

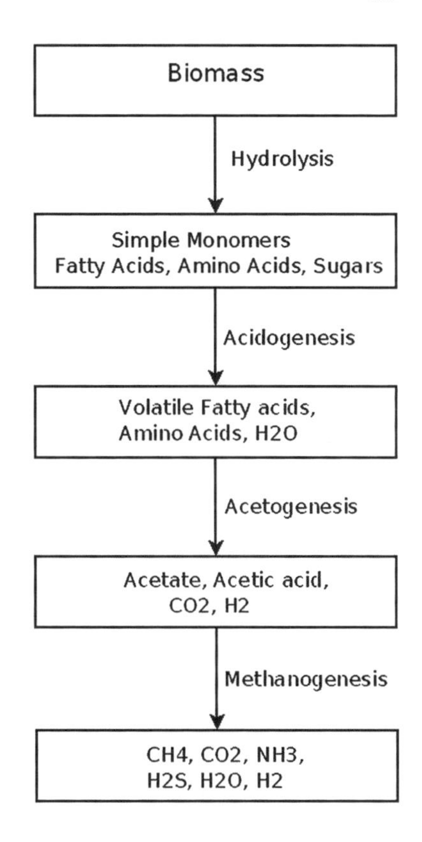

Fig. 1 Schematic flow diagram of anaerobic digestion [1]

biogas which mainly consists of methane is produced by the enzymatic reaction of the biogas forerunners. Figure 1 shows a schematic flow diagram of the biogas synthesis. Depending upon the choice of the Biomass, the gas composition may vary from 50–80 % Methane (CH_4), 20–50 % Carbon Dioxide (CO_2), 1–5 % Water Vapour (H_2O) and small amounts of Hydrogen Sulphide (H_2S), Ammonia (NH_3) and Hydrogen (H_2) [1].

Unprocessed biogas is a water vapour saturated mixture mainly of CH_4, CO_2, and traces of H_2S, NH_3 and H_2. The amount of water content is dependent on the temperature. The water in the gas can condense in the pipeline blocking the pipeline, and along with other constituents can also attribute to the formation of corrosive acidic solutions.

H_2S is a very poisonous gas whose maximum concentration at workplace shall not exceed 5 ppm in the EU [2]. Aqueous solution of H_2S is acidic and promotes corrosion of the metals in the pipeline and equipments. Moreover sulphur is a very well known catalyst poison. Sulphur reacts with the metals in the catalyst to form metal sulphide which are highly stable thus reducing the catalytic activity of the metals. Sulphur can also chemisorb onto active catalyst thus preventing reactant access. As the Reformer and the fuel cell have their own catalyst, desulphurisation of the biogas to ppb levels is absolutely necessary to prevent catalyst poisoning.

Although CO_2 has no significant part in environmental impact, it lowers the calorific value of the gas.

The component of our interest in Biogas for driving the fuel cell system is methane. In order to render Biogas compatible for a fuel cell system, the unwanted and poisonous components must be removed and thus enrich methane in various processes briefly discussed below.

2.1.1 Filtration and Drying

The solid particles and liquid part of the gas are removed from the biogas by filtration to prevent the dirt and oil clogging and limiting the efficiency of the next processes. The drying of biogas can be accomplished with the help of adsorption with silica gel and activated carbon, condensation and absorption by glycol. Adsorptive Silica gel process can adsorb water up to 40 wt% and is relatively economic. Condensation process can bring dew point to 2 °C and the absorption by glycol can separated higher hydrocarbons along with water. The choice of water separation process depends upon the end-use of the gas.

2.1.2 Desulphurisation

Although desulphurisation is a very important process, the choice of the process depends on the end-use of the biogas. The various types of desulphurisation are biological, chemical adsorption and sorption catalyst processes.

In biological process, microorganisms which oxidise H_2S are used to reduce the sulphur content at a gross level. This process, for example, can reduce the sulphur content to allowable limits for engine applications.

Reaction of H_2S with iron salts to precipitate sulphide, with iron oxides and subsequent oxidation to produce elemental sulphur, catalytic oxidation on with activated carbon and reaction with zinc oxide to form zinc sulphide are chemical desulphurisation processes to name a few.

A most important chemical process is with reaction with zinc oxide due to the tits desulphurisation capacity to under ppb ranges which is suitable for fuel cell system application [3].

2.1.3 CO_2 Separation and Methane Enrichment

Although in fuel cell application CO_2 acts as an inert chemical gas, its presence in the gas mixture substantially reduces the heating value of the bio gas and increases the specific heat capacity. Depending upon the gas mixture a CO_2 separation may or may not be necessary.

One of the common processes for CO_2 separation is Pressure Swing Adsorption (PSA). The principle of this process is the fact that, some gases tend to adsorb more readily and strongly on to solid surfaces than others. (e.g. CO_2 gets more readily adsorbed on special adsorbent like Zeolites than CH_4.) Higher the pressure higher is the adsorption affinity of the gases. After the adsorbent reaches near saturation with the gas, the process then swings to low pressure to desorb the gas from the adsorbent. Industrial practice is to have more than one pressure swing units so that while one unit is desorbing another unit can adsorb gases to give a near-continuous process.

Another most widespread process is high pressure water scrubbing. The difference in dissolubility of various gases in water is the principle of this process. The acidic and basic components like CO_2, H_2S and NH_3 are more soluble in water than non-polar, hydrophobic hydrocarbons. The dissolubility increases with the increase in pressure. The cleaned biogas is water vapour saturated at the absorption temperature and needs to be dried. Compared to other processes, high pressure water scrubbing requires only water absorption column along with compressors and heat exchangers and no expensive chemicals and this makes it relatively economic.

Other processes for CO_2 separation are membrane separation, Selexol process and cryogenic process only to name a few.

It is well practiced and documented that methane can be enriched up to 98 % in biogas [3].

2.2 *Fuel Cell System*

A fuel cell system consists of mainly the fuel cell itself with other components such as Reformer, heat exchangers, off-gas burners and controlling systems. Fuel cell converts the chemical energy of the reaction to electrical energy whereas the rest of the system perform the important functions such as reformer processing the fuel to feed the required amount of the anode gas, heat exchanger recovering energy from the hot exhaust gas back to the system and regulating energy before each component and electronic system controlling the various valves and temperatures of the whole system. The functionality of the important components is described briefly below.

2.2.1 Reformer

Reforming in general, is a chemical process to convert higher hydrocarbons to lower hydrocarbons. However reforming fuel for fuel cells is a process of splitting the hydrocarbons to produce H_2 as the main product. Reforming can be classified into steam reforming (SR), partial oxidation reforming (POX) and auto thermal reforming (ATR).

As the name indicates, SR splits the hydrocarbon with the help of steam. It is an endothermic process. The generic equation of SR is

$$C_xH_y + xH_2O \rightarrow xCO + (\frac{y}{2} + x)H_2 \tag{1}$$

With it, an associated water gas shift reaction also takes place namely,

$$CO + H_2O \xrightarrow{-41\,\mathrm{kJ/mol}} CO_2 + H_2 \tag{2}$$

This means that the overall gas will constitute of H_2, CO and CO_2. From Le Chatelier's principle we know that keeping the pressure of the reaction low will favour the forward reaction in Eq. 1 and pressure has no effect on the Eq. 2. The endothermic nature of the Eq. 1 means that the forward reaction is favoured at higher temperatures. Care must be taken so as to avoid carbon formation. Carbon formation can take place by pyrolysis,

$$C_xH_y \rightarrow xC + \frac{y}{2}H_2 \tag{3}$$

and Boudouard reaction

$$2CO \rightarrow C + 2CO_2 \tag{4}$$

Increasing the steam at the feed we can reduce carbon formation by reduction of partial pressure of CO and promoting carbon gasification reaction (5).

$$C + H_2O \rightarrow CO + 2H_2 \tag{5}$$

The minimum amount of steam required to avoid carbon formation may be calculated. It is an industrial practice to use an SCR of 2–3 as to avoid carbon deposition with a margin of safety [4].

Partial oxidation on the other hand is exothermic and the generic reaction is as follows,

$$C_xH_y + \frac{x}{2}O_2 \rightarrow xCO + \frac{y}{2}H_2 \tag{6}$$

Partial oxidation can be carried out at high temperatures (around 1,200–1,500 °C) and without catalysts, making the removal of sulphur compounds not necessary for the process although when used in a fuel cell system, in the subsequent processes it might be necessary to remove sulphur. POX can reform heavier petroleum fractions than catalytic processes and widely used in diesel and residual fraction reforming. If the temperature is reduced and a catalyst employed the process would be know as catalytic POX (CPOX). It can be seen from Eqs. 1 and 6 that POX reforming yields less hydrogen that SR.

Auto thermal reforming, also commonly used in fuel processing is neither endothermic nor exothermic. In this process both steam and oxygen (air) is fed with the fuel to a catalytic reactor. Thus it can be considered to be a combination of both steam reforming and partial oxidation. At a balanced state the exothermic POX provides the energy to the endothermic steam reforming. ATR also has an intermediate hydrogen yield between that of SR and POX [4].

In this study an ATR is chosen due its advantage over SR at being non endothermic and over POX at yielding more hydrogen.

2.2.2 Water Gas Shift Reaction

To increase the yield of hydrogen in the reformate gas a two stage Water Gas Shift (WGS) reactor is incorporated. The WGS reaction is given by the Eq. 2. The thermodynamics of the reaction are so as to favour the backward reaction at higher temperatures. Therefore cooling the reformate gas before passing it through a catalytic reactor will promote the forward reaction. The iron-chromium catalyst is found to be effective at temperatures of 400–500 °C. This could be called a High Temperature Shift Reaction (HTS) and further cooling it to ranges of 200–250 °C and passing through a reactor with a copper catalyst will further reduce the CO content of the reformate to under 0.5 % which is allowable for a High Temperature Proton exchange Membrane (HT PEM) Fuel Cell. Although the HT PEM membranes operating above 150 °C can withstand a CO concentration of up to 1 % [5], it was taken as an experimental parameter to keep the CO under 0.5 % so as not to irreversibly damage the system. It is necessary to mention that the shift reactor catalysts are easily poisoned by sulphur.

2.2.3 HT PEM FC

In this study a 1 kW_{el} HT PEM fuel cell is used. The basic principle of a proton exchange membrane (PEM) fuel cell is that the membrane separating cathode and anode gas chambers allow only proton transfer and the electron split from the hydrogen atoms have to travel the electric circuit to the anode.

Figure 2 shows a schematic diagram of PEM fuel cell. An HT PEM fuel cell's operating temperature range is 100–200 °C which is clearly above the operating temperature of a normal PEM fuel cell. The voltage of a fuel cell is usually very small, under 1 V, which means to produce a useful voltage the fuel cell have to be connected in series. Such a series of fuel cell is known as a Stack [4].

3 Modelling the System with Matlab/Simulink

The whole system is modelled using Matlab/Simulink. A schematic diagram of the Fuel cell system modelled in Matlab/Simulink is shown in Fig. 3. A brief description of the subsystems of the model is given below.

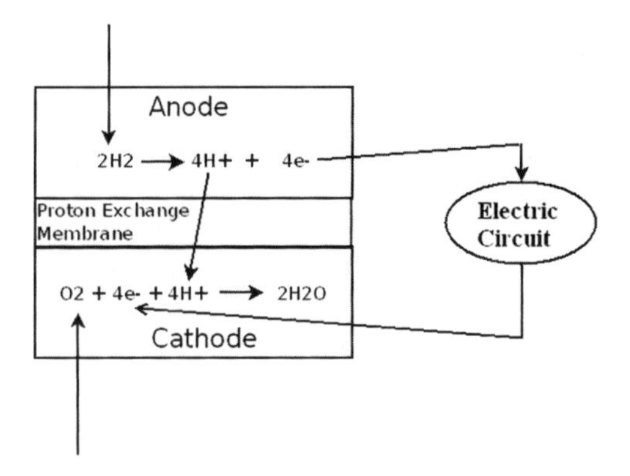

Fig. 2 Schematic diagram of a proton exchange membrane

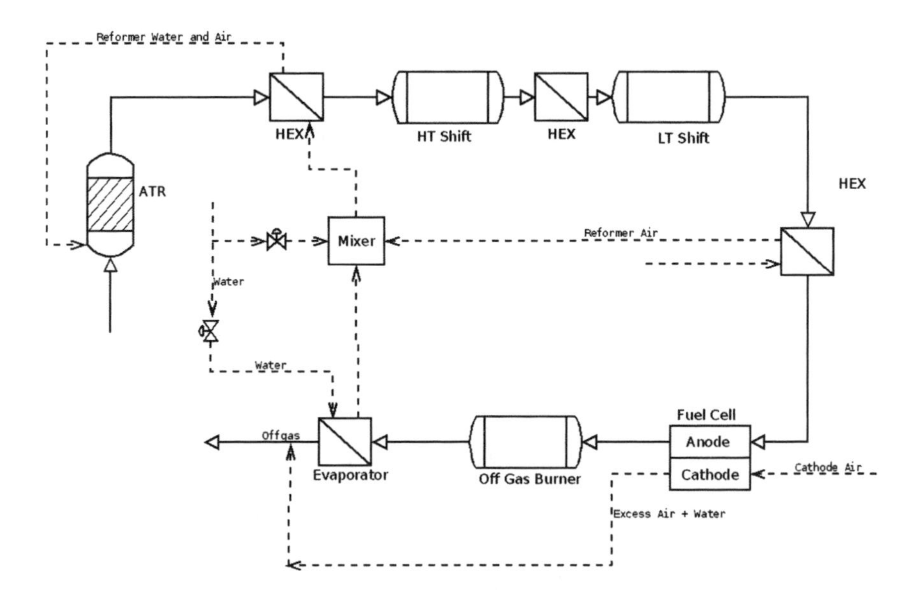

Fig. 3 Schematic flow diagram of the fuel cell system

3.1 Auto Thermal Reformer (ATR)

It is assumed that at the operating temperature the reaction proceeds is adiabatic and reaches the equilibrium. To model the ATR in Simulink, basics of minimization of Gibbs' free energy of a reaction (NASA Method) is used. It is assumed that the mixture of the gas with air and steam at equilibrium at the operating temperature and pressure. The composition of the mixture is calculated at this point. The Gibbs free energy of a mixture is given by

$$\frac{G}{RT} = \sum_{i=1}^{n} \left[\frac{x_i G_i^0}{RT} + x_i \ln \frac{x_i}{\sum x_i} + x_i \ln p \right] \quad (7)$$

G_i^0 is the molar free energy at 1 atmosphere of the species i
R is the Universal gas constant
n is the number of species
T is the temperature and
p is the pressure

at equilibrium *G/RT* is at minimum if the elemental composition is fixed.

The elemental constraint equation is given by

$$\sum_{i=1}^{n} a_{ij} x_i - b_j = 0 \quad (8)$$

a_{ij} is the number of atoms of element j in species i and
b_j is the total molar concentration of atoms of element j. this is given by the reactant concentrations
j ranges from 1 to m (m = number of elements)

Minimisation of *G/RT* with the elemental constraint equation is done using method of Lagrange multipliers λ_j (j = 1 to m) and the new function F defined as

$$F = \sum_{i=1}^{n} \left[\frac{x_i G_i^0}{RT} + x_i \ln \frac{x_i}{\sum x_i} + x_i \ln p \right] + \sum_{j=1}^{m} \lambda j \sum_{i=1}^{n} \left(a_{ij} x_i - b_j \right) \quad (9)$$

The solution for the function is found when F is a minimum, in other words all the partial derivates of F are null.

$$\left. \begin{aligned} \frac{\partial F}{\partial x_i} &= \frac{G_i^0}{RT} + \ln \frac{x_i}{\sum x_i} + \ln p - \sum_{j=1}^{m} \lambda_j a_{ij} \\ \frac{\partial F}{\partial \lambda_i} &= \sum_{i=1}^{n} a_{ij} x_i - b_j \end{aligned} \right\} = 0 \quad (10)$$

This is a system of $(n + m)$ non-linear equations with $(n + m)$ variables which can be solved iteratively to find the equilibrium composition of the reaction.

3.2 Water Gas Shift Reactor

Equation 2 shows the main reaction in the water gas shift reaction. In this model the methane formation is not considered although it is shown in the lab experiments. The WGS reaction is done in two stages, high temperature and low temperature. Literatures show different temperature ranges for high and low temperature shift reaction [6, 7]. It can be safely assumed to that high temperature range is 573–673 K and low temperature range is 423–573 K. The water gas shift reactor is slightly exothermic. This means that as the reaction proceeds the temperature increases and thus decreasing the equilibrium constant K_{eq}.

Equilibrium constant K_{eq} as a function of temperature for this reaction is given by [8]

$$K_{eq} = \exp\left(\frac{4577.8\,\mathrm{K}}{T} - 4.33\right) \quad (11)$$

A validated empirical rate expression for the consumption of CO can also be found in the literature [8] as

$$r_{CO} = 2.96 \times 10^5 \exp\left(-\frac{47{,}400\,\mathrm{J/mol}}{RT}\right)\left(p_{CO}p_{H_2O} - \frac{p_{CO_2}p_{H_2}}{K_{eq}}\right) \quad (12)$$

The WGS reactor is modelled as a plug flow reactor with 100,000 control volumes (CV) along the length. Such a high number of CV is chosen so as to avoid the reaction progressing backwards. Once the equilibrium is reached the gases do not react with each other and thus it is safe to break the loop. Figure 4 shows the effect of too less number of CV (1,000) on the exit temperature of the CVs. The graphs show temperature at the exit of each CV for same composition of the gas mixture with n = 1,000 for the left and n = 100,000 for the right graph. It clearly

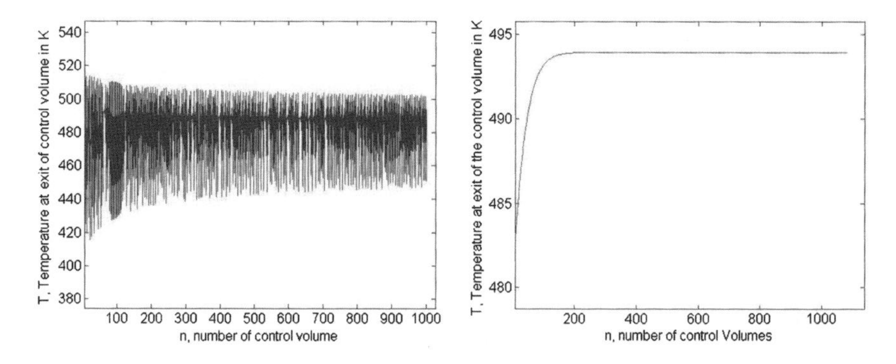

Fig. 4 Temperature fluctuations in a LTS reactor model due to insufficient number of CV

shows that the reaction in the left reactor model proceeds intermittently forward and backward never reaching equilibrium even at the end of the reactor whereas the right reactor model reaches equilibrium at 1,079th CV and thus breaking the loop.

3.3 Fuel Cell

The fuel cell modelled in this system is a High Temperature Proton Exchange Membrane FC which indicates that this FC works well above the 100 °C temperature. Unlike normal PEM FC which works under 100 °C, there is gas humidification necessary for HT PEM FC and it can tolerate up to 1 % CO and 10 ppm SO_2 allowing simpler fuel processing system [5].

The electromotive force EMF or reversible open circuit voltage of a hydrogen fuel cell is given by

$$EMF = \frac{-\Delta g_f}{2\,\mathrm{F}} \tag{13}$$

where Δg_f is the molar Gibb's Free Energy difference
and F is the Faraday's Constant 96,485 C.

The theoretical EMF which can be achieved assuming reversibility thus for a hydrogen fuel cell at 200 °C ($\Delta g_f = -220$ kJ) is 1.14 V. In practice the voltage is lower than this value due to the various impedances in the fuel cell. Activation losses due to the slowness of the reaction, ohmic losses due to the internal resistance to the flow of electrons through the electrode as well as the resistance to the flow of ions through the electrolyte and mass transport losses due to change in concentration of the reactants are the major types of losses found in a PEM fuel cell.

The output voltage of a cell, considering all these losses can be then put as

$$V_{cell} = E_{Nernst} - \eta_{act} - \eta_{ohm} - \eta_{trans} \tag{14}$$

From literature [9] the temperature dependent Nernst potential is given by

$$E_{Nernst} = 1.229 - 0.00085(T - 298.15\,\mathrm{K}) + \frac{RT}{2\,\mathrm{F}} \ln\left(p_{H_2}(p_{O_2})^{0.5}\right) \tag{15}$$

The same literature also gives the values of the Activation potential and ohmic losses as

$$\eta_{act} = \frac{1}{b} \arcsin h \left(\frac{1}{2 i_0 \left(\frac{1}{i} - \frac{1}{i_L} \right)} \right) \tag{16}$$

where $b = 26.32\,\mathrm{V}^{-1}, i_0 = 5\,e^{-6}\mathrm{A/cm^2}, i_L = 1.35\ \mathrm{A/cm^2}$ and i is the current density.

$$\eta_{ohm} = i\left(0.0269i^2 - 0.0637i + 0.1753\right) \tag{17}$$

The temperature of the system is simulated on grounds of the energy balance of the system.

$$mc_p\frac{dT}{dt} = \dot{Q}_r - P_{stack} - \dot{Q}_{cooling} \tag{18}$$

where

$$\dot{Q}_r = \dot{n}_{H_2}\,\Delta h_r \quad \text{is the change reaction enthalpy} \tag{19}$$

$$P_{stack} = V_{stack} \times I_{stack} \quad \text{is the net electrical power drawn form the fuel cell} \tag{20}$$

$\dot{Q}_{cooling}$ is energy dissipated to the environment.

The cooling power can alternatively be used for heating purposes which would increase the combined overall efficiency of the system.

3.4 Heat Exchanger

Plate heat exchangers (HEX) are modelled and used in this system. The P-NTU method was used to model the basic idea of modelling a HEX.

For an overall heat transfer coefficient, U (in W/m^{2}K) and an average temperature difference, ΔT_m (in K) and a n overall heat transfer area, A (in m^2) the heat transfer, $\dot{Q}$ (in W) in a HEX can be written as

$$\dot{Q} = UA\Delta T_m \tag{21}$$

where

$$U = \frac{1}{\left(1/\alpha_1 + 1/\alpha_{plate} + 1/\alpha_2\right)} \tag{22}$$

α_1 and α_2 (in W/m^{2}K) are the heat transfer coefficients of the fluids 1 and 2 respectively and α_{plate} is that of the plate. For the plate with a thickness of w (in m) and a thermal conductivity of k, (in W/mK) α_{plate} is given by

$$\alpha_{plate} = k_{plate}/w \tag{23}$$

Since the fluids are in motion heat transfer is convective and for a fluid with a thermal conductivity of k (in W/mK) and a flow characterized by the Nusselt's Number Nu (dimensionless) and hydraulic Diameter D_e (in m), the convective heat transfer coefficient α is given by

$$\alpha_i = k_i Nu_i/D_e \quad i = 1\, and\, 2 \tag{24}$$

The method for calculation of Nu is elaborately described in part Gb7 of VDI heat atlas [10]. The HEX was modelled with 100 % efficiency and the ATR reactor and Off Gas Burner were modelled with 90 % efficiency making up for the losses also on the heat exchangers.

4 Experimental Setup

The Fig. 3 shows a schematic flow diagram of the fuel cell system. A laboratory scale 1 kW electrical power fuel cell system was developed based on the flow diagram. It was found from simulation and from previous experience that 750 °C is the trade-off temperature for maximum H_2 yield and catalyst and reactor longevity. The maximum HT WGS reactor temperature was set at 350 °C and that of LT WGS reactor to 250 °C. The maximum fuel cell inlet temperature was set to 180 °C. The maximum allowed CO in anode gas as per the manufacturer of the fuel cell used in this system is 0.5 %. Simulations to achieve this with the least possible Steam to Carbon Ratio (SCR) were done. A series of start-stop operation was experimented to find the suitability of the system to mobile application. Long term operation tests were also undertaken. Since only the gas composition of the anode gas is of importance for the fuel cell a reference test with propane (C_3H_8) was used as the fuel. The model was validated against experimental results. The composition of the anode gas with C_3H_8 was then compared with that of methane enriched biogas.

5 Results and Conclusion

Although an auto thermal reforming with SCR 1.5 was tested and documented [11], it runs the risk of soot formation on the catalyst and thus increasing the possibility of irreversible damage of the catalyst. Moreover from the simulation tests it was found out that at SCR 1.5 the CO concentration is above the permissible level. Figure 5 shows the simulated CO concentrations (dry) of the fuel processor, i.e. at the exit of the WGS reactor. In accordance with the simulation results an SCR value between 2 and 2.5 was chosen to run the fuel cell system at safe operating conditions.

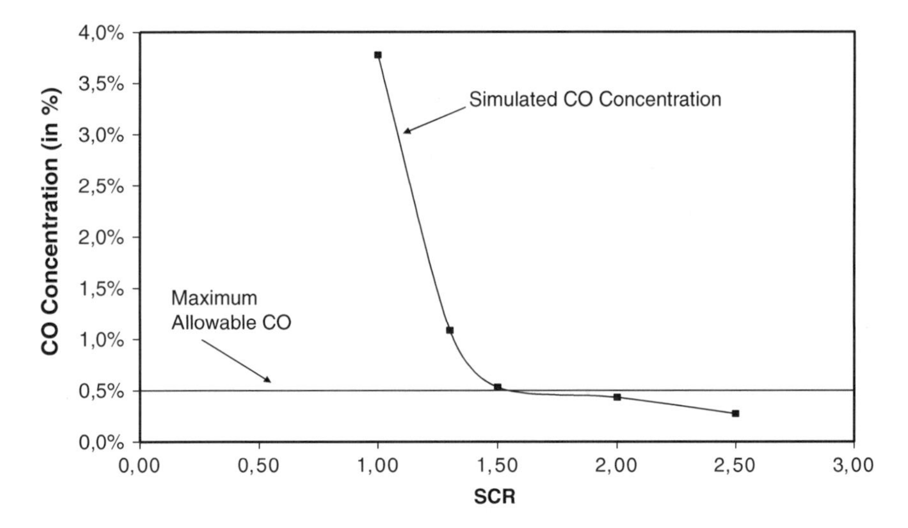

Fig. 5 CO concentration at different SCR values for propane at 750 °C

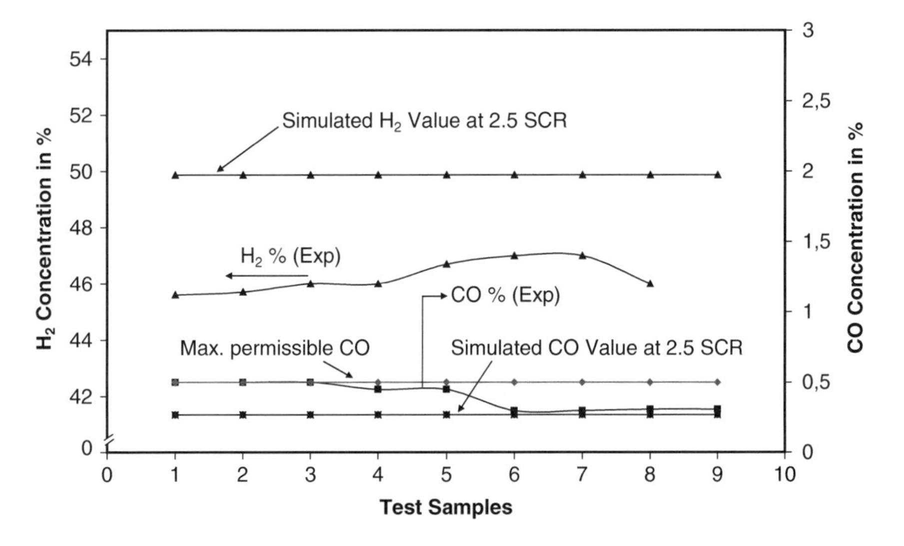

Fig. 6 Concentration of H_2 and CO (experimental and simulated)

Validation of the model was done with the reference experiment with acceptable tolerance taking into account of the fact that exothermic methane formation in the WGS reactor is not considered in the simulation. Figure 6 shows the simulated and experimental values of the CO and H_2 concentration in the reformate gas. The simulated results of H_2 concentration is at around 50 % (dry) whereas the experimental values are at around 46 %.

The fuel cell system was tested for its reforming in terms of percent yield of H_2 and CO and its reliability for cyclic operation. Figure 7 shows intermittent start and

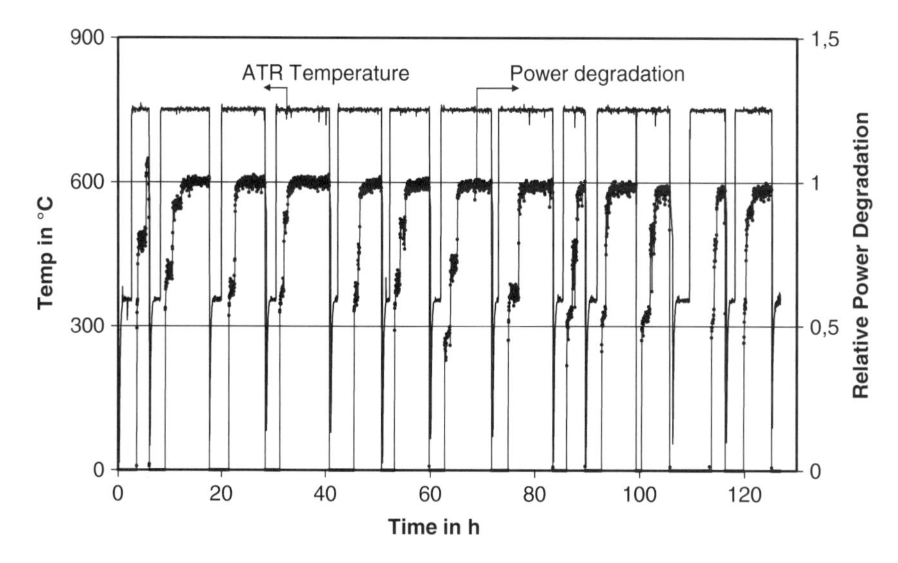

Fig. 7 Cyclic start and stop operation of the system

stop functioning of the fuel cell system for 13 cycles with the ATR temperature on the primary Y-axis (at the controlled 750 °C at each cycle) and the power degradation relative to the first cycle on the secondary Y-axis and Time in hours on the X-axis. The power degradation is shown to be lesser than 3 % at the thirteenth cycle constant voltage and current drawn at each cycle. The reason for the degradation of the power was leakage of the anode gas into the cathode and thus damaging the fuel cell.

Figure 8 shows the cold start of the fuel cell system in the experiment. The Fuel cell was 500 W_{el} and was run on partial load of 300 W. It takes around 1 hour for

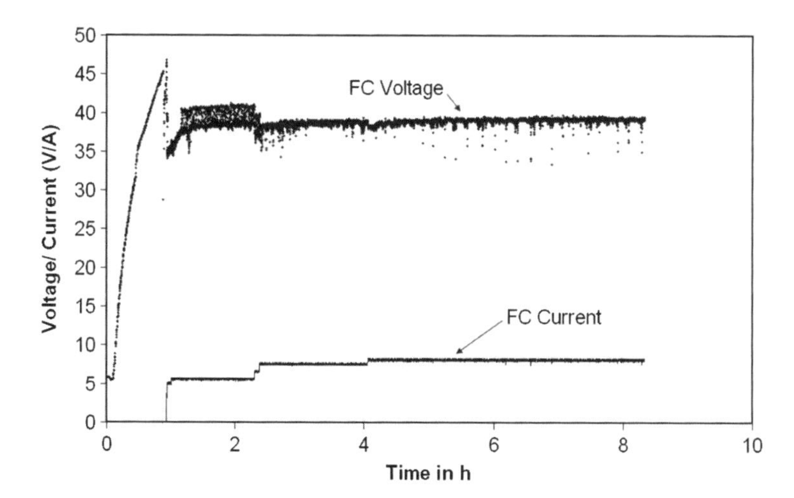

Fig. 8 One cycle operation of the fuel cell system

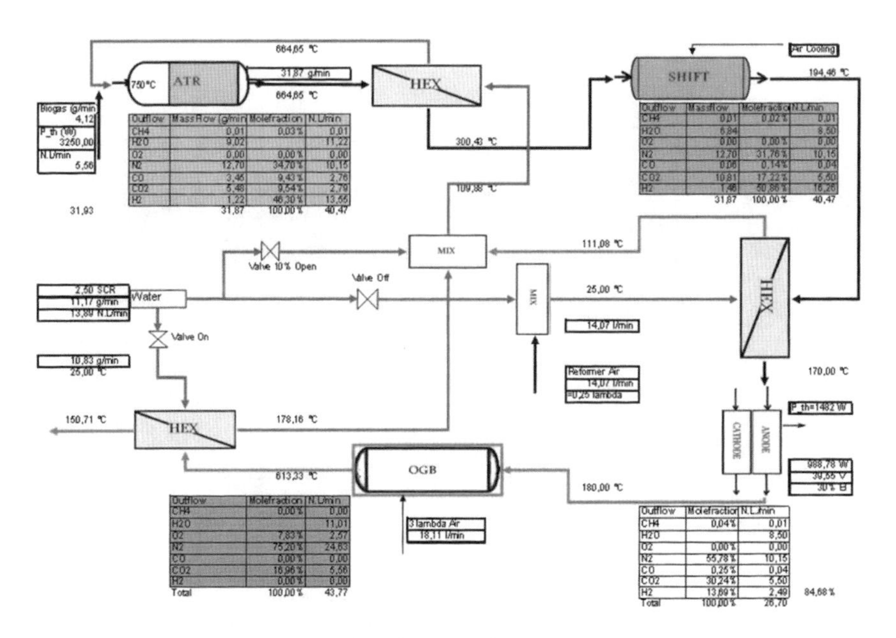

Fig. 9 Output of the model for 2.5 SCR 3,700 W biogas

the fuel cell system to warm up so that current could be drawn. This is not relatively good for a cold start but the experimental test was not optimized for quick cold start up. Quicker cold start can be achieved with measures like higher pre-heating temperatures and running with higher thermal power.

According to the reference test with propane an electrical efficiency over 30 % can be achieved. Also with 98 % methane enriched biogas the efficiency is at around 30 %. Figure 9 shows the output of the Matlab simulation model of biogas fuel cell system. It shows the fuel cell system with 3,250 W_{th}, 2.5 SCR. the CO concentration in the anode gas is reduced down to 0.14 %. The Electrical efficiency is defined as the ratio of electrical power drawn form the system to the thermal power fed into the system. The thermal power of a fuel flow can be easily found out as the product of heating value of the fuel with molar flow. The overall efficiency of the System can be defined as the ratio of the sum of electrical and thermal energy at the output to the thermal input energy. The overall efficiency can be around 65 % for the system with around 1,400 W thermal power that can be used for room heating purposes.

The system was tested for around 800 h operation without noticeable degradation. From this study it is found out that Auto Thermal Reforming of biogas for HT PEM FC application is not only feasible but also an efficient way to produce 'green' electricity and heating. The cyclic operation of the system also provides a proof that this system can be used for mobile application where continuous electric supply is not necessary.

Acknowledgments The authors would like to thank BreNaRo (Brennstoffgewinnung aus nachwachsenden Rohstoffen), an NRW Research School for financing this study. The authors would also like to thanks to their colleagues at FC Power GmbH sincerely.

References

1. Handreichung Biogasgewinnung und- nutzung: FNR, Leipzig (2004)
2. Technische Regeln für Gefahrstoffe: http://www.baua.de/de/Themen-von-A-Z/Gefahrstoffe/TRGS/TRGS-900.html (2006). Accessed 6 May 2011
3. Klinski, S.: Einspeisung von Biogas in das Erdgasnetz. FNR, Leipzig (2006)
4. Larminie, J., Dicks, A.: Fuel Cells Systems Explained, 2nd edn. Wiley, England (2003)
5. Li, Q., Jensen, J.O., Savinell, R.F., Bjerrum, N.J.: High temperature proton exchange membranes based on polybenzimidazoles for fuel cells. Prog. Polym. Sci. **34**, 449–477 (2009)
6. Callaghan, C.A.: Kinetics and Catalysis of the Water-Gas-Shift Reaction: a Microkinetic and Graph theoretic Approach, Dissertation, Worcester Polytechnic Institute (2006)
7. Iyoha, O.U.: Hydrogen Production in Palladium and Palladium-Copper Membrane Reactors at 1,173 K in Presence of H_2S. University of Pittsburgh, Pittsburgh (2007)
8. Choi, Y., Stenger, H.G.: Water gas shift reaction kinetics and reactor modeling for fuel cell grade hydrogen. J. Power Sources **124**, 432–439 (2003)
9. Nguyen, V.P., Ladewig, B., Lapicque, F.: Modelling and Simuliation of Fuel cells for residential Combined heat and power systems. http://perso.ensem.inpl-nancy.fr/Olivier.Lottin/FDFC08/CD/Contributions/Nguyen.pdf. Accessed 8 Sept 2010
10. VDI-Wärmeatlas: VDI. Springer, Berlin (2006)
11. Wang, W., Turn, S.Q., Keffer, V., Douette, A.: Parametric study of auto thermal reforming of LPG. Prepr. Pap.-Am. Chem. Soc., Div. Fuel Chem. **49**(1), 143 (2004)

Local Dynamics and Statistics of Streamline Segments in Fluid Turbulence

P. Schaefer, M. Gampert and N. Peters

Abstract Based on local extreme points of the absolute value u of the velocity field u_i, streamlines are partitioned into segments as proposed by Wang (J. Fluid. Mech. 648:183–203, 2010). The temporal evolution of the arc length l of streamline segments is analyzed and associated with the motion of the isosurface defined by all points on which the gradient in streamline direction $\partial u/\partial s$ vanishes. This motion is diffusion controlled for small segments, while large segments are mainly subject to strain and pressure influences. Due to the non-locality of streamline segments, their temporal evolution is not only a result of slow but also of fast changes, which differ by the magnitude of the jump Δl that occurs within a small time step Δt. The separation of the dynamics into slow and fast changes allows the derivation of a transport equation for the probability density function (pdf) $P(l)$ of the arc length l of streamline segments. While slow changes in the pdf transport equation translate into a convection and a diffusion term when terms up to second order are included, the dynamics of the fast changes yield integral terms. The convection velocity corresponds to the first order jump moment, while the diffusion term includes the second order jump moment. It is theoretically and from DNS data of homogeneous isotropic decaying turbulence at two different Reynolds numbers concluded that the normalized first order jump moment is quasi-universal, while the second order one is proportional to the inverse of the square root of the Taylor based Reynolds number $Re_\lambda^{-1/2}$. It's inclusion thus represents a small correction in the limit of large Reynolds numbers. Numerical solutions of the pdf equation yield a good agreement with the pdf obtained from the DNS data. It is also concluded on theoretical grounds that the mean length of streamline segments scales with the Taylor microscale rather than with any other turbulent length scale, a finding that can be confirmed from the DNS.

P. Schaefer · M. Gampert · N. Peters (✉)
Institut Für Technische Verbrennung, RWTH-Aachen, Templergraben 64, 52056 Aachen, Germany
e-mail: n.peters@itv.rwth-aachen.de

M. Klaas et al. (eds.), *Fuels From Biomass: An Interdisciplinary Approach*,
Notes on Numerical Fluid Mechanics and Multidisciplinary Design 129,
DOI 10.1007/978-3-662-45425-1_10

1 Introduction

The turbulent motion of fluids is a highly complex and still unresolved problem. It owes its complexity mainly to the interaction of a wide range of spatial and temporal scales. In addition, the influence of pressure renders turbulent flows, due to its coupling of spatially distant points, non-local in nature. Today, even for the simplest of all turbulent flows, namely homogeneous isotropic decaying turbulence, no closed theory exists [1–3]. Due to its randomness, turbulent flows have to be described by means of statistics. These can be one-point statistics, which capture only local properties of the flow field or multipoint statistics [4, 5] which then also contain information about the non-local behavior of the field. Historically, profound insights into the statistical structure of homogeneous isotropic turbulence have mostly been gained by means of the two-point correlation function and its transport equation, the von Kármán-Howarth equation [6], or its alternative formulation, the Kolmogorov equation [7, 8]. However, in such an approach the local dynamics of the turbulent field have been averaged out. In addition, one may ask, whether the "artificial" cartesian frame in which such theories are developed can be replaced by a more "natural" frame based on the flow field itself. In 1971, Corrsin [9] asked the question *What types (of geometry) are naturally identifiable in turbulent flows*? In a first approach, vortex structures have been identified and analyzed for instance by She et al. [10] and Kaneda and Ishihara [11]. These were found to form tubes in regions of high vorticity, while a sheet-like structure was identified in regions of low vorticity. However, vortex tubes and sheets do not allow a unique decomposition of the flow field into unambiguous sub-ensembles. This problem was overcome by Wang and Peters [12] in their theory of dissipation elements, which allows the decomposition of turbulent scalar fields based on the concept of gradient trajectories which end in local extreme points of the scalar field. The work has its roots in early works by Gibson [13] who analyzed the role of extreme points in turbulent scalar mixing processes. For further details on the theory, see Schaefer et al. [14, 15] and for experimental results see Schaefer et al. [16]. Despite its success in describing turbulent scalar fields, the concept of dissipation elements cannot be used for the analysis of vector valued fields, such as the turbulent velocity field. To overcome this problem, Wang [1] proposed to study streamlines in turbulent velocity fields. The geometrical properties of streamlines in turbulence have for instance been studied by Rao [17] and Braun et al. [18] who geometrically parametrize streamlines by their curvature and torsion. Streamlines were also shown to play an important role in the problem of turbulent pair diffusion. Goto and Vassilicos [19] argue that there exists a frame, in which the so called streamlines persistence is maximized. Streamlines are considered persistant if their geometry changes slowly enough for a particle to approximately follow their path for a significantly long time. In that case, particles initially close to each will separate once they approach a stagnation point, where streamlines diverge. In an evolving turbulent velocity field $u_i(x_i, t)$, streamlines can be traced at any instant t in the frozen field from any point in space by following the normalized directional vector

$t_i = u_i/u$, where $u = (u_i u_i)^{1/2}$ denotes the absolute value of the velocity vector. Streamlines are, different from gradient trajectories in a scalar field, a priori infinitely long, unless they hit a stagnation point, in which all three velocity components vanish and they diverge. Wang [1] has proposed to divide streamlines into segments based on local extreme points of u along the streamline. Segments are then bound by two extrema, i.e. points where the projected velocity gradient $t_i \partial u/\partial x_i = \partial u/\partial s = 0$ vanishes. Wang [1] further characterized streamline segments as positive and negative ones, depending on the sign of the velocity difference Δu at their ending points. While within positive segments $\partial u/\partial s > 0$, within negative segments $\partial u/\partial s < 0$, thus the flow along positive segments is locally accelerated, while along negative segments it is decelerated. In order to describe streamline segments statistically, one may choose the arc length distance l between and the velocity difference Δu at the two extreme points. Then, most of the statistical properties are captured in the joint probability density function (jpdf) $P(l, \Delta u)$. Based on Bayes' theorem, we can relate the marginal pdf of the length of streamline segments to the joint and conditional pdfs yielding

$$P(l) = \frac{P(l, \Delta u)}{P_c(\Delta u|l)}, \tag{1}$$

where in Eq. 1 $P(l)$ denotes the marginal pdf and $P_c(\Delta u|l)$ the conditional pdf. The current work is concerned with the marginal pdf of the length l of streamline segments in homogeneous isotropic decaying turbulence. In Chap. "GIS-Based Model to Predict the Development of Biodiversity in Agrarian Habitats as a Planning Base for Different Land-Use Scenarios", we will discuss the local instantaneous dynamics of streamline segments and discern two physical effects, namely fast and slow changes of their length. The distinction between fast and slow changes will allows us to derive a transport equation for the normalized pdf $\tilde{P}(\tilde{l})$ in which the slow processes lead to drift and diffusion terms which contain the so called jump moments of l, while the fast changes yield integral terms. In Chap. "The Cellulolytic System of Cyst Nematodes", we identify all points on which $\partial u/\partial s = 0$ vanishes as an isosurface containing all ending points of streamline segments. The slow changes are then attributed to the motion of the isosurface whose dynamics are treated based on a level set approach and related to the jump moments. In Chap. "New Pathways for the Valorization of Fatty Acid Esters", the jump moments are evaluated from the DNS data of homogeneous isotropic decaying turbulence at two different Reynolds numbers and the results are compared with the theory. The pdf transport equation is then solved numerically using approximation ansätze for the jump moments and the result is compared to the pdf obtained from the DNS. Concluding remarks are given in Chap. "Soluble Organocalcium Compounds for the Activation and Conversion of Carbon Dioxide and Heteroaromatic Substrates".

2 Dynamics of Streamline Segments

To compute the pdf $P(l)$, streamlines are calculated in the frozen turbulent field from homogeneously distributed (equal distanced) points in space in direction of the vector t_i and in its opposite direction $-t_i$. Then, streamline segments are identified along the streamline and based on their arc length l, the pdf $P(l)$ is calculated. However, the dynamics of streamline segments cannot be obtained from the analysis of a frozen turbulent field. We thus, instead of analyzing a frozen turbulent field, will conceptually imagine streamline segments passing through fixed points in space at many subsequent time-steps. This shift in perspective from the statistics of the frozen turbulent field to a time series observation will allow us to identify the mechanisms responsible for the temporal evolution of the length of streamline segments and derive a transport equation for the marginal pdf $P(l)$, which is linked to the local dynamics of the turbulent field itself.

In three dimensional space, the condition $\partial u/\partial s = 0$ defines an isosurface, which divides space into two regions, namely a region, in which $\partial u/\partial s > 0$ containing all positive segments to be denoted with (+) and a region with $\partial u/\partial s < 0$ containing all negative segments to be denoted with (−). Thus, streamlines starting from an arbitrary point in space will intersect the isosurface thereby entering alternatingly into regions denoted with (+) and (−) signs. The situation is illustrated in Fig. 1a where the isosurface as well as a streamline is shown for a turbulent flow field calculated by DNS in three-dimensional space. Regions of positive and negative $sgn(\partial u/\partial s)$ are denoted with (+) and (−) signs, respectively. The streamline enters the box from the top left corner and intersects the corrugated isosurface five times before leaving the box at the lower right corner. The five intersections define the boundaries for four segments along the streamline. Figure 1b shows the corresponding variation of u along the same streamline in a one-dimensional plot, where the five segments are demarked with dotted lines whose locations correspond to the intersections shown in Fig. 1a. The variation is plotted over the arc length s, so that the horizontal distance between two dotted lines corresponds to the length l of the segment.

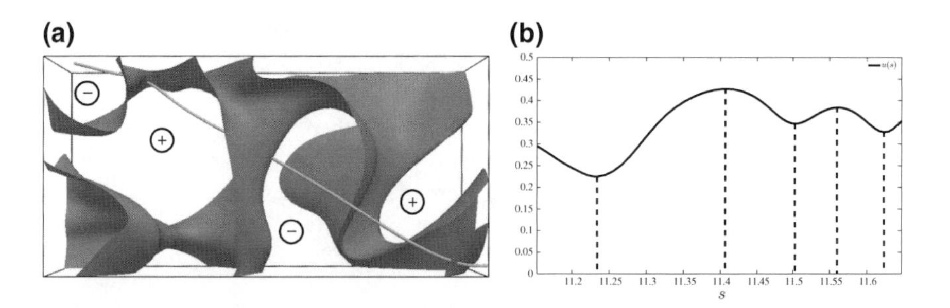

Fig. 1 **a** Isosurface defined by $\partial u/\partial s = 0$ and streamline from DNS (case 1). **b** Variation of u along the streamline over the arc length (arbitrary units)

We will in the following discern two different effects that will influence the evolution of the length $l(t)$ over time. The above differ by the magnitude of the jump Δl that they entail within Δt. Firstly, there will be continuous changes, to which we will refer as slow changes. These occur, when the streamline passing through a fixed point reaches approximately the same position on the isosurface, the latter has not changed in an abrupt manner (such that a new intersection of the streamline and the isosurface is formed on the way) and the streamline geometry remains approximately the same, i.e. the directional field has not changed dramatically. For this case, jumps in the streamline length Δl are considered to be small (as compared to the value of l itself) so that $\mathcal{O}(\Delta l) \ll l$. Such changes can be attributed to the continuous evolution of the isosurface as well as the directional field by convection and diffusion. Secondly, however, there will be discontinuous, fast changes, where the value of l will abruptly change by a value Δl, which can be of the order of the instantaneous length itself (of course the lower bound $l \geq 0$ cannot be violated) so that $\mathcal{O}(\Delta l) = l$. Such changes will occur, if either the isosurface, previously already close to the streamline along its way to the intersecting point now intersects at an intermediate point, or the directional field changes thus leading to an intersection which is not close to the old one any more. In the variation of u along the streamline such a fast process corresponds to the disappearance or new formation of an extreme point. Fast changes can further be discerned by the sign of the jump Δl, where we refer for $sgn(\Delta l) < 0$ to it as a cutting process, while for $sgn(\Delta l) > 0$ the process is referred to as an attachment process. We will discuss the physical picture behind the fast process based on the two-dimensional (2D) illustration of a cutting process shown in Fig. 2. (Note that an attachment process would correspond to the same picture with reversed time.) The X marks the fixed grid point through which at two consecutive time steps a streamline passes (line with arrow head). The solid lines denote the situation at time t_0 while the dotted lines correspond to the situation at $t_0 + \Delta t$. As can be observed, while the right boundary of the segment has changed little, the left boundary has abruptly changed due to a new intersection of the streamline with the isosurface, which has shifted in space. The illustration on

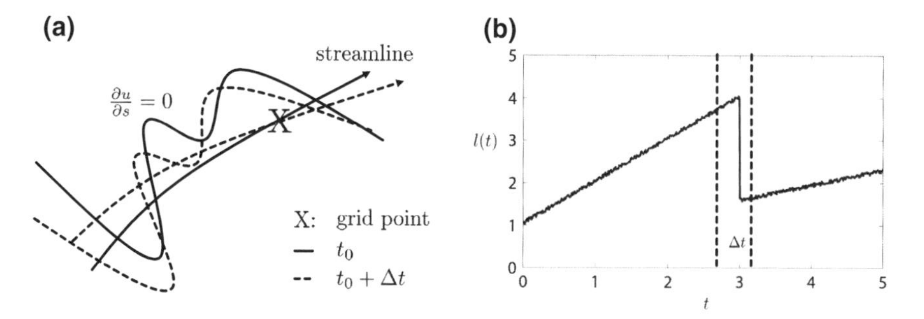

Fig. 2 **a** Illustration of a scenario that leads to a cutting. **b** Time series of $l(t)$ displaying the fast change Δl when a cutting occurs

the right shows the corresponding abrupt jump in the time series of the streamline segment length. Note, that such fast changes are not an artifact of finite time steps in numerical simulations or alike, but rather are an intrinsic feature resulting from the non-local dynamics of streamline segments. It is in addition remarkable to note that such fast dynamics are present, although the underlying fields ($\partial u/\partial s$ and t_i) evolve continuously in time as they are diffusion controlled at the small scales.

3 Stochastic Processes with Slow and Fast Changes

Although the physical picture behind streamline segments and dissipation elements is quite different, the modeling of the random cutting events as well as the slow processes coincided in form with the one proposed by Wang and Peters [12], so that we will not go into further details and adapt their derivation of a transport equation for the pdf of the length of streamline segments, yielding in normalized coordinates (normalized with the mean length l_m)

$$\begin{aligned}\frac{\partial\tilde{P}(\tilde{x},\hat{\tau})}{\partial\hat{\tau}}+\frac{\partial}{\partial\tilde{x}}\left[(\tilde{a}_1(\tilde{x})-\frac{\tilde{x}}{\tau_{lm}a_\infty})\tilde{P}(\tilde{x},\hat{\tau})\right]=&\frac{1}{2}\frac{\partial^2}{\partial\tilde{x}^2}\left[\tilde{a}_2(\tilde{x})\tilde{P}(\tilde{x},\hat{\tau})\right]\\&+\Lambda_c\left(2\int_0^\infty\tilde{P}(\tilde{x}+\tilde{z},\hat{\tau})d\tilde{z}-\tilde{x}\tilde{P}(\tilde{x},\hat{\tau})\right)\\&+2\Lambda_a\left(\int_0^{\tilde{x}}\frac{y}{x}\tilde{P}(\tilde{x}-\tilde{y},\hat{\tau})\tilde{P}(\tilde{y},\hat{\tau})d\tilde{y}-\tilde{P}(\tilde{x},\hat{\tau})\right),\end{aligned} \tag{2}$$

where $\Lambda_c=(\lambda l_m)/a_\infty$ is a normalized cutting frequency per unit length and time, while $\Lambda_a=\mu/a_\infty$ denotes a normalized attachment frequency per unit time. Note, that the value of Λ_c is an eigenvalue of the problem to be obtained from the normalization condition $\int\tilde{x}\tilde{P}d\tilde{x}=1$, while in Chap. "The Cellulolytic System of Cyst Nematodes" we will show that Λ_a is related to the slope of the pdf at the origin. $\tilde{a}_1$ and $\tilde{a}_2$ are the normalized first and second order jump moments and correspond to drift and diffusion of l in phase space. Note, that different from [12] $\tilde{a}_2$ has been included in the present work.

In the following, we will derive functional expressions for the drift velocity and the diffusion term based on the local instantaneous dynamics of streamline segments in turbulent flows.

4 Jump Moments of the Length of Streamline Segments

4.1 First Order Jump Moment

Let us derive the transport equation governing the scalar $u_s = \partial u/\partial s = t_i \partial u/\partial x_i$. Taking the derivative with respect to time, yields.

$$\frac{\partial u_s}{\partial t} = t_i \frac{\partial}{\partial t}\frac{\partial u}{\partial x_i} + \frac{\partial u}{\partial x_i}\frac{\partial t_i}{\partial t} = \underbrace{\frac{\partial}{\partial s}\frac{\partial u}{\partial t}}_{1} + \underbrace{\frac{\partial u}{\partial x_i}\frac{\partial t_i}{\partial t}}_{2}. \tag{3}$$

We will treat the motion of the isosurfaces defined by $u_s = 0$ using a level set method [20], and thus we are only interested in the condition on the isosurface itself. DNS calculations show that on the isosurface on average the first term on the r.h.s. of Eq. 3 is much larger than the second. Physically, this finding can be attributed to the fact that in incompressible flows all streamlines have zero Gaussian curvature $\kappa = \partial t_i/\partial x_i = 0$ when passing through the isosurface, meaning that locally all streamlines are parallel to each other [1]. In other words, the flow field in the vicinity of the isosurface is quite regular. It can then be expected, that the region around isosurfaces are quite stable yielding a locally small $\partial t_i/\partial t$. Thus, neglecting the term labeled 2 in Eq. 3, we obtain on the isosurface

$$\frac{\partial u_s}{\partial t} + u_i \frac{\partial u_s}{\partial x_i} = -\frac{\partial^2 p}{\partial s^2} + \nu u^{-1} \frac{\partial}{\partial s}\left(\left(\frac{\partial u}{\partial x_i}\right)^2 - \left(\frac{\partial u_i}{\partial x_j}\right)^2\right) + \nu \frac{\partial}{\partial s}\left(\frac{\partial^2 u}{\partial x_i^2}\right). \tag{4}$$

We write the last term on the right hand side as

$$\frac{\partial}{\partial s}\left(\frac{\partial^2 u}{\partial x_i^2}\right) = \frac{\partial^2 u_s}{\partial x_i^2} - 2\frac{\partial t_j}{\partial x_i}\frac{\partial^2 u}{\partial x_i \partial x_j} - \frac{\partial^2 t_j}{\partial x_i^2}\frac{\partial u}{\partial x_j}, \tag{5}$$

and note that the second two terms on the r.h.s. of Eq. 5, as well as the first two terms on the r.h.s. of Eq. 3 act as source/sink terms for u_s, of which only the pressure term is not proportional to viscosity. We combine all those source/sink terms proportional to viscosity to one term and write

$$\Pi_\nu = \nu\left[u^{-1}\frac{\partial}{\partial s}\left(\left(\frac{\partial u}{\partial x_i}\right)^2 - \left(\frac{\partial u_i}{\partial x_j}\right)^2\right) - \left(2\frac{\partial t_j}{\partial x_i}\frac{\partial^2 u}{\partial x_i \partial x_j} + \frac{\partial^2 t_j}{\partial x_i^2}\frac{\partial u}{\partial x_j}\right)\right], \tag{6}$$

to obtain the following transport equation of u_s valid on the isosurface

$$\frac{\partial u_s}{\partial t} + u_i \frac{\partial u_s}{\partial x_i} = -\frac{\partial^2 p}{\partial s^2} + \nu \frac{\partial^2 u_s}{\partial x_i^2} + \Pi_\nu. \tag{7}$$

Equation 7 is a convection—diffusion equation with additional source terms, for which Peters [21] has derived the velocity with which isosurfaces move as

$$\frac{\partial \mathbf{x}_0}{\partial t} = \mathbf{u} + \mathbf{n} s_D, \tag{8}$$

where $\mathbf{n} = \nabla u_s / |\nabla u_s|$ is the unity normal vector of the isosurface. The displacement speed s_D due to diffusion and the source/sink terms reads

$$s_D = \frac{-\frac{\partial^2 p}{\partial s^2} + \nu\left(\kappa|\nabla u_s| + \frac{\partial^2 u_s}{\partial n^2}\right) + \Pi_\nu}{|\nabla u_s|}, \tag{9}$$

where $\kappa = \nabla \mathbf{n}$ is the curvature of the isosurface. The first contribution is due to the pressure term, while the two terms in brackets are directly proportional to the local curvature and due to normal diffusion across the isosurface, respectively. Finally, the last contribution is due to the source/sink terms proportional to viscosity. Equation 8 will be the basis for the derivation of the drift velocities of streamline segments. Projection along t_i, taking the difference and averaging yields for the average rate of change of the streamline segment length

$$\left\langle\frac{\Delta l}{\Delta t}\right\rangle = \left\langle \Delta\left(\mathbf{t}\cdot\frac{\partial \mathbf{x}_0}{\partial t}\right)\right\rangle = \langle \Delta u | l\rangle + \langle \Delta(\cos\alpha s_D)\rangle, \tag{10}$$

where $\cos(\alpha) = n_i t_i$ denotes the angle between the normal vector of the isosurface and the tangent vector to the streamline. For large segments, viscous terms in Eq. 9 can be neglected, so that in the limit $l \to \infty$ we write

$$\left\langle\frac{\Delta l}{\Delta t}\right\rangle\Big|_{l\to\infty} = \langle \Delta u | l\rangle - v_P(l), \tag{11}$$

with

$$v_P(l) = \left\langle \Delta\left(\cos\alpha \frac{\partial^2 p/\partial s^2}{|\nabla u_s|}\right)\right\rangle. \tag{12}$$

The first term on the r.h.s of Eq. 11 corresponds to the conditional mean strain rate due to the velocity difference at the ending points of segments, while the second term is due to the pressure related source term in Eq. 4.

For small segments, however, the velocity difference will be small and the displacement speed will be viscous dominated and similar in form to the one proposed in [12]. With the normalization introduced in Chap. "Feasibility Study of Auto Thermal Reforming of Biogas for HT PEM Fuel Cell Applications", we arrive at the normalized drift velocity

$$\tilde{v}_D(\tilde{l}) = -c_e \frac{\nu}{l_m^2 a_\infty}\frac{1}{\tilde{l}} = -\frac{\nu_e}{\tilde{l}}, \tag{13}$$

where we have absorbed the proportionality constant c_e in the normalized viscosity ν_e, which appears as a model parameter of order unity in the pdf transport equation that takes complex three-dimensional effects of the isosurface into account that have been neglected so far [22]. Note that Eq. 13 shows that the diffusion controlled drift of small segments is singular at the origin, leading to a linear rise of the pdf at the origin. In addition, Eq. 13 allows us to express the normalized attachment frequency in terms of the slope of the pdf at $\tilde{l} \to 0$ as

$$\Lambda_a = v_e \frac{\partial \tilde{P}}{\partial \tilde{x}}|_{\tilde{x}\to 0}, \tag{14}$$

cf. Wang and Peters [12] for details.

From Eqs. 11 and 13 we conclude, that the macroscopic drift of streamline segments consists of three contributions, namely a viscous dominated drift for small segments, that becomes singular at the origin, while large segments are dominated by the strain and pressure induced motion of the isosurface. We thus write in normalized coordinates

$$\tilde{a}_1 = \tilde{v}_D + \widetilde{<\Delta u|l>} + \tilde{v}_P, \tag{15}$$

where all three contributions are a function of the normalized length $\tilde{l}$ and their functional dependence will be determined from the DNS data.

4.2 Second Order Jump Moment

The diffusion term a_2 cannot easily be derived from first principles but will rather be discussed in the following based on scaling arguments and the general picture of turbulent flows as being a multi length-, multi time-scale problem. Physically, a_2 describes small scale fluctuations superimposed on the mean compression or elongation of segments. For the following discussion, we distinguish between two regions in which a_2 scales differently. For $l \gg \eta$ a streamline segment will be embedded in eddies of integral size and small scale fluctuations will be due to the motion of much smaller eddies, which must be of the order of the Kolmogorov scale $\eta = (\nu^3/\varepsilon)^{1/4}$. For this region we propose to write

$$a_2(l) = < \tfrac{\Delta l^2}{\Delta t} > \; = < \tfrac{\Delta l}{\Delta t} \Delta l > \; \propto <|\tfrac{\Delta l}{\Delta t}|> \eta \propto |a_1(l)|\eta. \tag{16}$$

The underlying assumptions for the above reasoning is that for segments larger than the Kolmogorov scale the value of Δl becomes statistically decorrelated from $\Delta l/\Delta t$. This yields for the ratio of the normalized functions for $l \gg \eta$

$$\frac{\tilde{a}_2(\tilde{l})}{|\tilde{a}_1(\tilde{l})|} \propto \frac{\eta}{l_m} \propto Re_\lambda^{-1/2}, \tag{17}$$

where the proportionality to the Taylor based Reynolds number $Re_\lambda = \lambda u'/\nu$ with the Taylor length $\lambda = \sqrt{10k/\varepsilon\nu}$ arises from the fact that the mean length of streamline segments scales with the Taylor length, which will be shown in the course of this work. However, for segments much smaller than the Kolmogorov length the segments will only experience viscous effects and the two ending points will be so close to each other, that locally the flow field can be considered laminar. This means, that fluctuations must vanish in the limit $l \to 0$, so that $a_2(0) = 0$. From Eq. 16 we conclude that a_2 exhibits three distinct regions: for $l \gg \eta$ it will scale linearly with l following the linear scaling of a_1, for $l \ll \eta$, $a_2 \to 0$ while for intermediate values of $l = \mathcal{O}(\eta)$, where a_1 is dominated by the drift velocity $|v_D| \propto \nu/l$ a steep rise of a_2 will be present. Based on the above reasoning, we conclude that for all values of l, $a_2 \ll a_1$ in the limit of large Reynolds-numbers, which justifies a truncation of the Taylor series at $\mathcal{O}(2)$ in the derivation of the pdf transport equation. The scaling derived here on phenomenological grounds will be verified based on the DNS data in Chap. "Planar, Stereoscopic, and Holographic PIV-Measurements of the In-Cylinder Flow of Combustion Engines".

5 Numerical Validation

In the course of this chapter, we present basic information about our direct numerical simulations (DNS) of decaying homogeneous isotropic turbulence at two different Reynolds numbers. We will use the data to identify streamline segments at two consecutive time steps to calculate the first and second order jump moments and compare their scaling with the theoretical considerations in the above chapter. Using approximation ansätze for the two functions, we numerically solve the proposed pdf transport equation and compare the result with the pdfs obtained from the DNS.

5.1 DNS of Homogeneous Isotropic Decaying Turbulence

We have performed two DNS calculations, where we have solved the Navier-Stokes equations numerically in a cubic box of size 2π with periodic boundary conditions employing a pseudo-spectral method in space and a second-order Adam-Bashforth method in time. Aliasing errors are eliminated by isotropic truncation. Computations were performed on an IBM BlueGene/P machine at the research center Jülich and the calculation used 16,384 processors. The initial velocity field is random and isotropic and is generated so that it satisfies a prescribed energy spectrum. The initial energy spectrum is taken from Mansour and Wray [23].

The most important parameters characterizing the DNS are summarized in Table 1.

Table 1 Parameters of the different DNS cases

Case	1	2
No. of grid cells	$1,024^3$	$1,024^3$
Reynolds number Re_λ	50	116
Viscosity ν	5.00×10^{-4}	1.00×10^{-4}
Kinetic energy k	4.85×10^{-2}	3.44×10^{-2}
Dissipation ε	1.25×10^{-2}	5.89×10^{-3}
Integral time τ	3.88	5.84
Kolmogorov scale η	1.00×10^{-2}	3.61×10^{-3}
Resolution $\Delta x/\eta$	0.61	1.70
Kolmogorov time t_η	0.20	0.13
Taylor scale λ	1.39×10^{-1}	7.64×10^{-2}

5.2 Drift Velocity and Diffusion from DNS Data

Starting from fixed grid points, streamlines are traced in direction of t_i and $-t_i$ until they intersect with the isosurface. From this, the length (as the arc length of the streamline segment) $l(t)$ and the velocity difference at the ending points Δu attributed to each of the grid points is obtained. This procedure is repeated at two consecutive times. From this data we calculate the drift and diffusions terms as $a_1 = <\Delta l/Deltat>$ and $a_2 = <\Delta l^2/Deltat>$. However, as the latter are attributed to slow processes only, large jumps have to be excluded. To this end values where $\Delta l/\Delta t$ is larger than a threshold are excluded from the sampling, where the threshold is taken to be a few multiples of the mean $<|\Delta l/\Delta t|>$ (including fast processes). As large jumps due to fast processes are by at least an order of magnitude larger than slow processes, the procedure is rather insensitive to the value of the threshold. Apart from physically relevant fast changes, artificial large jumps are introduced to the data when the isosurface swipes over the grid point, so that the latter does not remain within a region of same $sgn(\partial u/\partial s)$. These, too, have to be excluded from the sampling. To reduce the number of such exclusions and to resolve the drift velocity of very small segments, the temporal separation between the two fields was chosen to be approximately $\Delta t = 0.2t_\eta$ for both cases, which yields an exclusion of about 0.2 % of the sample points.

Based on Eq. 15 we expect three contributions to the macroscopic drift of streamline segments. While the conditional mean velocity difference $<\widetilde{\Delta u|l}>$ as well as the sum of all contributions are easily accessible from the DNS data, the drift velocity $\tilde{v}_D$ and the pressure induced velocity $\tilde{v}_P$ are more difficult to obtain. We thus propose to calculate the sum $(\tilde{v}_D + \tilde{v}_P)$ from the balance of Eq. 15. Figure 3 shows the normalized conditional mean velocity difference and the above mentioned sum calculated from the balance. As can be observed in normalized coordinates (for details of the values used for the normalization see Table 2) the curves for the two cases collapse in the entire range.

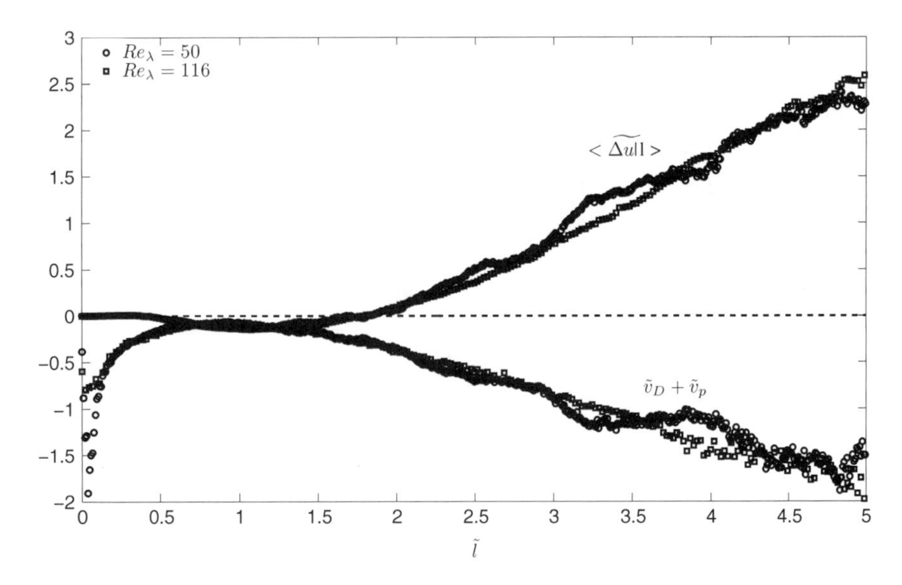

Fig. 3 $<\widetilde{\Delta u|l}>$ and $\tilde{v}_D(\tilde{l}) + \tilde{v}_P(\tilde{l})$ over the normalized length for case 1 and 2

Table 2 Statistics of the streamline segments for DNS cases 1 and 2

Case	1	2
l_m	1.09×10^{-1}	4.66×10^{-2}
l_m/λ	0.78	0.61
τ_{l_m}	9.69	16.68
a_∞	1.2	1.5
c_e	−2.5	−2.5
v_e	0.09	0.08

The conditional mean velocity difference shows a plateau, where it is nearly zero, for small length $\tilde{l} < 0.5$, followed by a negative branch in the region $0.5 < \tilde{l} < 1.5$. For values of $\tilde{l} > 1.5$ we observe a linear scaling. Based on this linear regime the slope a_∞ for the normalization was determined. Physically, the velocity difference at the ending points of streamline segments thus has a zero net effect for small segments, while intermediate segments are compressed and large segments are subject to extensive strain. It is most interesting to note, that obviously the scaling of the two-point velocity difference along streamlines based on their ending points does not follow a classical Kolmogorov scaling as $<\Delta u> \propto (\varepsilon l)^{1/3}$ but rather a linear scaling. This finding will be analyzed in more detail at a different place. A similar result however has been reported for the two-point velocity difference along gradient trajectories, where again for small separation distances a negative branch and for large separation distances a linear scaling was found.

For details on the observations and a theoretical discussion, see Wang [24] and Gampert et al. [25].

The second curve in Fig. 3 shows the sum of the diffusion induced drift of small segments and the effect of the pressure related term in Eq. 15. While $\tilde{v}_P(0) = 0$, as it is a two-point difference, we cannot draw a final conclusion on its form for intermediate segments, as here the influence of the diffusion induced drift may not be negligible. However, from theoretical considerations in chapter "GIS-Based Model to Predict the Development of Biodiversity in Agrarian Habitats as a Planning Base for Different Land-Use Scenarios" we know that the diffusion term $\tilde{v}_D$ must vanish for large segments, so that $\tilde{v}_P$ also scales linearly, though with a negative slope, for large separation distances. This contribution thus counteracts the stretching effect of the mean velocity difference for large segments.

Figure 4 shows $\tilde{a}_1$ as a function of the normalized coordinate $\tilde{l}$ for both the high and the low Reynolds number case. We observe, as could be expected from the above discussion, a very good collapse of the two curves. Three distinct regions can be distinguished, in which $\tilde{a}_1(\tilde{l})$ scales differently: First, in the region $0 < \tilde{l} < 0.5$ we observe the viscosity dominated drift of small segments, which leads to their annihilation and whose scaling confirms Eq. 13. Second for $0.5 < \tilde{l} < 2.5$ a region in which the drift is negative and almost a constant. Third for $\tilde{l} > 2.5$ a region in which the drift scales linearly with $\tilde{l}$. The inset shows a zoom into the viscous dominated region where a higher resolution was used, and which shows the dimensional drift velocity in compensated form $(v_D(l)l/v)$. Apart from a region very close to the origin, where due to numerical resolution inaccuracies the curve tends

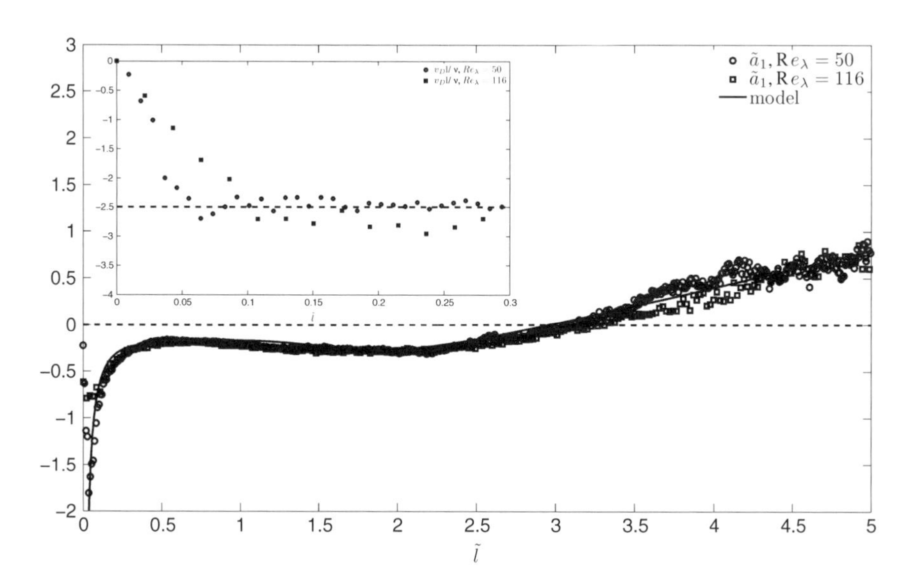

Fig. 4 $\tilde{a}_1(\tilde{l})$ and $<\widetilde{\Delta u|l}>$ over the normalized length for case 1 and 2

to zero, we observe a wide plateau which confirms the theory derived in Sect. 4.1. The value of the proportionality constant in Eq. 13 turns out to be $c_e \approx 2.5$ for both cases, which yields a value of $v_e = 0.09$ and $v_e = 0.08$ for case 1 and 2 respectively. These values however have to be treated with caution and rather represent a lower bound, as due to the singularity of $\tilde{v}_D$ at the origin large contributions to the diffusion dominated drift cannot be resolved with the DNS data separated by a finite time step Δt. Physically, the diffusion induced drift at the origin leads to the anhiliation of very small elements. Intermediate and larger elements ($\tilde{l}<3$) are on average compressed due to the combined action of the pressure and velocity induced motion of the isosurface, while only large elements ($\tilde{l} > 3$) are on average stretched.

In order to numerically solve the pdf transport equation, Eq. 2, we need to provide an analytical expression for $\tilde{a}_1$. We propose to model each of the three different contributions separately. For the diffusion induced drift we write

$$\tilde{v}_D(\tilde{l}) = -\tfrac{v_e}{\tilde{l}}\left(1 - c(1 - \exp(-2\tilde{l}))\right), \tag{18}$$

where the first factor is based on the theoretical derivation in chapter "GIS-Based Model to Predict the Development of Biodiversity in Agrarian Habitats as a Planning Base for Different Land-Use Scenarios", while the second factor is based on considerations in Wang and Peters [12] and extends the theoretically accessible limit $\tilde{l} \to 0$. The constant c is determined during the numerical procedure and ensures that the net effect of the drift velocity is zero on average.

For $\tilde{v}_P$ and $<\widetilde{\Delta u|l}>$ we propose

$$\begin{aligned} <\widetilde{\Delta u|l}> &= \left[(\tilde{l}-1)^4 + 1\right]^{1/4} - 1.19, \\ \tilde{v}_P &= -0.6\left[(\tilde{l}-0.6)^4 + 1\right]^{1/4} + 0.62. \end{aligned} \tag{19}$$

Such an ansatz ensures the correct linear scaling for large values of $\tilde{l}$. The additive constant follows from the fact that both functions must vanish at the origin as they are two-point differences. The ansatz is shown as the solid black line in Fig. 4 and is in good agreement with the DNS data.

Figure 5 shows the normalized function $\tilde{a}_2$ over the normalized length $\tilde{l}$ for both cases. We observe the three different regions already outlined in, Chap. "The Cellulolytic System of Cyst Nematodes" (B): a thin layer at the origin where the boundary condition $\tilde{a}_2(0) = 0$ counteracts the steep rise due to the diffusion induced drift followed by a linear scaling that, following the reasoning in Sect. 4.2 corresponds to the linear scaling of $\tilde{a}_1$. In addition, we note that the order of magnitude analysis from Eq. 17 proofs to be correct for the strain as well as the diffusion dominated region as the scaling with $Re_\lambda^{-1/2}$ can be confirmed. This again justifies the truncation of the Taylor expansion in the derivation of the pdf transport equation at order $\mathcal{O}(2)$. The solid lines show an ansatz of the form

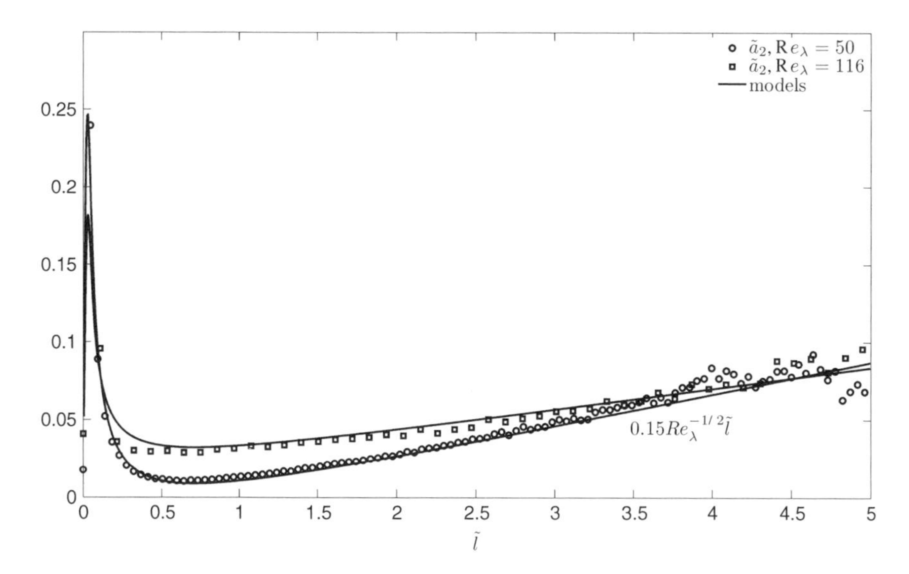

Fig. 5 $\tilde{a}_2(\tilde{l})$ over the normalized length for case2

$$\begin{aligned} \tilde{a}_2(\tilde{l}) &= Re_\lambda^{-1/2}\left(\frac{\nu_e}{\tilde{l}} + a\tilde{l} + \left(Re_\lambda^{1/6} - b\right)\right)f(\tilde{l}) \\ f(\tilde{l}) &= 1 - \exp(-(35\tilde{l}^3)), \end{aligned} \quad (20)$$

which takes the diffusion controlled scaling with $\tilde{l}^{-1}$ for small values of $\tilde{l}$ as well as the linear scaling for large values of $\tilde{l}$ into account. The function f has been introduced as a damping function to correctly ensure the return to zero at the origin. The values of the parameters used for the ansatz are $a = 0.15$ and $b = 2.1$. The solid lines in Fig. 5 show that the approximations are in good agreement with the DNS data.

5.3 The Marginal Pdf of Streamline Segments

Figure 6 shows the normalized marginal pdf of the length of all streamline segments $\tilde{P}(\tilde{l})$ over the normalized length $\tilde{l}$. The symbols correspond to the two DNS cases. Although the Reynolds-number differs by a factor of two, an almost perfect collapse of the pdfs can be observed. In addition, both pdfs show a steep, linear rise at the origin which is due to the diffusion induced drift velocity that leads to the annihilation of small segments. The maxima of both curves occur at approximately $\tilde{l} \approx 0.6$ with a value of the pdf of $\tilde{P}_{max} \approx 0.75$. For large values of $\tilde{l}$ we observe an exponential decay of the tail of the pdfs. Such an exponential tail is characteristic

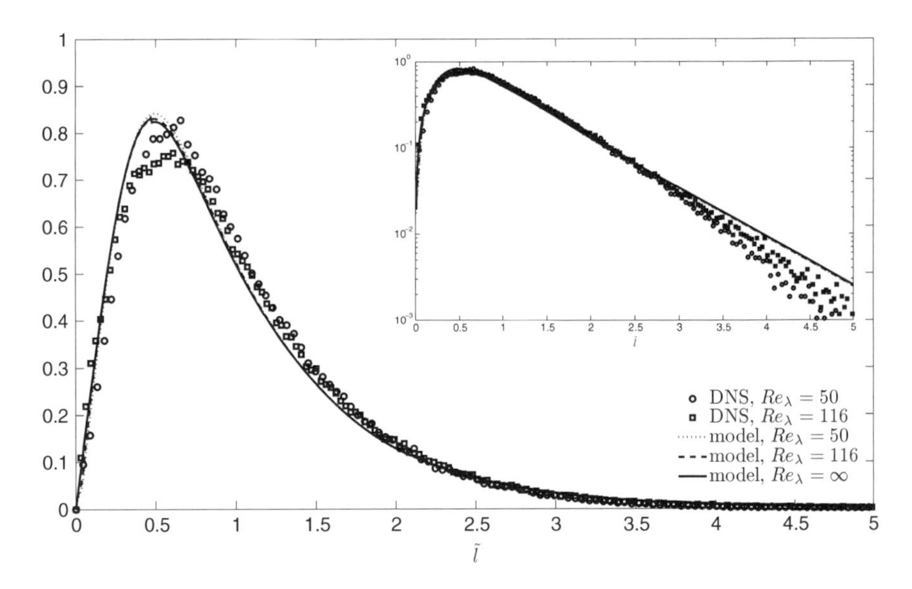

Fig. 6 $\tilde{P}(\tilde{l})$ from DNS and model with $v_e = 0.2$

for an underlying Poisson process, which has been taken into account for the modeling of the integral terms corresponding to the fast processes.

The solid and dotted lines correspond to the steady state solution of Eq. 2. The equation was solved using a finite difference scheme for the drift and diffusion term with the corresponding functions $\tilde{a}_1$ and $\tilde{a}_2$ chosen along the above introduced ansätze. Note that the value $v_e = 0.5$ was chosen for the numerical solution, which is larger than the value from the DNS analysis. However, as discussed above, the latter can only be seen as a lower bound. The value of the product of the two time scales $\tau_{lm}a_\infty$ was taken from the DNS data and turns out to be large for both cases so that its influence in the model, where the inverse appears, becomes small. The solution was iterated in time until the steady state was reached. The obtained pdf was normalized at each time step and the value of the eigenvalue Λ_a was obtained via Eq. 14, while the eigenvalue Λ_c was chosen so that the first moment of the normalized pdf becomes identical to one. The numerical values of the solution are given in Table 3. We first note, that the model seems to be rather insensitive to the Reynolds number showing that its influence is indeed a higher order effect. We have included the numerical solution for the limiting case of infinite Reynolds

Table 3 Parameters of the numerical solution of Eq. 2

Case	1	2	–
Re_λ	50	116	∞
Λ_a	1.247	1.358	1.371
Λ_c	2.343	2.541	2.554
c	1.822	1.856	1.904
v_e	0.5	0.5	0.5

numbers in Fig. 6, for which $\tilde{a}_2 = 0$. The product of the time scales for this case was chosen to be identical to the one of case 2. With increasing Reynolds numbers we only observe a slight shift of the maximum value of $\tilde{P}(\tilde{l})$, which for the considered Reynolds numbers is $\tilde{P}_{max} \approx 0.82$ and occurs at $\tilde{l} \approx 0.5$. The same trend is also observed in the DNS data, although the model slightly overpredicts the maximum of the DNS data, which has a value of about 0.8 for case 1 and 0.75 for case 2. Apart from the slight overshoot at the maximum the model is in quite good agreement with the DNS data, reproducing the linear increase for small values of $\tilde{l}$ due to the diffusion controlled drift of small segments as well as the exponential decay for large segments. The exponential tail for large values of $\tilde{l}$ is further highlighted in the log-lin inset in Fig. 6, where only for values of $\tilde{l} > 3$ the model overpredicts the exponential decay.

Finally, let us consider the scaling of the mean length l_m of streamline segments. Based on Eq. 13 and the constancy of v_e at the two different Reynolds numbers, we express the mean length as

$$l_m \propto (\nu a_\infty)^{1/2} \propto (\nu k/\varepsilon)^{1/2} \propto \lambda, \qquad (21)$$

where the proportionality to the Taylor microscale follows from the assumption that $a_\infty \propto \tau$, where $\tau = k/\varepsilon$ is the integral time scale.

6 Conclusion

Streamlines constitute natural and flow intrinsic geometries in turbulent flow fields. Different from gradient trajectories, they allow the analysis of vector valued fields, such as the turbulent velocity field itself. Based on an approach by Wang [1] we have considered streamline segments in homogeneous isotropic decaying turbulence. In order to model their length distribution $P(l)$ we have analyzed the dynamics which streamline segments are subject to. Due to their non-local nature, slow and fast changes were discerned, of which the latter yield large instantaneous jumps of the length of segments, while the first are responsible for continuous changes. The distinction of slow and fast changes has allowed the expansion of the transition probabilities associated with the slow changes into a Taylor series up to second order, yielding a convection and a diffusion term in the transport equation for the length distribution of streamline segments. The transition probabilities due to the fast changes remain in the equation yielding integral terms which were modeled based on a generic cutting/reconnection process, while the convection and diffusion terms could be associated with first and second order jump moments of l.

In order to derive expressions for the jump moments, we have identified all points defined by $\partial u/\partial s = 0$ as an isosurface, in which all streamline segments ending points lie. This isosurface divides space into two regions and streamlines pass alternatingly through these regions. Assuming the geometry of streamlines to be slowly changing as compared to the motion of the isosurface, which is treated by means of a level set method, the dynamics of streamline segments are mostly attributed to the relative motion of two neighboring isosurfaces, which define the boundaries of a streamline segment. Based on the governing equation for the isosurface two contributions to the dynamics of streamline segments were identified, namely a diffusion controlled region, which leads to the annihilation of very small segments and a strain/pressure dominated region yielding an extension of large segments. The second order jump moment is related to the first order one on theoretical grounds and it turns out that for large Reynolds numbers it is of higher order small.

Both the first and second order jump moments are accessible from the DNS data of homogeneous isotropic decaying turbulence, when at two consecutive time steps streamline segments associated with homogeneously distributed grid points are identified and, excluding fast changes, the slow ones are analyzed. The results are examined based on the theoretical considerations and in the course of the analysis it turns out, that the conditional velocity difference at the ending points of streamline segments scales linearly with the separation distance. Using its slope a_∞, which constitutes the inverse of a time scale, as one of the normalization parameters, the curves for two different Reynolds numbers collapse well and the scaling for the second order jump moment can be confirmed. The unusual scaling of the conditional mean velocity difference, which does not follow the typical Kolmogorov $1/3$ scaling, will be the subject of future research.

When using ansätze to provide analytical expressions for the first and second order jump moments to numerically solve the pdf transport equation it turns out that the influence of the diffusion term is small, so that at leading order the normalized pdf $\tilde{P}(\tilde{l})$ becomes Reynolds number independent. This is shown to be true even at relatively low Reynolds numbers of $Re_\lambda = 50 - 100$.

The final conclusion we draw from our analysis, is that the mean separation distance between two neighboring extreme points of the absolute value of the velocity field along streamlines, which corresponds to the length of the streamline segment, scales with the Taylor microscale rather than with any other turbulent length scale.

Acknowledgments This work was funded by the NRW-Research School "BrenaRo", the Cluster of Excellence "Tailor-Made Fuels from Biomass", which is funded by the Excellence Initiative of the German federal state governments to promote science and research at German universities and by the Gauss Center for Supercomputing in Jülich.

References

1. Wang, L.: On properties of fluid turbulence along streamlines. J. Fluid Mech. **648**, 183–203 (2010)
2. Krogstad, P., Davidson, P.A.: Freely-decaying, homogeneous turbulence generated by multi-scale grids. J. Fluid Mech. **680**, 417–434 (2011)

3. Vassilicos, J., Valente, P.C.: The decay of homogeneous turbulence generated by a class of multi-scale grids. J. Fluid Mech. (to be published)
4. Oberlack, M., Rosteck, A.: New statistical symmetries of the multi-point equations and its importance for turbulent scalign laws. Discrete Continous Dyn. Syst. Ser. S **3**, 451–471 (2010)
5. Rosteck, A.M., Oberlack, M.: Lie algebra of the symmetries of the multi-point equations in statistical turbulence theory. J. Nonlinear Math Phys **18**, 251–264 (2011)
6. von Kármán, T., Howarth, L.: On the statistical theory of isotropic turbulence. Proc. Roy. Soc. Lond. A **164**, 192–215 (1938)
7. Kolmogorov, A.N.: The local structure of turbulence in an incompressible viscous fluid for very large Reynolds numbers. Dokl. Akad. Nauk SSSR **30**, 301–305 (1941)
8. Kolmogorov, A.N.: Dissipation of energy under locally isotropic turbulence. Dokl. Akad. Nauk SSSR **32**, 16–18 (1941)
9. Corrsin, S.: Random geometric problems suggested by turbulence. In: Rosenblatt, M., van Atta, C. (eds.) Statistical Models and Turbulence, volume 12 of Lecture Notes in Physics, pp. 300–316. Springer, Berlin (1971)
10. She, Z.S., Jackson, E., Orszag, S.A.: Intermittent vortex structures in homogeneous isotropic turbulence. Nature **344**, 226–228 (1990)
11. Kaneda, Y., Ishihara, T.: High-resolution direct numerical simulation of turbulence. J. Turbul. **7**, 1–17 (2006)
12. Wang, L., Peters, N.: The length scale distribution function of the distance between extremal points in passive scalar turbulence. J. Fluid Mech. **554**, 457–475 (2006)
13. Gibson, C.H.: Fine structure of scalar fields mixed by turbulence i. zero gradient points and minimal gradient surfaces. Phys. Fluids **11**, 2305–2315 (1968)
14. Schaefer, P., Gampert, M., Goebbert, J.H., Wang, L., Peters, N.: Testing of different model equations for the mean dissipation using Kolmogorov flows. Flow Turbul. Combust. **85**, 225–243 (2010)
15. Schaefer, P., Gampert, M., Gauding, M., Peters, N., Trevi, C.: *ñ* o. The secondary splitting of zero gradient points in a turbulent scalar field. J. Eng. Math. **71**(1), 81–95 (2011)
16. Schaefer, L., Dierksheide, U., Klaas, M., Schroeder, W.: Investigation of dissipation elements in a fully developed turbulent channel flow by tomographic particle-image velocimetry. Phys. Fluids **23**, 035106 (2010)
17. Rao, P.: Geometry of streamlines in fluid flow theory. Def. Sci. J. **28**, 175–178 (1978)
18. Braun, W., De Lillo, F., Eckhardt, B.: Geometry of particle paths in turbulent flows. J. Turbul. **7**, 1–10 (2006)
19. Goto, S., Vassilicos, J.C.: Particle pair diffusion and persistent streamline topology in two-dimensional turbulence. New J. Phys. **6**, 65 (2004)
20. Sethian, J.A.: Level Set Methods and Fast Marching Methods. Cambridge Monographs on Applied and Computational Mathematics (1999)
21. Peters, N.: Turbulent Combustion. Cambridge University Press, Cambridge (2000)
22. Wang, L., Peters, N.: Length scale distribution functions and conditional means for various fields in turbulence. J. Fluid Mech. **608**, 113–138 (2008)
23. Mansour, N.N., Wray, A.A.: Decay of isotropic turbulence at low Reynolds number. Phys. Fluids **6**, 808–814 (1993)
24. Wang, L.: Scaling of the two-point velocity difference along scalar gradient trajectories in fluid turbulence. Phys. Rev. E **79**, 046325 (2009)
25. Gampert, M., Goebbert, J.H., Schaefer, P., Gauding, M., Peters, N., Aldudak, F., Oberlack, M.: Extensive strain along gradient trajectories in the turbulent kinetic energy field. New J. Phys. **13**, 043012 (2011)

Planar, Stereoscopic, and Holographic PIV-Measurements of the In-Cylinder Flow of Combustion Engines

T. van Overbrüggen, I. Bücker, J. Dannemann, D.-C. Karhoff, M. Klaas and W. Schröder

Abstract The experimental analysis of the highly unsteady three-dimensional flow in internal combustion (IC) engines requires measurement techniques that are able to capture the velocity field with high temporal and spatial resolution. Among other techniques, particle-image velocimetry (PIV) is used to measure the flow in combustion engines. Depending on the specific goal of a measurement series, either standard 2D-2C PIV, stereoscopic PIV (2D/3C) or fully three-dimensional PIV methods (3D/3C) can be used. In this study, the fundamentals of particle-image velocimetry (PIV) are explained in detail, with a special focus on the application of this measurement technique to internal combustion engines. As far as the generation of the particle images is concerned, this paper describes the special characteristics of seeding particles for use in IC engines including the generation process followed by a short introduction into different light sources and light-sheet generation methods. With regard to image acquisition and processing, digital imaging devices and image evaluation methods are described. Moreover, three component two dimensional and three dimensional PIV measurement techniques, namely stereoscopic-PIV and holographic-PIV, are concisely explained. Hereafter, two-component PIV measurements in several planes, three-component-PIV measurements in a set of planes and holographic-PIV measurements in the whole volume of the intake flow of a four valve piston engine at 160° crank angle are analyzed. The results of the stereoscopic PIV measurements show the highly three-dimensional propagation of the engine flow. Furthermore, the feasibility of holographic PIV for the analysis of engine flows is confirmed.

T. van Overbrüggen (✉) · I. Bücker · J. Dannemann · D.-C. Karhoff · M. Klaas · W. Schröder
Institute of Aerodynamics RWTH Aachen University, Wüllnerstr. 5a, 52062 Aachen, Germany
e-mail: t.van-overbrueggen@aia.rwth-aachen.de

M. Klaas et al. (eds.), *Fuels From Biomass: An Interdisciplinary Approach*,
Notes on Numerical Fluid Mechanics and Multidisciplinary Design 129,
DOI 10.1007/978-3-662-45425-1_11

1 Introduction

The analysis of flow fields in technical applications, e.g., engines, turbines compressors, and so forth, is of great importance for the further development of these technical systems. While measurement techniques like hot-wire anemometry and laser-doppler anemometry (LDA) are often used to analyze the flow in one single measurement point with a high temporal resolution, the analysis of a complete flow field using theses techniques, is very time consuming as the flow has to be measured point by point. To gain a broader impression of the spatial development and expansion of the flow structures, it is advantageous to use whole-field measurement techniques like particle-image velocimetry.

First planar PIV measurements were performed by Guibert et al. [17]. They introduced the PIV measurement technique for in cylinder flows and presented first digital measurements of a two stroke engine in motored conditions.

Reuss et al. [33] accomplished two-dimensional analog PIV measurements in the burned and unburned gas of a firing, reciprocating internal combustion engine at either 2 or 4 crank angle before top dead center (btdc). The showed the velocity, filtered velocity, vorticity, and strain-rate distribution for this crank angle to analyze the interaction of the flame and the flow. They present a strong correspondence between the flow direction of coherent structures and the flame wrinkling, although the flame appears to propagate around, rather than through eddy structures.

Konrath et al. [24] performed first holographic PIV (H-PIV) measurements for a light sheet in the center plane of a test engine for 120°, 180°, and 240° after top dead center (atdc). They showed the velocity distribution for all three velocity components in this plane. Furthermore, they state that whole three-dimensional measurements are necessary to fully understand the formation and development of the in-cylinder flow.

Dannemann et al. [6–8] performed multi-planar flow measurements in a motored four-valve combustion engine for several crank angles of intake and compression. They showed the propagation of the tumble vortex, as well as a ring vortex pair below the inlet valves. Furthermore, the turbulent kinetic energy was used, to visualize small scale turbulent structures inside the engine flow.

Bücker et al. [2] accomplished multi-planar stereoscopic PIV measurements for 14 axial planes in a motored direct-injection spark-ignition (DISI) engine with tumble intake ports at 15 crank angles during intake and compression strokes. They showed a c-shaped tumble vortex and two ring vortices beneath the intake valves. Furthermore, their temporal analysis presented the evolution of the mean kinetic energy which is conserved until late compression through the tumble motion.

This paper presents the fundamentals of particle-image velocimetry (PIV) with a special focus on the application of PIV to internal combustion engines, i.e., the measurement equipment and the experimental setup as well as exemplary results of different PIV setups in two single-cylinder IC engines. A brief description of the basic features of this measurement technique will be followed by a short overview of seeding, light sources, light-sheet optics, image recording hardware, and image post-processing. Next, three different PIV setups for measurements in IC engines, namely, planar 2C/2D PIV, stereoscopic PIC (S-PIV, 3C/2D) and holographic PIV (H-PIV, 3C/3D) will be explained. The paper will conclude by experimental results illustrating the flow field inside the cylinder of two different motored four-valve combustion engines.

2 Principles of Particle-Image Velocimetry

Particle-image velocimetry is a non-intrusive, whole field, (laser) optical measurement technique that determines the velocity components of a flow field, based upon the tracking of seeding particles that are added to the flow. The experimental setup of particle-image velocimetry typically consists of five basic elements. Figure 1 shows a typical PIV setup in a wind tunnel. Tracer particles, which were

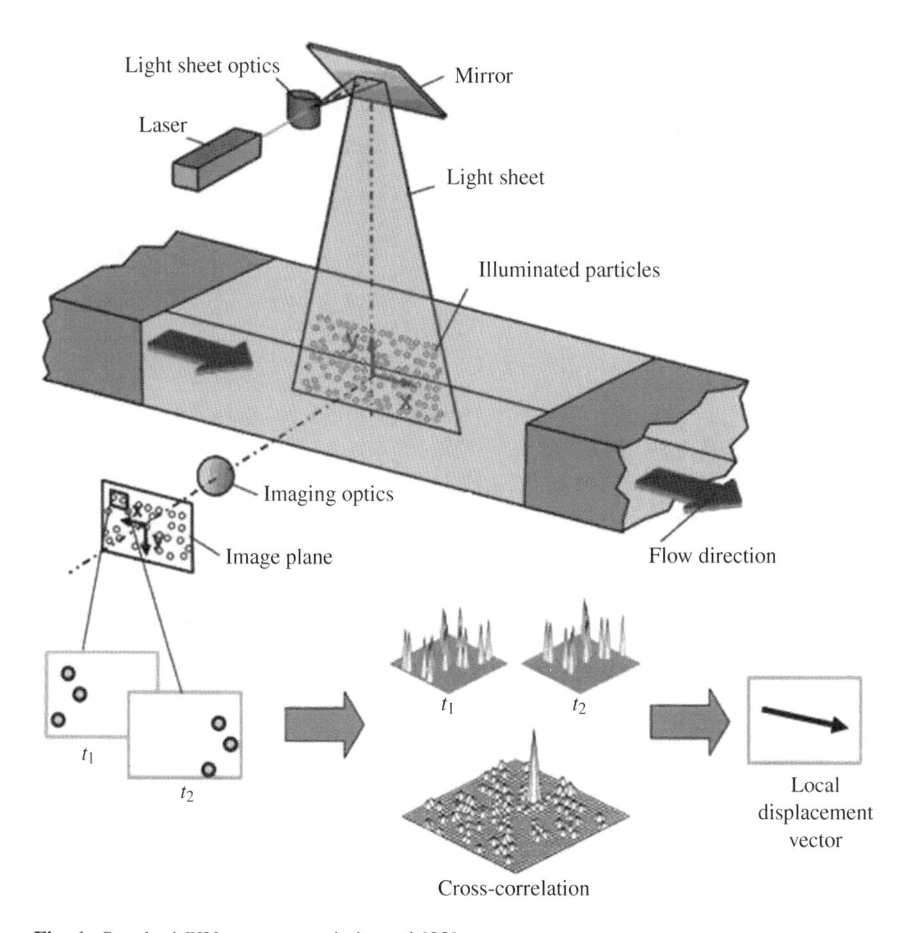

Fig. 1 Standard PIV setup at a windtunnel [32]

added to the flow (seeding) are illuminated by means of a laser light sheet at least twice within a short time interval (light sheet generation and flow illumination). A camera records the scattered light on different frames (recording). Finally, a software calculates the displacement of the particle images between the different light pulses and a post-processing is performed to eliminate and replace spurious velocity vectors [32] (processing/post-processing). To calculate a vector velocity map the images are divided in small subareas, so called “interrogation windows”. Cross-correlation algorithms are used to calculate the displacement between the different illuminations of the different interrogation windows. The velocity is calculated from the displacement and the time between the different illuminations.

Compared to measurement techniques like hot-wire anemometry or velocity measurements by pressure tubes PIV does not require probes that might disturb the flow through its mere presence. As a whole-field measurement technique PIV is capable of measuring the flow in a large area. In addition, the measurement of all three velocity components is possible. For this purpose, the experimental setup has to be changed either to a stereoscopic (3C-PIV) [32] or to a holographic (Holo-PIV) [19, 23, 31] or tomographic (Tomo-PIV) [10] PIV setup. The last two setups are also capable of measuring not only the flow in a light sheet but also inside a whole measurement volume.

Since PIV is an indirect measurement technique, not the velocity of the flow itself but rather the velocity of the tracer particles is measured. Hence, tracer particles, that do not follow the flow perfectly, can disturb the measurement. Therefore, one has to carefully test the particle characteristics if they fit to the specific flow problem. Moreover, the seeding density has to be adjusted. A flow with not enough particles generates no correlation peaks in some interrogation windows, whereas too many particles lead to ambiguous correlation peaks due to noise.

2.1 Tracer Particles

As mentioned above, the measurement of the velocity is indirect due to the fact that the velocity of the tracer particles is measured instead of the flow velocity itself. Therefore, it is important to determine the mechanical properties of the tracer particles as well as the light scattering behavior to find the best suited particles for a given flow.

2.1.1 Fluid Following Behavior

One main source of errors in PIV measurements is the influence of inertia forces that act upon the seeding particles. On the one hand, the density of the particles might differ from the density of the fluid and the particle diameter is not infinitely small. Hence, the mass of the particles and therefore the diameter affects the inertia of a particle. This leads to the conclusion that the particles should be as small as possible and that their density should be identical to that of the fluid. On the other hand, a minimum particle diameter is necessary to scatter enough light [32]. Therefore, a compromise must be found between high light intensity and good flow tracking capability. To calculate the influence of acceleration forces Ruck [34] determines the sedimentation speed u_g as a function of particle density, provided that the flow around the particle can be assumed laminar around a sphere and therefore STOKES drag can be applied, as follows:

$$u_g = \frac{1}{18}\frac{\rho_p \cdot d^2}{\rho_f \cdot \nu}\left(1 - \frac{\rho_f}{\rho_p}\right) \cdot g \tag{1}$$

where u_g is the particle velocity due to gravitational forces, ρ_p is the density of the tracer particle, ρ_f the density of the fluid, ν the kinematic viscosity, d the diameter of the particle and g the acceleration due to gravity. In analogy to Eq. (1) the influence of general acceleration on a particle can be calculated by

$$u_p = \frac{1}{18}\frac{\rho_p \cdot d^2}{\rho_f \cdot \nu}\left(1 - \frac{\rho_f}{\rho_p}\right) \cdot a \tag{2}$$

where u_p is the particle velocity due to general acceleration and a is the general acceleration. Moreover, the ability of particles to react to accelerations is of great importance. Therefore, Ruck [34] calculates the relaxation time τ_s

$$\tau_s = \frac{1}{18}\frac{\rho_p \cdot d^2}{\rho_f \cdot \nu} \tag{3}$$

This value determines the time a particle of given size needs to adapt to a new flow condition. Therefore, a low relaxation time corresponds to a good flow tracking behavior. Figure 2 shows the sedimentation speed and the relaxation time of Di-Ethyl-Hexyl-Sebacat (DEHS) droplets, titanium dioxide and Expancel Microspheres in air. It is evident that Microspheres have a much lower sedimentation speed and relaxation time than DEHS and TiO_2 due to their much lower density. At not constant fluid acceleration or if STOKES drag can not be applied the calculation of the flow velocity becomes more difficult [32]. Nonetheless, τ is an eligible indicator to determine the flow following behavior.

2.1.2 Light Scattering Behavior

The amount of scattered light depends on polarization, observation angle [32] and the difference in refractive indices of the particle and the surrounding medium. For spherical particles of the size of approximately $0.1\,\mu m \leq d_p \leq 100\,\mu m$, light is scattered according to the rules the rules of the Mie-theory. For particles with diameters significantly smaller than the wavelength of light, the light scattering

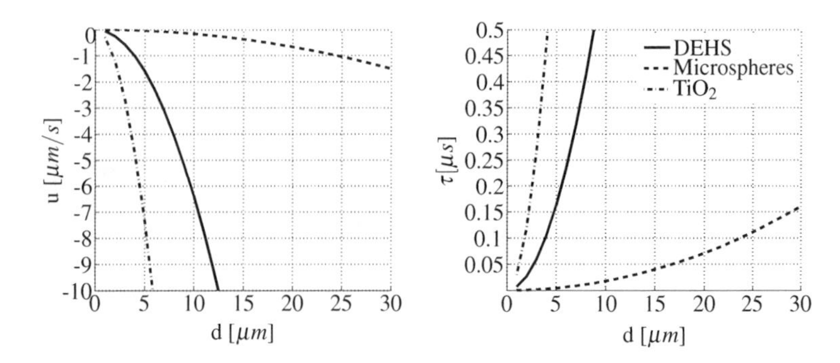

Fig. 2 Sedimentation speed and relaxation time of DEHS, TiO_2 and Microspheres for different diameters in air

behavior can be described using the Rayleigh approximation. Larger particles scatter the light by means of the rules of geometric optics. PIV particles usually are in the size range mentioned above an thus Mie scattering theory can be applied. A detailed description of Mie scattering is given by Born [3], Born and Wolf [4] and van de Hulst [43].

Figure 3a–f show the light scattering behavior of DEHS, titanium dioxide (TiO_2) and Expancel Microspheres particles with different diameters for light with a wavelength of $\lambda = 532$ nm in air. With increasing particle diameter the intensity of scattered light increases rapidly. Moreover, the intensity of light for different particle diameters is characterized by oscillation effects if only one observation angle is taken into account. If different observation angles for one particle size are observed, one can see that the highest intensity is given by forward scatter. Nevertheless, a recording at 90° is most often used, due to the limited depth of field at forward scatter configuration.

Furthermore, the light is spread in all directions by the particle. Therefore, particles inside a light sheet are not only illuminated by direct illumination but also by the scattered light from other particles in the light sheet. Hence, a larger quantity of particles increases the intensity of every single particle. Nevertheless, too high particle concentrations are disadvantageous due to the background noise that will increase with an increasing number of particles. For measurements inside a piston engine, normally DEHS with diameters of about $d = 0.5$ μm is used. Furthermore, for holographic PIV measurements, Microspheres are used, due to their high light scattering behavior.

2.1.3 Particle Generation

As explained in Sect. 2.1.1, it is important for PIV measurements to know the diameter of the tracer particles to determine the error due to velocity lag of the particles. If solid particles like TiO_2 or Microspheres are used, the diameter is defined by the production process. However, for liquid particles in air flows or

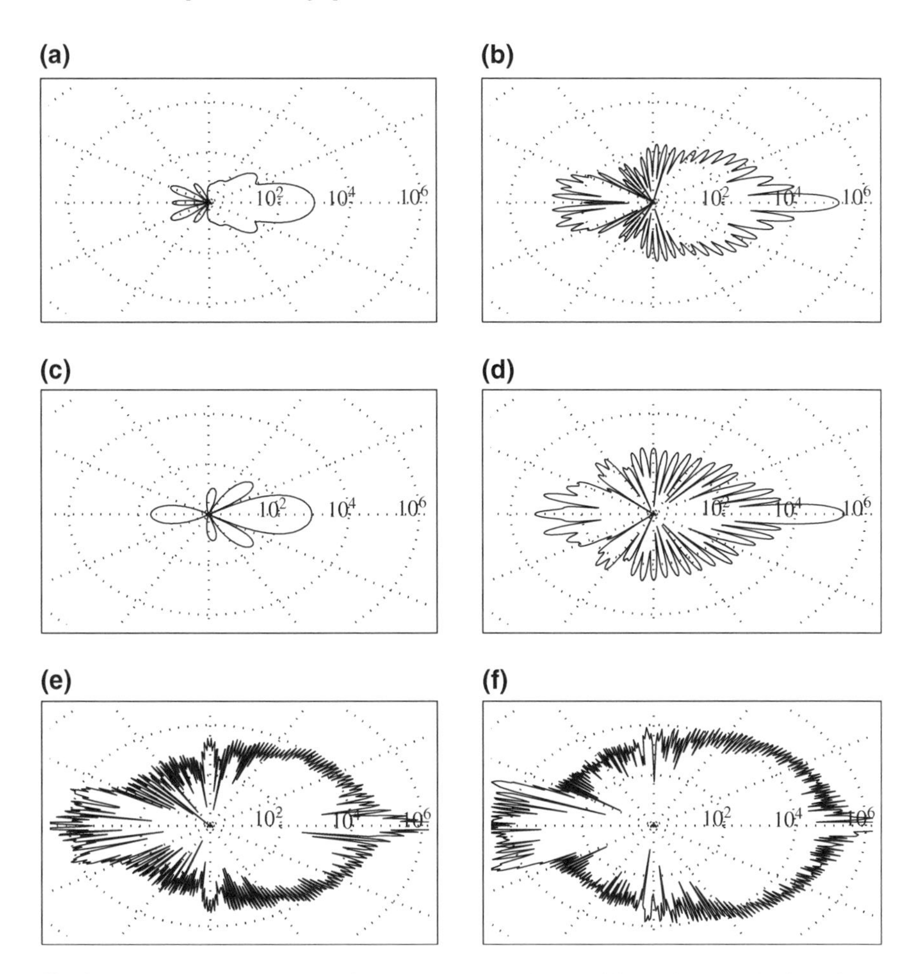

Fig. 3 Light scattering behavior of DEHS (*top*) and TiO_2 (*middle*) and expancel microspheres (*bottom*) particles in air for light with a wavelength of $\lambda = 532$ nm. **a** DEHS: $\varnothing = 1$ µm, **b** DEHS: $\varnothing = 5$ µm, **c** TiO_2: $\varnothing = 1$ µm, **d** TiO_2: $\varnothing = 5$ µm, **e** Microspheres: $\varnothing = 20$ µm, **f** Microspheres: $\varnothing = 40$ µm

gaseous bubbles in water flows special seeding generators have to be used. To seed air flows with oil droplets, a seeding generator with laskin nozzles, as shown in Fig. 4, is used. The seeding generator consists of a closed container, two air supplies, and an outlet for the seeding particles. One of the air supplies is connected to several laskin nozzles, while the other one is used to generate a bypass airflow. A laskin nozzle consists of a pipe that is closed at its end. Shortly above the end, four holes with a diameter in the order of a millimeter are circularly drilled inside the pipe. Above the holes a circular plate is attached to the pipe. The plate also has four holes, which are at the same radial position like the holes inside the pipe. To generate seeding particles, compressed air, with an overpressure of about one bar with respect to the outlet pressure, is applied to the laskin nozzles. The air leaves the

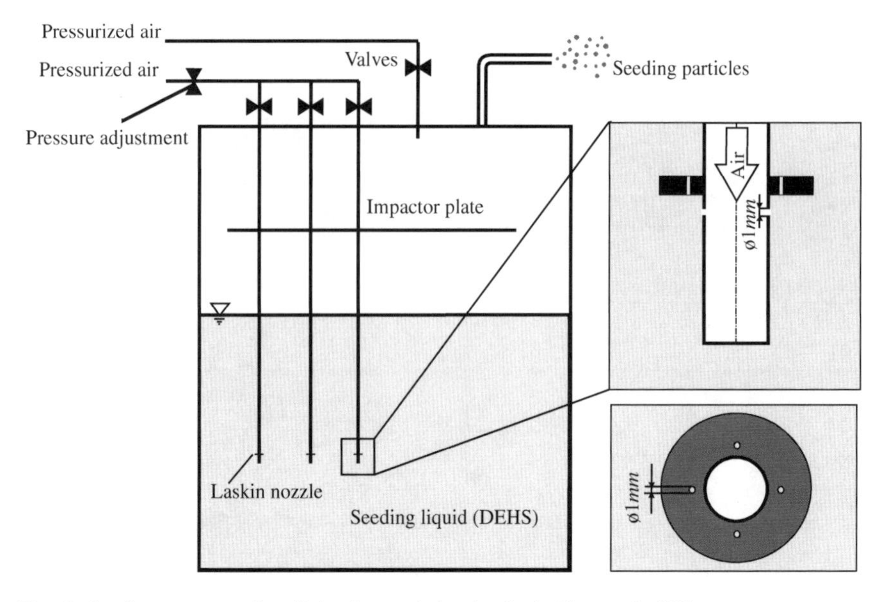

Fig. 4 Seeding generator for oil droplets and sketch of a laskin nozzle [32]

nozzles through the holes and small droplets are generated by the shear stress between the air and the oil. The air bubbles carry the droplets out of the oil. The impactor plate retains bigger droplets inside the generator, whilst smaller droplets leave the generator through the outlet. To control the amount of generated seeding, the number of nozzles and the overpressure can be varied.

For liquid flows, solid particles or hollow glass spheres are used. However, for applications, where measurements are disturbed through sedimented particles at walls or the model, air bubbles are an alternative. To generate air bubbles Schröder et al. [35] used a micro bubble generator, that is able to generate air bubbles of the size of 5–30 μm with a mean diameter of 12.7 μm.

Another alternative of generating micro bubbles is the generation of hydrogen and oxygen bubbles through electrolysis. This process was used by Große et al. [15] and Soodt et al. [38] to measure the flow inside a realistic model of the first bifurcations of the human lung. The tracer diameter depends on the shear flow rate, the distance between the cathode and the anode of the electrolysis unit, and the voltage applied to the unit.

2.2 Light Sources

Light sources for PIV applications have to fulfill several requirements. At first, a homogeneous illumination of the flow using a thin light sheet with high light intensity is desirable. Hence, light sources with small diverging angles are best suited. For this reasons, mostly lasers are used as PIV light source. Laser beams can easily be bundled into thin light sheets for illuminating tracer particles, their light has a high intensity and is monochromatic and laser light is coherent. The latter being of special interest for holographic PIV. A detailed description of the fundamentals of laser systems can be found in [9, 28, 36, 40, 42] and is beyond the scope of this article.

New investigations from Willert et al. [47] show that light emitting diodes (LEDs) can be used as a light source for PIV measurements. LEDs generate a high energy output with a high efficiency factor. Furthermore, it is possible to run LEDs in pulsed operation with pulse distances of nanoseconds and very high repetition rates [11]. Hence, they can be driven either at decreased heat radiation, or at a higher pulse current and therefore at higher light intensity. LEDs are a cost-efficient alternative for high-speed lasers, which are, until now, the most expensive part of a high-speed PIV system. One drawback of LEDs is the difficulty to bundle the light to thin light sheets. The standard light sheet optics are not usable for LEDs. Therefore, Willert et al. [47] used a fiber optic bundle to catch the light. For measurements inside a piston engine, standard laser light is used since thin light sheets are important.

2.3 Light Sheet Optics

The essential element of standard light sheet optics is a cylindrical lens. If a laser with a low divergence angle and a small beam diameter is used, e.g., an Ar^+ laser, one cylindrical lens can be sufficient to generate a light sheet. Lasers with worse beam conditions, e.g., Neodym-doped Yttrium Aluminium Garnat (ND:YAG)-lasers, need a combination of different lenses to generate thin light sheets with high intensity [32]. For this purpose, different lens combinations can be used.

Figure 5a shows a combination of three cylindrical lenses. The first lens is used to generate the light sheet, whereas the second lens has been added to collimate the light to generate a light sheet of constant height. The third lens focuses the light sheet to an appropriate thickness. The biconcave lens was chosen because focal lines should be avoided. For high power lasers, the air near the focal line might be ionized.

The setup in Fig. 5b is used to make the light sheet optics more flexible. The height of the light sheet is defined through the focal length of the planoconvex cylindrical lens in the middle. The light sheet thickness can be adjusted by changing the distance between the two spherical lenses.

For applications where it is important to change the light sheet height and thickness independently, the configuration shown in Fig. 5c is feasible. Since it is not possible to change both parameters with a configuration where spherical lenses are used, only cylindrical lenses are employed. The focal length of the first lens defines the light sheet height, whereas the width of the light sheet is defined by the last two lenses. Additionally this configuration is capable of generating thinner light sheets than the beam diameter of the laser beam [32].

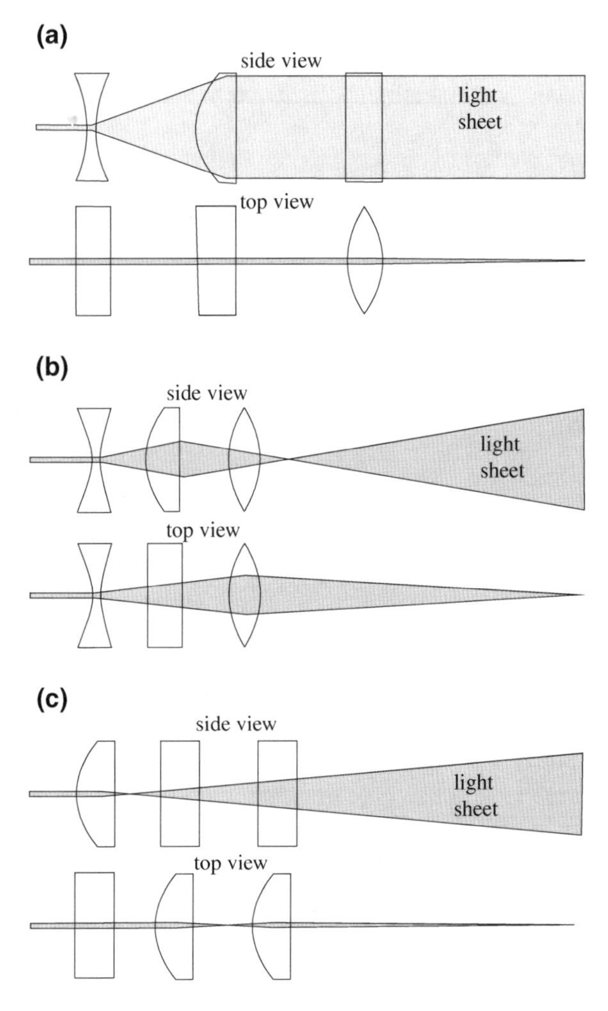

Fig. 5 Light sheet optics using three cylindrical lenses (one of them with negative focal length) (**a**), light sheet optics using two spherical lenses (one of them with negative focal length) and one cylindrical lens (**b**) and light sheet optics using three cylindrical lenses (**c**) [32]

2.4 Image Recording

The first PIV setups worked with analog cameras. These cameras worked mostly with double exposure, such that two PIV pictures where captured on one image. Therefore, a determination of the flow direction was not possible. In modern PIV investigations, mostly digital cameras are used. To capture images, either charged-coupled device (CCD) or complementary metal–oxide–semi-conductor (CMOS)-chips are used. The general function of a CCD-chip as well as the function of a CMOS-chip is well described in [16, 22, 32]. The choice of the photo sensor depends on the application. A CMOS-chip can be read out very fast, due to the possibility to control every pixel separately, whereas in a CCD-chip the pixels can only be controlled row by row. Therefore, CMOS-chips are better suited for high speed measurements. Furthermore, the CMOS-chip has some other benefits. Its manufacturing size is smaller than that of a CCD, its dynamic range is higher, and the power consumption is less. Nevertheless, it also has some drawbacks: the light sensitivity is lower than that of a CCD and its random noise is slightly increased. Therefore, CMOS-chips are mostly used in high-speed PIV measurements. In contrast, CCD-chips are used for applications when very low noise and high light intensity is desirable.

2.5 Digital Image Processing

In the early (analogue) years of PIV, image processing was based upon auto-correlation algorithms, due to the double exposure of the images. Nowadays, where digital cameras are used for image recording, cross-correlation algorithms are used. The cross-correlation function determines the similarity of two signals at two different time steps. In the context of PIV, the cross-correlation possesses its maximum where the first image and the shifted second image have the maximum consistency. The shifted distance determines the particle displacement.

To generate a vector map from the PIV images, they are divided in small interrogation windows. To increase spatial resolution, these windows may overlap each other. A cross-correlation algorithm then calculates the displacement between one interrogation window of image one and the corresponding window of image two. A fast and stable cross-correlation algorithm is shown in Fig. 6. Both images are Fourier transformed and a complex conjugated multiplication of the two 2D complex matrices generates a cross-power spectrum of the two images. Using a Fourier transform, this cross-power spectrum can be transformed into the cross-correlation of the two images. The peak of the cross-correlation then represents the displacement of the particles of this pair of windows. This process is then repeated for all windows of an image, thus generating a vector map.

In some cases, the cross-correlation algorithm fails to determine the correct displacement and velocity, respectively, and creates spurious vectors which must be filtered using several filter methods. Among other filters, the most common filters are the global velocity filter, which filters every vector, that lies not inside a defined velocity range or the local velocity filter, which filters all vectors, whose velocity

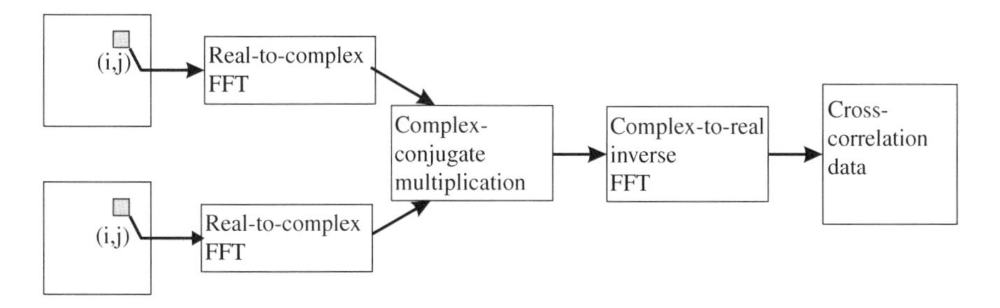

Fig. 6 Implementation of cross-correlation using fast Fourier transforms [32]

diverges too much from the mean velocity of the surrounding vectors. The spurious vectors usually are replaced by interpolation. After generating a complete vector map from the filtered data, subsequent adaptive cross-correlations can be used to refine the measurement results. For this, the interrogation window size is reduced. Furthermore, the displacement of the interpolated vector fields is taken into account. Hence, the interrogation window of the second image is shifted. This allows a finer determination of the velocity. After this, further filtering and interpolating of invalid vectors can be done. An iterative refining of the velocity distribution is possible. A detailed description of the PIV algorithm including details like window shifting, window deformation, and multi-pass cross-correlation can be found in Raffel et al. [32] and Westerweel [45, 46] and is beyond the scope of this article.

2.6 Three-Component PIV Measurements

2.6.1 Stereo-PIV

The standard PIV setup is able to measure the two velocity components in the light sheet plane, i.e., the measurement plane (in-plane velocity). However, this light sheet possesses a certain thickness. For highly three-dimensional flows this may lead to a measurement error due to a velocity component perpendicular to the light sheet as it is illustrated in Fig. 7. Here, d is the light sheet thickness, dx the particle displacement in the x-direction, dz the particle displacement in the z-direction, x_1 the measured particle displacement in the x-direction and x_2 the true displacement in the x-direction. This error can be minimized through a large viewing distance with large focal lenses. Nevertheless, the knowledge of the third velocity component is of great importance for a wide field of PIV investigations. To measure third velocity component (out-of-plane velocity), several investigators used a second camera to get a stereoscopic view of the flow field [1, 5, 13, 29, 30, 37, 44]. Both cameras are tilted with respect to the measurement plane (Fig. 8). The two cameras

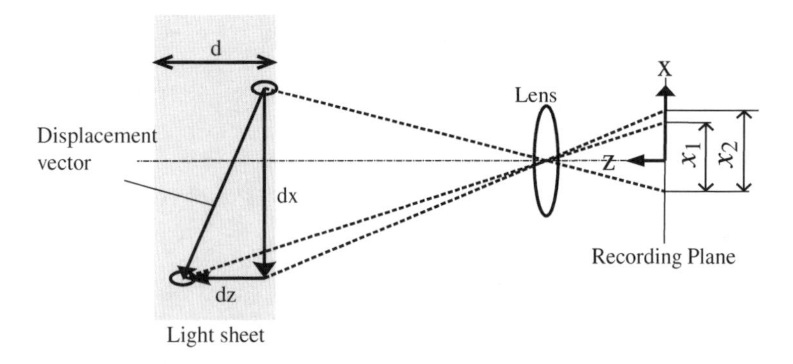

Fig. 7 Displacement error through third velocity component in 2C-PIV [32]

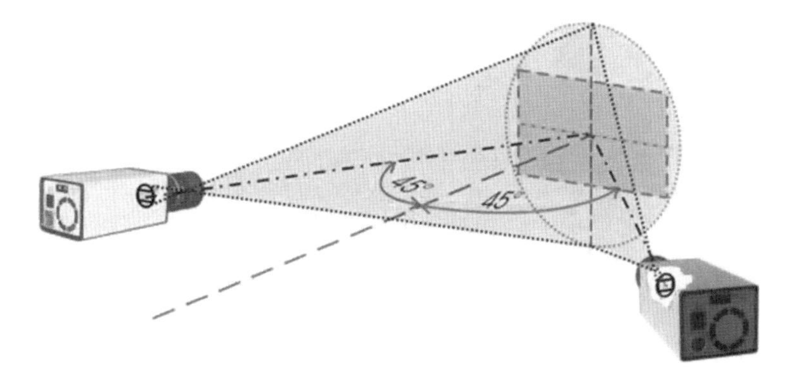

Fig. 8 Camera configuration for stereoscopic PIV

record a slightly different image of the flow and therefore a different displacement for the same interrogation windows. The true velocity components are calculated out of the displacements of both cameras. The Eqs. (4)–(10) calculate the true displacement vector for the single interrogation windows [32]

$$\tan\alpha = \frac{x_i(t+\Delta t)}{z_0} \tag{4}$$

$$\tan\beta = \frac{y_i(t+\Delta t)}{z_0} \tag{5}$$

$$u_r = \frac{x_i(t+\Delta t) - x_i(t)}{M\cdot\Delta t} \tag{6}$$

$$v_r = \frac{y_i(t+\Delta t) - y_i(t)}{M\cdot\Delta t} \tag{7}$$

$$u = \frac{u_r\cdot\tan\alpha_l + u_l\cdot\tan\alpha_r}{\tan\alpha_r + \tan\alpha_l} \tag{8}$$

$$v = \frac{v_r\cdot\tan\beta_l + v_l\cdot\tan\beta_r}{\tan\beta_r + \tan\beta_l} \tag{9}$$

$$w = \frac{u_l - u_r}{\tan\alpha_r + \tan\alpha_l} \tag{10}$$

The quantities α and β are the angles between the z-axis and the ray from the tracer particle through the lens center to the recording plane, whereas α lies inside the XZ-plane and β inside the YZ-Plane. The velocity components u, v and w are in the X-, Y-, and Z-direction, respectively. The quantity M defines the magnification of the used lenses. The variables x_0, x_i, y_0 and y_i define the position of the tracer particle in the X- and Y-direction at time step t and t′ respectively. These formulas are usable for every opening angle as long as the viewing axes are not collinear.

One problem that occurs with angular viewing with standard camera configurations is that not the complete light sheet plane is in the focal plane of the camera. Therefore, regions away from the focal plane are blurred during the recording. This problem appears for every image that does not satisfy the Scheimpflug condition. This condition states that the image plane, the lens plane, and the object plane of a camera have either to intersect in one point, or to be parallel. For standard PIV measurements the second condition is fulfilled. However, for stereoscopic PIV a Scheimpflug imaging arrangement is necessary. This arrangement is realized by tilting the lens with respect to the camera [32].

2.6.2 Holographic PIV

Holographic PIV uses the imaging technique of the holography to record and analyze velocity distributions in a three-dimensional measurement volume rather than in a light sheet. To get a volume information, not only the light intensity but also the phase of the light wave has to be recorded. A principle to measure the intensity as well as the phase was first presented by Gabor [12]. He described a method based on the interference of two wave fronts to get a replica of the original object wave. However, at this time, light sources with high coherence length where not available. Leith and Upatnieks [26] firstly demonstrated the principle of reconstructed wavefronts and the name holography was introduced. They used light from a laser that has not only a specific wave length, but also a high coherence length.

For recording a hologram two different techniques are possible. For in-line holography, a measurement volume is illuminated by laser light. The light is scattered by the seeding particles. Furthermore, some of the light does not hit a particle and therefore leaves the measurement volume not scattered. After the light has passed through the measurement volume, both parts of the light interfere and form an interference pattern. This pattern is recorded by a photosensitive emulsion either on glass or on a triacetat (TAC) film carrier. After the development process a permanent absorption grating is formed. If this grating is illuminated by light of the same wave length as it has been needed to record the image, a real image of the recorded flow pattern is produced. Off-axis holography works with two beams. The first beam illuminates the measurement area. This light is scattered by the particles. A second beam, the reference beam, is expanded and illuminates the whole recording material. The scattered light and the reference beam interfere and form an interference pattern that also generates a real image of the recorded flow, after the development process. The principle of recording a hologram is described by Herrmann [19], Konrath [23], and Kreis [25] in detail.

3 Results and Discussion

To illustrate the measurement results of the different PIV techniques two-component/two-dimensional (2C/2D) PIV measurements, three-component/two-dimensional (3C/2D) PIV measurements, and three- component/three-dimensional (3D/3D) PIV measurements at 160° crank angle are presented for two test engines. The first and the last measurement were performed in an engine that is based on a Suzuki DR 750 (SUZI) motorcycle four-stroke single-cylinder engine with a bore of D = 105 mm and a stroke of 84 mm. The engine is driven by an electrical 55 kW engine (Fig. 9). The original cylinder was replaced by a transparent liner made of Plexiglas that is placed between the original cylinder head of pent roof geometry and an extended lower iron liner, which keeps the piston properly aligned. The flat piston crown consists of Plexiglas to assure optical access. The engine is operated at a mean revolution speed of $\bar{n}$ = 1,500 rpm without fuel injection and combustion. The typical progression of the engine speed n is shown in Fig. 10 by the decay due to compression and the valve lift curves for the inlet and outlet valves over three full engine cycles.

The second measurement was conducted in a test engine built by FEV Motorentechnik with a bore of D = 75 mm and a stroke of 82.5 mm (Fig. 11 (TINA)). This optical engine is a motored model of a direct-injection spark-ignition (DISI) engine. The engine is equipped with a tumble intake port, a pent-roof combustion chamber, and a flat piston crown. Optical access is provided via the quartz-glass piston crown and the full-stroke quartz-glass liner. The injector and spark plug are mounted in the symmetry plane, the former between the intake valves and the latter

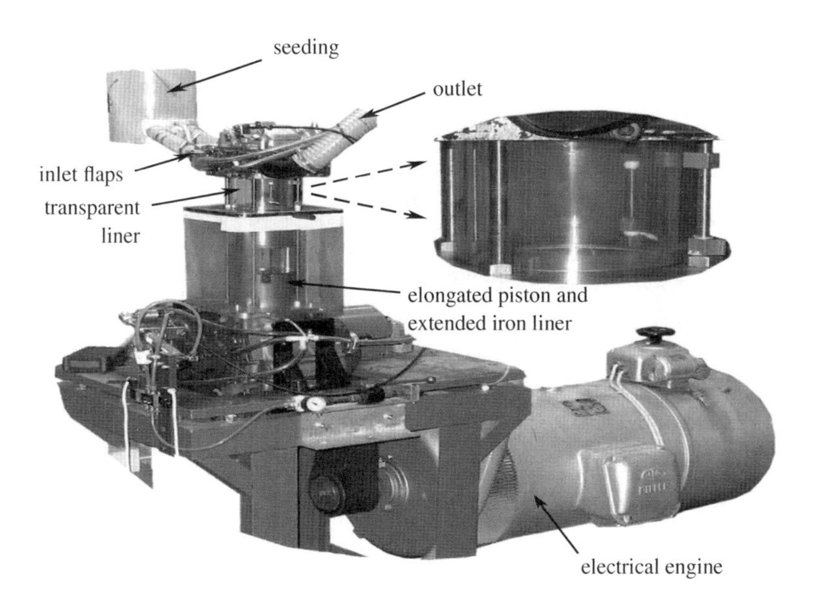

Fig. 9 First test engine "SUZI"

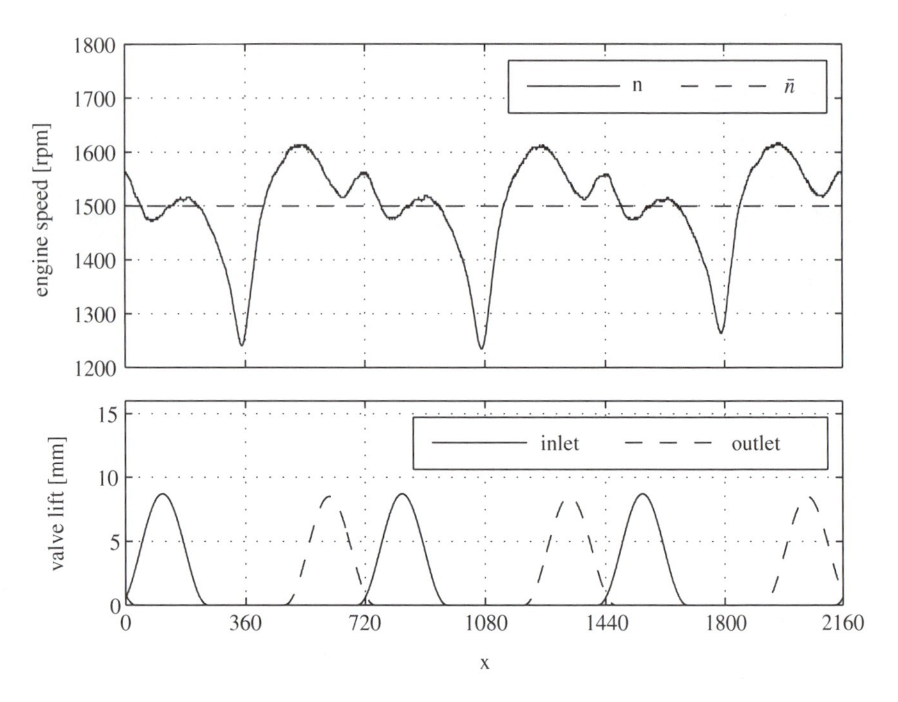

Fig. 10 Progression of engine speed and valve lift curves of the "SUZI"-engine

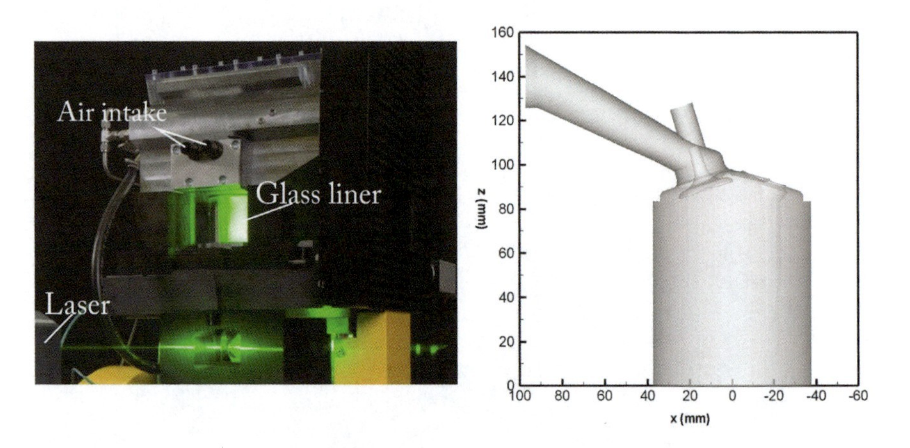

Fig. 11 Optical setup (*left*) and tumble port (*right*) of the "TINA" engine

between the exhaust valves. The engine is driven by a 30 kW electrical engine. Measurements have been conducted at 1,500 rpm. The intake valve lift and the piston speed curves are shown in Fig. 12. The main characteristics of both test engines are summarized in Table 1:

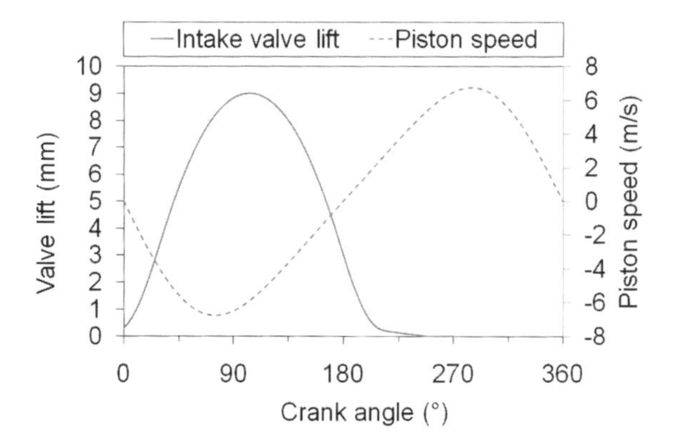

Fig. 12 Intake valve lift and piston speed of the "TINA"-engine

Table 1 Main characteristics of the test engines

Engine data	Suzuki DR750	FEV test engine
Bore (mm)	105 mm	75.0 mm
Stroke (mm)	84 mm	82.5 mm
Cylinder displacement (cm^3)	728 cm^3	364 cm^3
Compression ratio	12:1	7.4:1
Valve lift intake/exhaust (mm)	8.7/8.5 mm	9/9 mm
Valve diameter intake/exhaust (mm)	40/34 mm	27.1/23 mm
Valve angle towards vertical axis (° CA)	20°	17.5°
Exhaust valves close (1 mm lift) (° before TDC)	3° before TDC	20° before TDC
Intake valves open (1 mm lift) (° after TDC)	7° after TDC	10° after TDC
Intake and exhaust valve event duration (°)	211°	186°

3.1 Data Analysis

The ensemble averaged mean velocity component is defined by (Heywood [20] and Tennekes and Lumley [41])

$$\bar{U}_{i,j} = \frac{1}{N}\sum_{n=1}^{N} U_{i,j,n}, \tag{11}$$

where $U_{i,j,n}$ is the total velocity of the cell i,j in the n-th particle image and N is the number of particle images recorded for the specific crank angle. The ensemble-averaged variance u'^2 is defined by (Heywood [20] and Tennekes and Lumley [41])

$$\bar{U}'^2 = \sigma_u^2 = \frac{1}{N}\sum_{n}^{N} = 1u_i'^2 = \frac{1}{N}\sum_{n=1}^{N}(U_{i,j,n} - U_{i,j}), \tag{12}$$

with u′ being the velocity fluctuation and σ_u the standard deviation of u. These fluctuations are the sum of turbulent fluctuations and cyclic variations. Furthermore, turbulence is important for internal combustion engines, as it accelerates the propagation of the flame front. Turbulence can be quantified by the turbulent kinetic energy k (Tennekes and Lumley [41])

$$k = \frac{1}{2}\left((\bar{u})'^2 + (\bar{v})'^2 + (\bar{w})'^2\right). \tag{13}$$

Note that for the 2D/2C measurements, the corresponding out of plane velocity component $\bar{u}'^2$ or $\bar{v}'^2$ is zero. Since the turbulent kinetic energy also includes the cyclic variations of the engine flow, an uncertainty range with respect to the turbulent structures exists. Nevertheless, the general character of the flow field with respect to the turbulent kinetic energy can be qualitatively investigated. Flow structures are analyzed using the vortex center identification criterion Γ_1 (Graftieaux [14])

$$\Gamma_1 = \frac{1}{s}\sum_{i=1}^{s} \frac{r \times v}{\|r\|\|v\|}. \tag{14}$$

The quantities s is the number of grid points used to calculate Γ_1, r the radius vector, and v the velocity vector at point i.

3.2 2C-PIV in Different Planes

3.2.1 Experimental Setup

Particle-Image Velocimetry System

A schematic of the experimental setup is shown in Fig. 13. The PIV system consists of a Spitlight600 laser, a PCO Sensicam QE double shutter PIV camera, and a Nikon lens with a focal length of 50 mm and a minimum f-number of 1.8. The maximum laser energy is 400 mJ per double pulse and the maximum frequency is 10 Hz. A system of 4 lenses produces the light sheet at a thickness of approximately 1 mm whereas the light sheet covers the whole cylinder stroke. DEHS at a mean diameter of 0.5 μm is used as tracer particles. The seeded air is stored in an air reservoir upstream of the inlet flaps. The flaps open shortly before the measurement to reduce a contamination of the transparent liner. The PCO Sensicam QE has a resolution of 1,376 × 1,040 pixels at a maximum frame rate of 4 Hz. The frame rate is set to 3.125 Hz. Hence, every fourth intake stroke is measured. The laser pulse separation is set to Δt = 10 μm. The positions of the measurement planes are shown in Fig. 14.

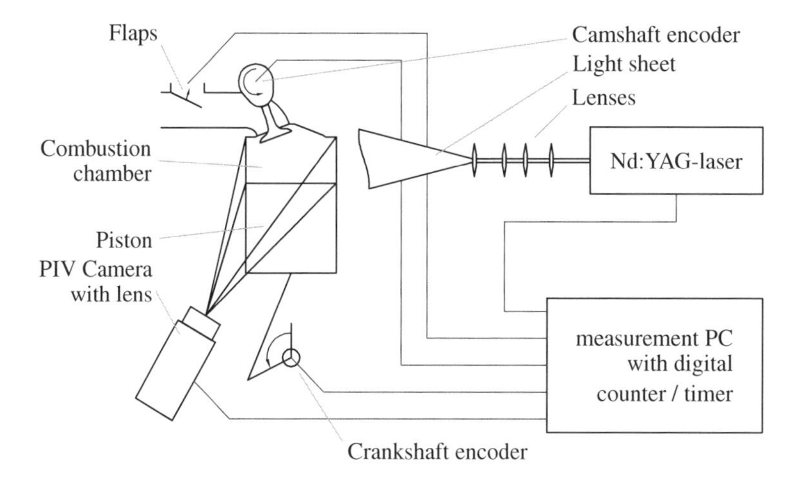

Fig. 13 Schematic of the experimental setup of the "SUZI" engine

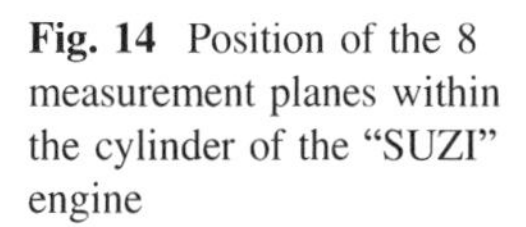
Fig. 14 Position of the 8 measurement planes within the cylinder of the "SUZI" engine

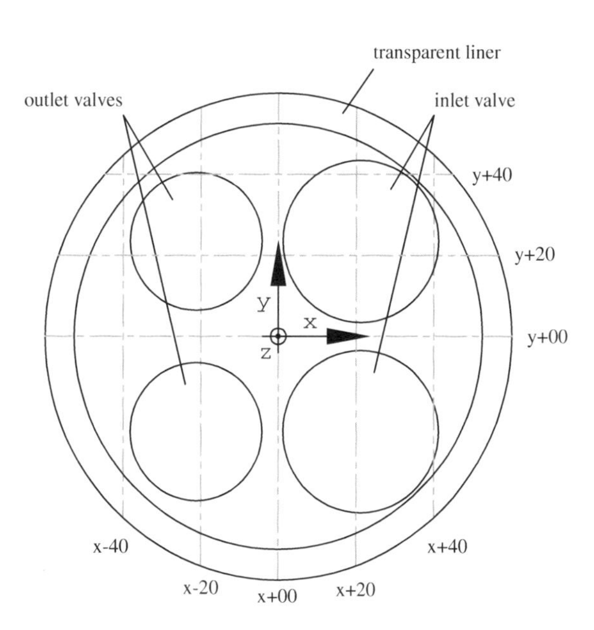

To yield ensemble-averaged data, about 280 double images where recorded for each measurement plane. Stansfield et al. [39], Huang et al. [21], and Li et al. [27] used 46, 100, and 120 double images, respectively.

Image Post-processing

The PIV post-processing of the recorded images is realized with the commercial software VidPIV (ILA GmbH, Germany). The evaluation is carried out using adaptive cross-correlation techniques with window shifting and deformation. The final interrogation window size was 32 × 32 pixel with an overlap of 50 %. The particle concentration was set to a minimum of 3–5 particle per interrogation window (32 × 32 pixel) for all measurement planes. The final vector field resolution is approximately $\Delta x = 1.5\,\mathrm{mm}$, i.e., $\Delta x/D = 0.014$. Optical distortions are compensated by means of a non linear mapping function, which maps the pixel coordinates of the camera to physical coordinates within the cylinder. For this purpose, a calibration grid is placed inside the cylinder and a calibration image is recorded for every measurement plane.

3.2.2 Results

In the following, the flow fields in the different measurement planes are described and discussed in detail. Figure 18 shows the ensemble averaged flow inside the cylinder at a crank angle of 160° atdc, whereas Figs. 15a, b and 16a–c show the results in the y-z planes and Fig. 17a–c show the results in the x-z planes respectively. The velocity fields are color-coded by the turbulent kinetic energy. The red lines show the streamlines of the flow.

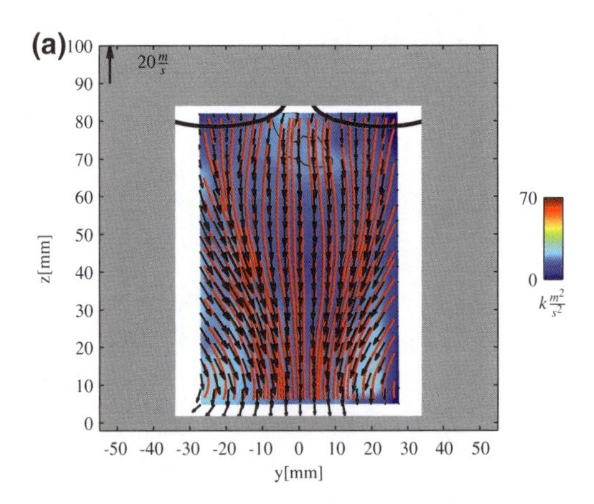

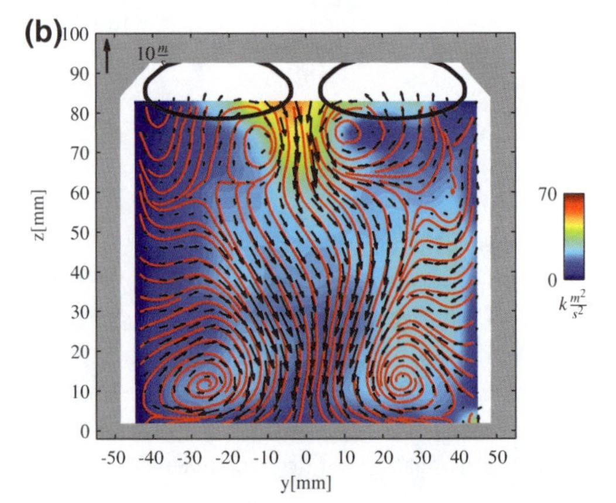

Fig. 15 Velocity vectors (*every 3rd arrow*), streamlines (*red lines*), and turbulent kinetic energy (*color code*) at 160° atdc for two negative x-planes (x = −40 mm (**a**), x = −20 mm (**b**))

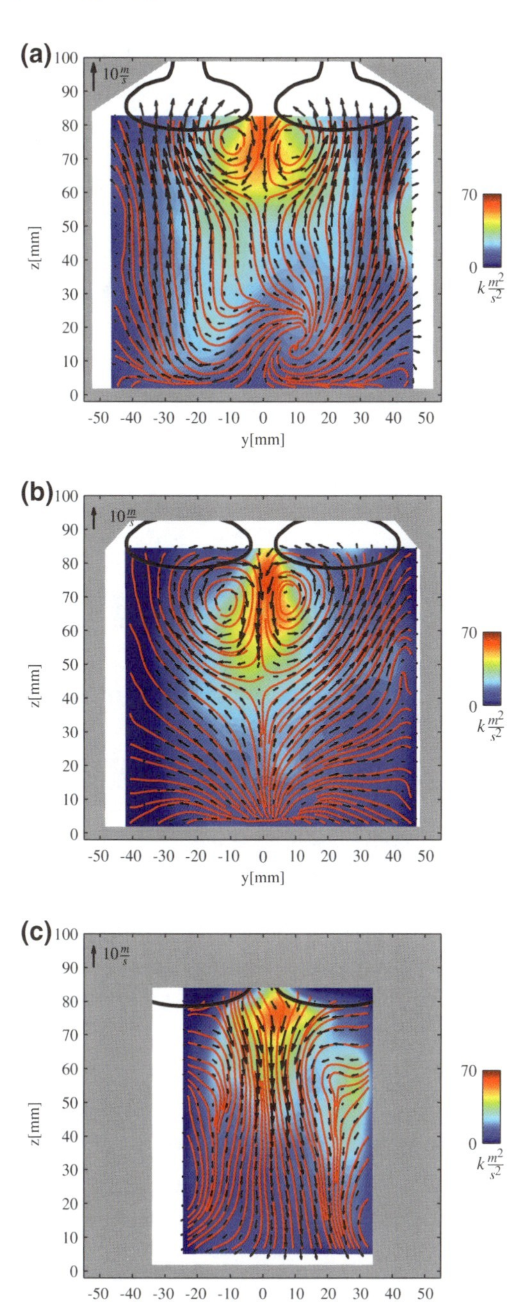

Fig. 16 Velocity vectors (*every 3rd arrow*), streamlines (*red lines*), and turbulent kinetic energy (*color code*) at 160° atdc for the y-z symmetry plane and two positive x-planes (x = 0 mm (**a**), x = 20 mm (**b**), x = 40 mm (**c**))

The dominant flow structures in the y-z-planes are the two counter-rotating vortices beneath the inlet valves. These vortices occur in measurement planes x = −20 mm, x = 0 mm, and x = 20 mm. Furthermore, two counter-rotating vortices

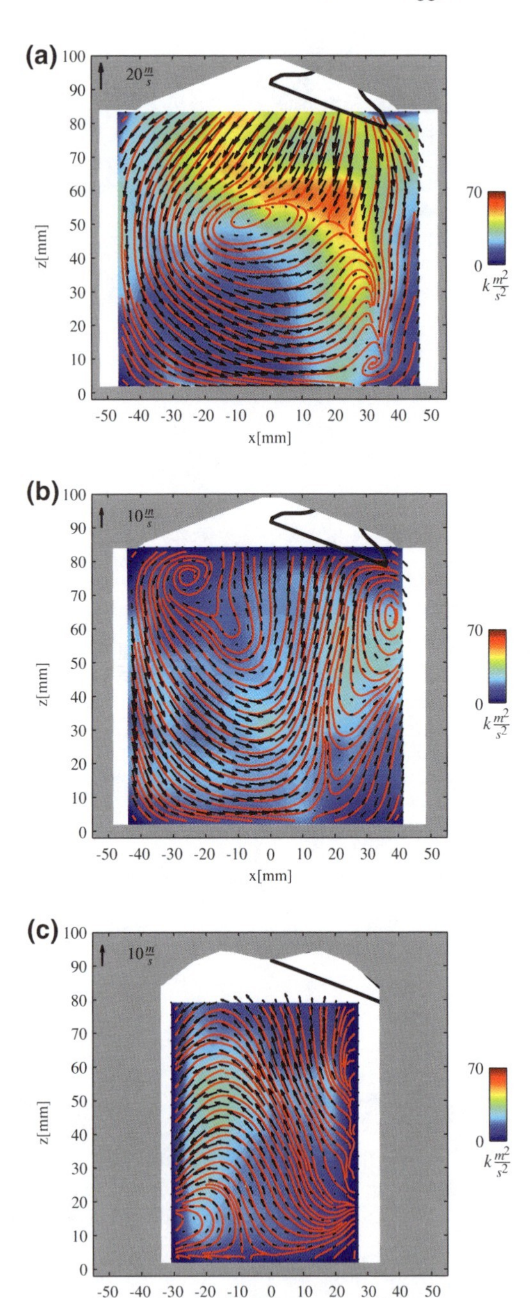

Fig. 17 Velocity vectors (*every 3rd arrow*), streamlines (*red lines*), and turbulent kinetic energy (*color code*) at 160° atdc in three y-planes (y = 0 mm (**a**), y = 20 mm (**b**), y = 40 mm (**c**))

in the lower part of the cylinder can be seen in the measurement plane x = −20 mm. The flow in the measurement plane x = −40 mm, as well as in measurement plane x = 40 mm is dominated by a strong downward velocity. In the measurement plane x = 0 mm a single vortex at (y, z) = (10, 15 mm) can be seen. Apart from this, the whole flow in these planes is nearly axis-symmetric with respect to the y = 0 mm axis.

Due to the fact, that the flow field in the x-z plane is nearly symmetric with respect to the y = 0 mm plane, no measurements are performed in the y = −20 mm and y = −40 mm planes. The main flow structure in the y = 0 mm plane is the tumble vortex at (x, z) = (−10, 50 mm). The left part of the inlet flow leaves the inlet ducts, is reflected by the cylinder walls and the piston, and forms the tumble. The right part of the flow is reflected by the right cylinder wall and the piston. Both flows form a small vortex at (x, z) = (30, 10 mm). The main flow structure in the y = 20 mm plane are two counter-rotating vortices. This plane is approximately located in the center of an inlet valve (y = 23.5 mm). The left streaming jet is deflected by the pent roof and the cylinder wall and finally by the piston towards the center of the cylinder. The right streaming jet is redirected through the cylinder wall and the piston. Both flows merge and a roll up and two counter-rotating vortices can be observed. The flow at y = 40 mm is dominated by an upward flow and a small vortical structure at (x, z) = (−25, 15 mm).

To gain a better understanding of the three-dimensional character of the flow field inside the cylinder, the single flow fields, already shown in Figs. 15, 16 and 17 are merged to generate a three-dimensional illustration as it is shown in Fig. 18. The

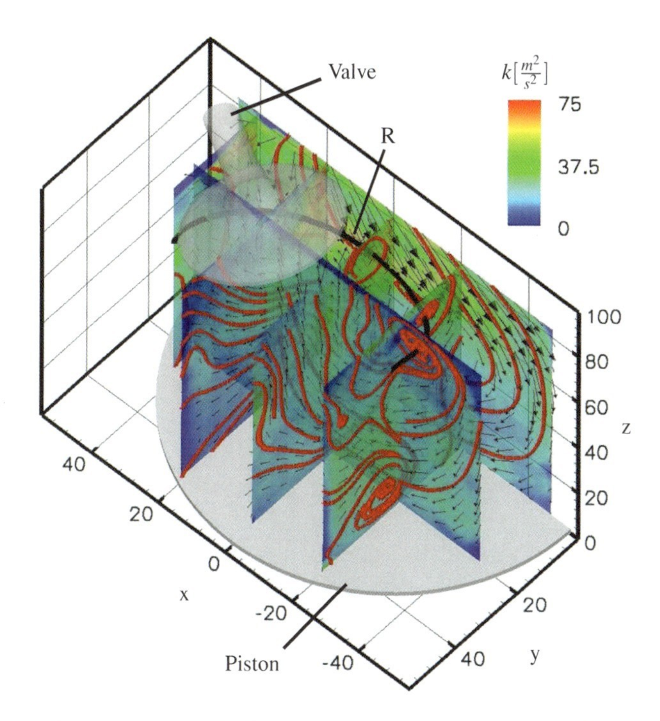

Fig. 18 Three-dimensional view of velocity vectors (*every 4th arrow*), streamlines (*red lines*), and turbulent kinetic energy (*color code*) inside the cylinder at 160° atdc

figures show the in-plane velocity vectors (black arrows). Only every 4th measured vector is shown for clarity. Furthermore, the instantaneous streamlines (red lines) and the turbulent kinetic energy (color code) are visualized. Because of symmetry reasons, which are already shown in Figs. 15, 16 and 17, only one halve of the engine is shown. The main flow structure is the ring vortex, which is generated through the inlet valves. Furthermore, the tumble vortex can clearly be seen.

3.3 2D/3C PIV

3.3.1 Experimental Setup

Stereo-PIV System

The Stereo PIV system consists of a pulsed Nd:YAG laser New Wave "Solo 200XT" and two PCO Sensicam double shutter PIV cameras, each with a Nikon lens with a focal length of 105 mm, a sensor size of 1280 × 1024 pixels and lenses and cameras are arranged following the Scheimpflug condition. The number of illuminated pixels varies with crank angle. A large f-number of 11 is used to increase the diameter of the Airy Disk and thus the particle image diameter. The laser energy is set to 80 % of its maximum of 200 mJ per double pulse. The sampling rate is 2 Hz such that every seventh cycle is captured. The laser pulse separation is set to $\Delta t = 9\,\mu m$. To reduce reflections on the glass liner, the light sheet is inserted from below via a 45° mirror in the elongated slotted piston and via the quartz-glass piston crown, see Fig. 19. The light sheet possesses a thickness of approximately 1 mm and covers the complete cylinder stroke and 79 % of the width of the central planes. This limitation of the field of view is due to the union nut holding the glass piston crown on top of the piston. The in-cylinder flow is seeded with small droplets of Di-Ethylhexyl-Sebacat (DEHS) upstream of the intake manifold to create a homogeneous particle distribution.

The velocity fields shown in Fig. 20 are recorded in 14 axial planes parallel and perpendicular to the main tumble plane at chosen crank angles during the intake and compression stroke (40–320° crank angle (CAD) after top dead center at an

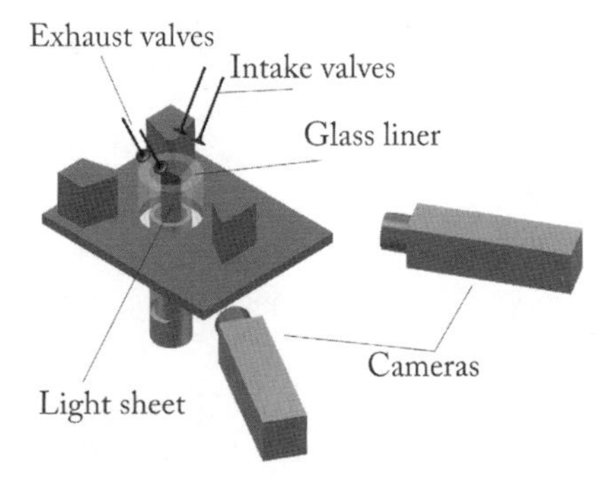

Fig. 19 Experimental stereo PIV setup at the "TINA" engine

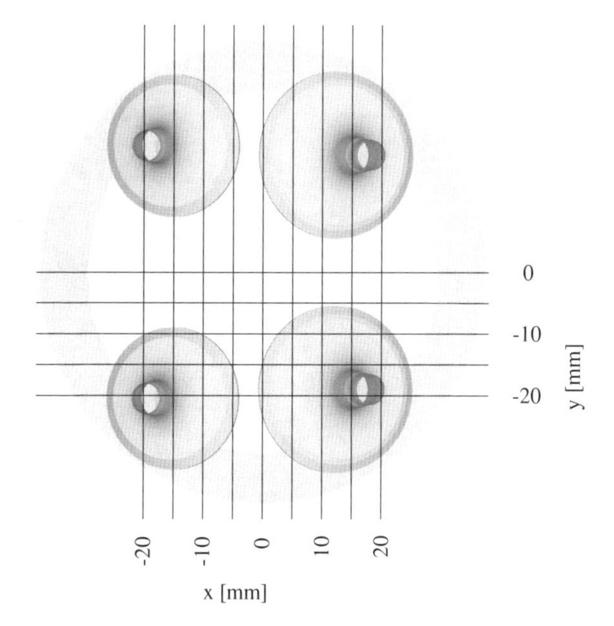

Fig. 20 Measurement planes and engine geometry. Side view (*left*) and top view (*right*)

increment of 20°). For each plane and crank angle, 70 samples are taken. Previous investigations have shown that 70 samples yield sufficient convergence of the ensemble average.

The cross correlation is realized by the commercial software VidPIV from ILA GmbH. The final interrogation window size is 32 × 32 pixels with 50 % overlap. Thus, the velocity vector cell size is 1.4 mm. The quartz-glass liner possesses a curved wall of 20 mm thickness, leading to optical aberrations of the particle images in the radial direction. Additional perspective distortion is introduced by the angular displacement stereoscopic system system. For the correction of these deviations, a calibration target with equally spaced markers is placed inside the cylinder and recorded prior to each measurement. The mapping and de-warping of the particle images is done by a quadratic mapping function algorithm. The de-warped calibration images have a residual distortion of approximately 1 pixel on the center line and up to two pixels towards the left and right edge of the field of view respectively. The resolution of the raw particle images before de-warping is 12 × 8 pixels/mm (axial × radial). After de-warping it is 11 × 11 pixels/mm.

3.3.2 Results

Stereoscopic, Quasi-Volumetric Results at 160° CA

The set of cross-tumble-plane measurements and tumble plane measurements is shown in Fig. 21a and b, respectively. The velocity fields are color-coded by the Γ_1 -criterion, where $\Gamma_1 > 0$ and $\Gamma_1 < 0$ denote the center of a counter-clockwise and clockwise rotating vortex, respectively. The vortex centers are detected in all

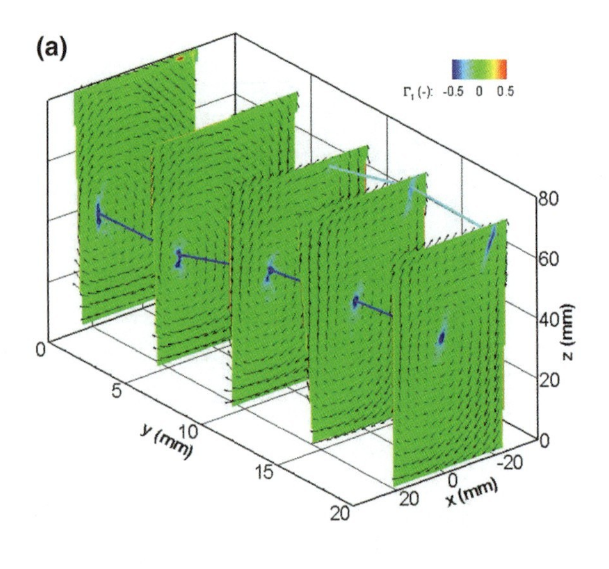

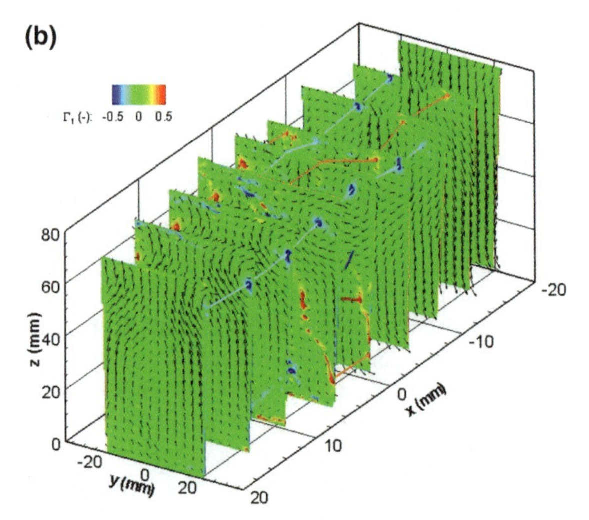

Fig. 21 Vortices in the tumble planes (**a**) and cross-tumble planes (**b**) in the "TINA" engine

measurement planes between x = +15 and x = −10 mm offset from the cross-tumble plane and are connected by straight colored lines. The dominant flow structure is the main tumble (red line). Furthermore, two ring vortices beneath the inlet valves are detected. The inner half (orange line) of the ring vortices is positioned higher than the outer half (blue line) and both orange and blue lines have an inclination due to the strong influence of the tumble flow. Figure 22a gives an overview of the three-dimensional vortical structures. The colors of the lines correspond to the colors in Fig. 21a and b. The black line connects the measured vortex centers of the left ring vortex. The planar velocity fields in the y + 00 plane and the x + 00 plane are color-coded by the ensemble averaged total velocity U_{EA}. The tumble vortex has a c-shape

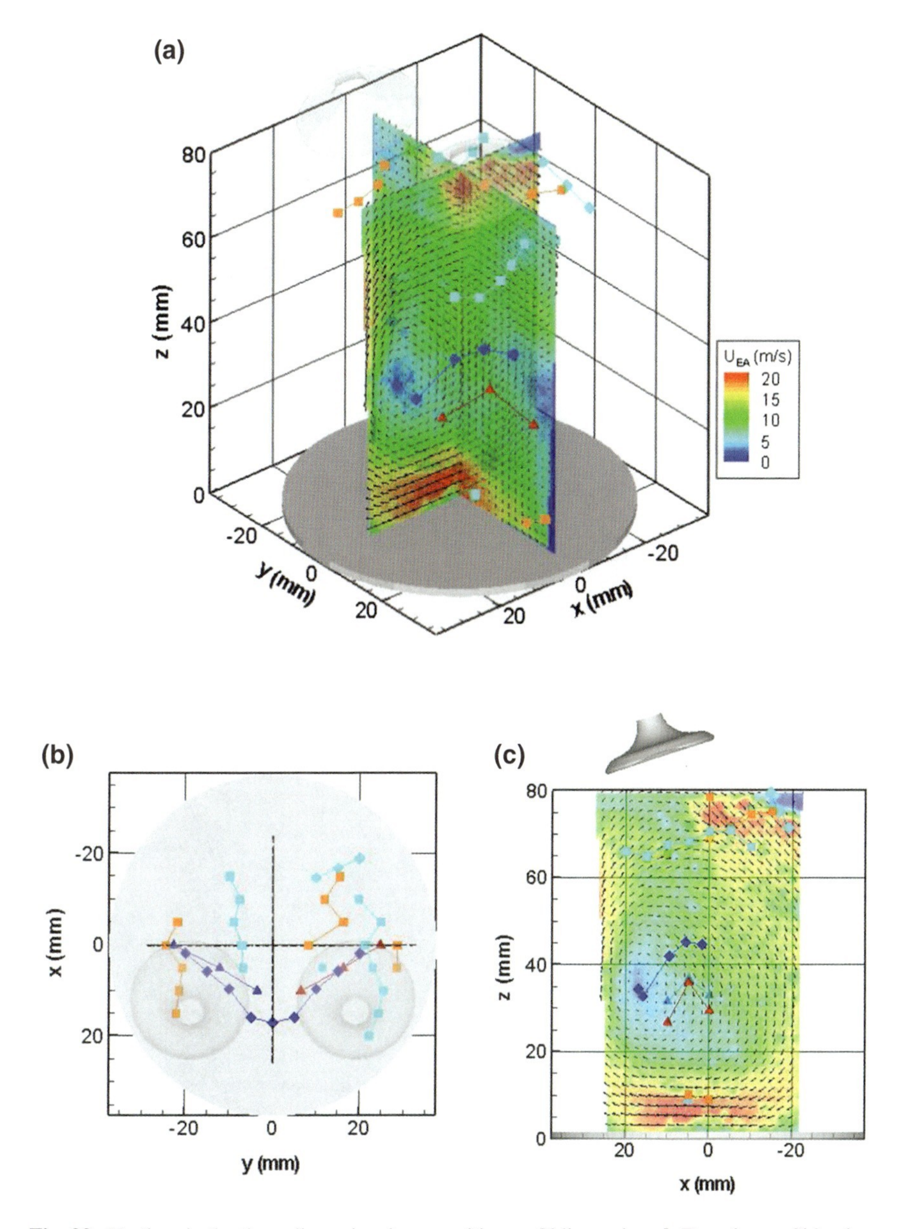

Fig. 22 Vortices in the three-dimensional composition. **a** Oblique view. **b** Top view. **c** Side view

which follows the curved cylinder wall. Because of the symmetry, the tumble was mirrored along the symmetry plane. The diamond-shaped blue symbols represent the tumble-vortex centers detected in the tumble planes while the triangle-shaped blue and red symbols represent the the tumble-vortex centers detected in the cross-tumble planes. In both planes the tumble is detected in the same x and y position. Due to the limited field of view of about 79 % of the bore of the cylinder, the end points of the tumble vortex can not be visualized but could be extrapolated to end at the cylinder walls. Figure 21b shows the top view of the combustion chamber. The ring vortices are stretched towards the exhaust valves due to the influence of the tumble. Figure 21c shows a side view of the symmetry plane with the in-plane velocity vectors of the tumble flow and the total velocity U_{EA} in the color-code.

3.4 Volumetric HPIV

3.4.1 Experimental Setup

The Holographic PIV system is based on a system by Konrath et al. [23, 24] and was developed further to perform fully three-dimensional flow measurements within the cylinder of a four-valve single-cylinder internal combustion engine. The system mainly consists of three optical setups: the image relay setup, containing the holographic film plate, the reference beam module, and the flow field illumination setup. A schematic of the PIV system is shown in Fig. 23. In the following, the

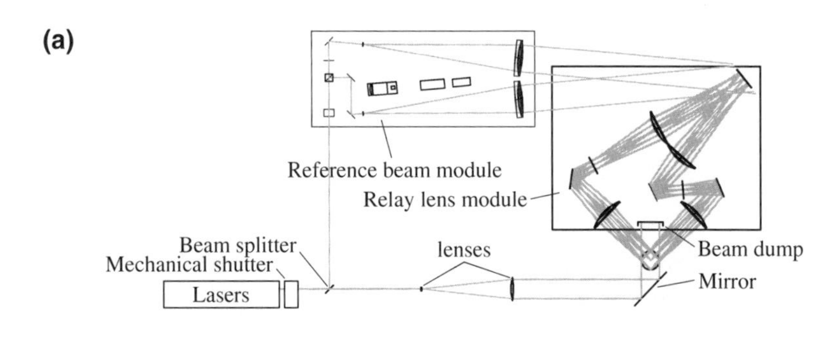

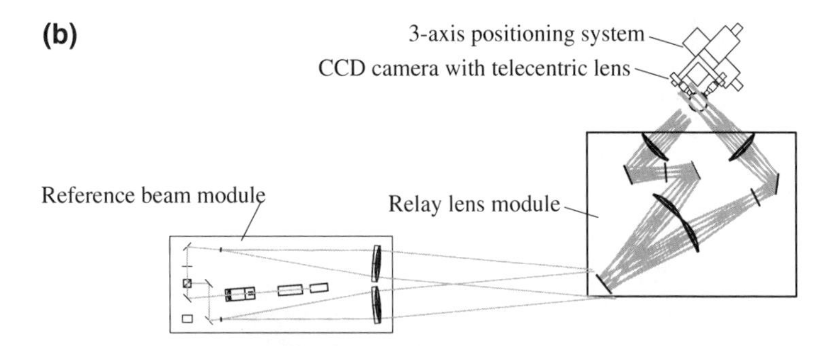

Fig. 23 Schematic of the recording and reconstruction setup of the holographic particle-image velocimetry system. **a** Recording configuration. **b** Reconstruction configuration

basic features as well as some modifications of the HPIV system are explained. A more detailed description of the system is given in [23, 24].

The recording of the holograms is done by two single cavity Nd:YAG lasers with a total laser pulse energy of about two times 400 mJ. To enlarge the coherence length of the laser beams in order to realize holographic recordings, both lasers are seeded, resulting in a coherence length of approximately 1.5 m, which is large enough to yield a sufficiently large contrast within the hologram plane. To separate the two laser pulses in the reference module, they are slightly shifted. That means, the beams have a local separation of about 7 mm at the separation mirror. Therefore, they can be divided without pockels cell. However, they hit the expansion lenses such that they illuminate the same part of the measurement volume. A beam splitter separates the two beams in a low energy reference- and a high energy object-beam. The path length of the reference and the object beam are aligned to increase the contrast of the interference patterns. The holograms are recorded on a 102×127 mm^2 green sensitive VRPM glass plate by Slavich.

To measure all three velocity components with equally high precision, the flow field is simultaneously recorded from two orthogonal directions. For each direction, two lenses and mirrors are used to relay the particle image in front of a single hologram plate, whereas both optical paths have the same length. The illumination directions of the relayed particle images are ±15° (horizontal) with respect to the normal of the hologram plate. The whole relay module is mounted on a stable plate that can be rotated by 180° to perform phase-conjugate reconstruction, ensuring the same geometry during hologram reconstruction and recording. Thus, image aberrations introduced by lenses are eliminated.

To derive the velocity vectors from the particle image, the particle field is captured twice at a certain time interval. A temporal separation of the single particle images is achieved using different reference beams. They illuminate the hologram from ±5° vertically and 53° horizontally with respect to the normal of the hologram plane. To ensure no overlap of direct and conjugate images of one of the four recorded images using the chosen angles, the maximum numerical aperture is set to 0.094 by placing aperture stops inside the focal (Fourier) planes of the relay lenses.

The reconstruction of the holograms is performed using a continuous wave (cw) Nd:YAG laser that has the same wave length as the recording laser. The cw-laser is mounted on the reference beam module. A laser modulator and a polarizing beam splitter is used to switch between the two reference beams. Distortions in radial direction and astigmatic aberrations of the plexiglas liner are cancelled, due to a half cylinder made out of the same material and with the same dimensions as the original liner. This cylinder is placed in the reconstruction setup on the same position, where the original cylinder was. A precise reversal of the reference beam angles during reconstruction is achieved using the alignment method described by Heflinger et al. [18]. The remaining reference beam alignment error is compensated using the method described by Konrath et al. [23]. The holographic images are scanned and digitalized using a Guppy CCD camera and a telecentric lens mounted on a three axis positioning system. Both views are scanned separately and the images are post processed using in-house adaptive cross correlation algorithms. Spurious velocity vectors are filtered after every iteration step by different filter algorithms [32].

3.4.2 Results

In the following, the flow fields in different planes of the measured volumes are described and discussed in detail. Figures 25–27 show the flow inside the cylinder of the "SUZI" engine at a crank angle of 160° atdc, whereas Figs. 25a, b and 26a–c show the results in the y-z planes and Fig. 27a–c show the results in the x-z planes, respectively. The velocity fields are color coded with the out-of-plane velocity component. The red lines show the streamlines of the flow. For better visualization of the large scale flow structures, two hologram measurements are averaged and the velocity components are slightly smoothed. The measurement volume is shown in Fig. 24. Due to slightly different beam profiles of the two Spitlight lasers, the sectional plane of the two volumes is oval. Nevertheless, it was possible to analyze a wide range of the flow inside the cylinder.

The dominant large scale structure in the y-z planes are, like in the 2D measurements in Sect. 3.2.2, the two counter-rotating vortices below the inlet valves. These vortices can be seen in all y-z measurement planes. They result in the intake flow through the inlet valves and a roll up of the flow under the intake valves. Furthermore, in measurement plane x = 0 mm and x = 10 mm a second counter-rotating vortex pair at the bottom of the cylinder with the vortex centers at

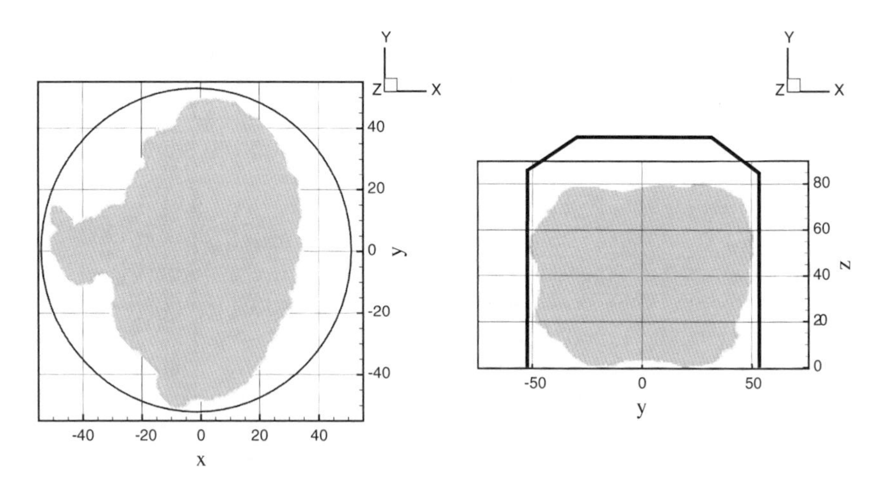

Fig. 24 Illuminated measurement volume for the holographic PIV measurements

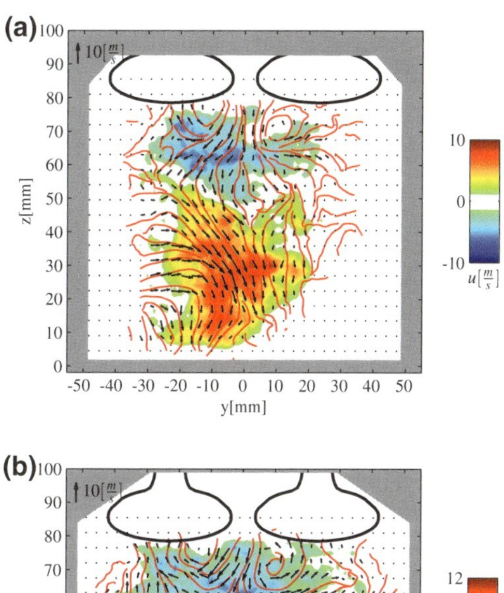

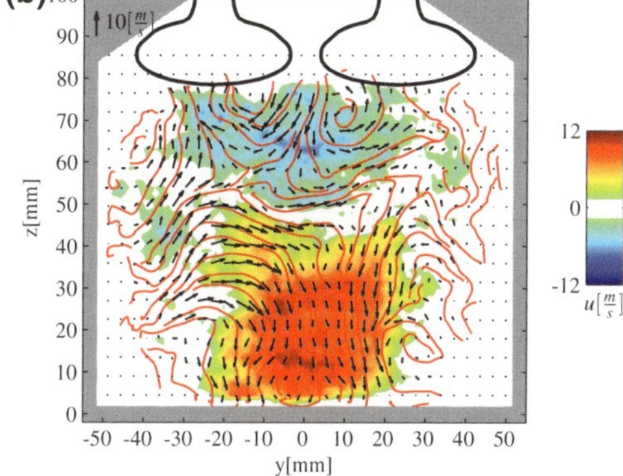

Fig. 25 Velocity vectors (*every 3rd arrow*), streamlines (*red lines*), and out-of-plane velocity component (*color code*) at 160° atdc for two negative x-planes (x = −20 mm (**a**), x = −10 mm (**b**))

(y, z) = (−20, 20 mm) and (y, z) = (25, 20 mm), can be seen. Furthermore, the whole flow is nearly axis-symmetric with respect to the y = 0 mm axis. The x = −20 mm plane is dominated by a downward velocity, whereas the flow field in the x = 20 mm plane possesses a strong upward velocity component. In all measurement planes, a strong positive out-of-plane velocity can be seen at the lower half of the cylinder, whereas the out-of-plane component in the upper half of the cylinder is positive. This results from the tumble vortex, which is also described in the next paragraph.

Since the flow in the x-z plane is nearly axis-symmetric with respect to the y = 0 mm plane, only measurements in the y = 0 mm, y = 10 mm and y = 20 mm planes are shown. In the y = 0 mm plane, the fluid enters the combustion chamber through the gaps between the inlet valves and the pent roof. The flow direction possesses the same inclination than the inlet valves. After entering the combustion chamber, the flow is redirected through the cylinder walls and the piston. This forms the main flow structure in the y = 0 mm plane, which is the tumble vortex. Its center can be seen at (x, z) = (−10, 50 mm). The right part of the flow was not

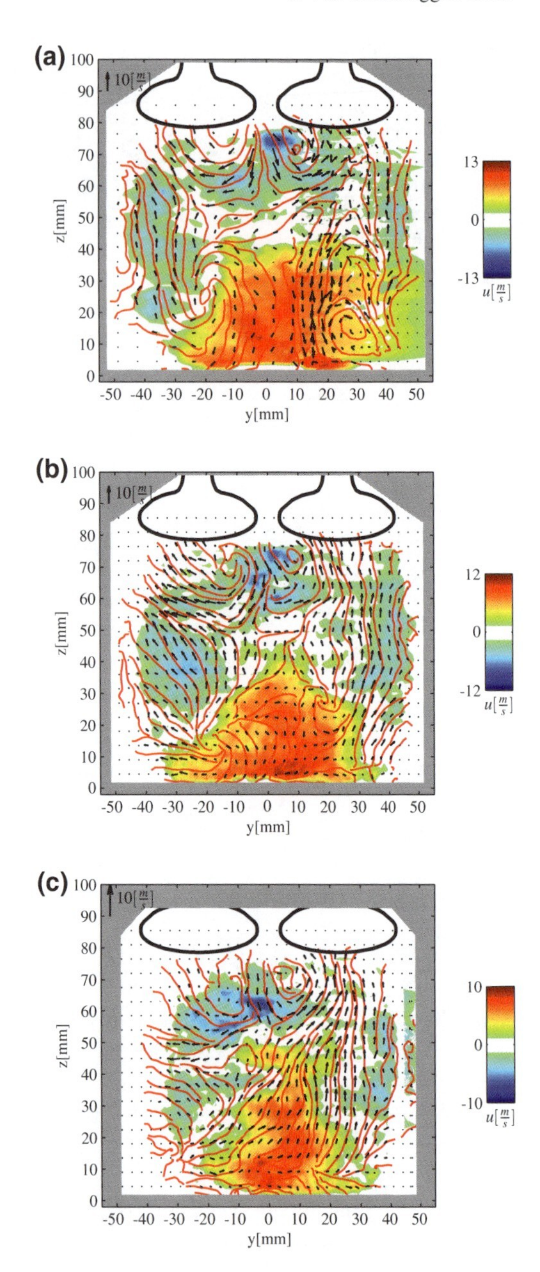

Fig. 26 Velocity vectors (*every 3rd arrow*), streamlines (*red lines*), and out of plane velocity component (*color code*) at 160° atdc for the y-z symmetry plane and two positive x-planes (x = 0 mm (**a**), x = 20 mm (**b**), x = 40 mm (**c**))

analyzed, because of the sightly different beam profiles, which is shown in Fig. 24. Figure 28 shows the three-dimensional view of the flow in the whole measurement area. The lines show the streamtubes, which visualize the flow direction. The streamtubes are color coded with the absolute velocity. Here, the tumble, which where already shown in Fig. 27a–c can clearly be seen. Two regions with a high

Fig. 27 Velocity vectors (*every 3rd arrow*), streamlines (*red lines*), and out of plane velocity component (*color code*) at 160° atdc in three y-planes (y = 0 mm (**a**), y = 10 mm (**b**), y = 20 mm (**c**))

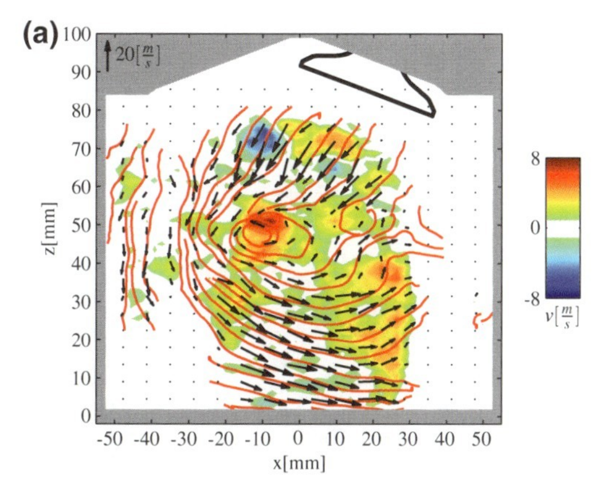

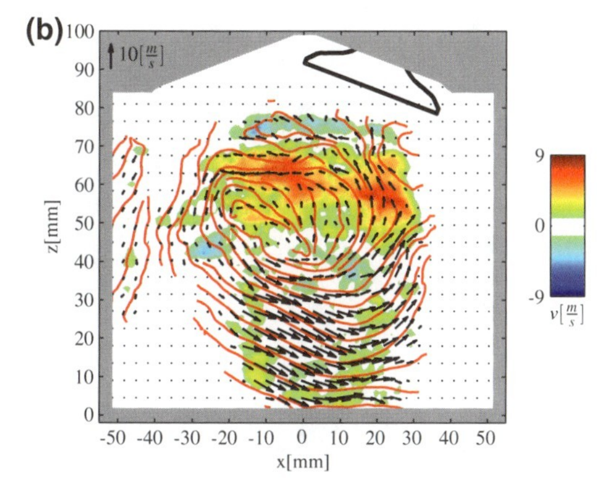

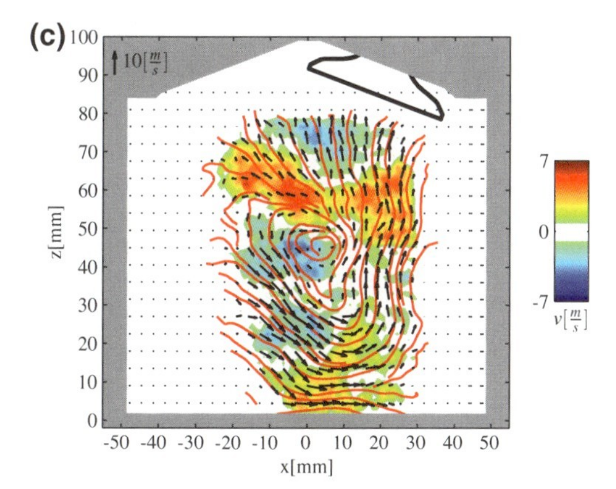

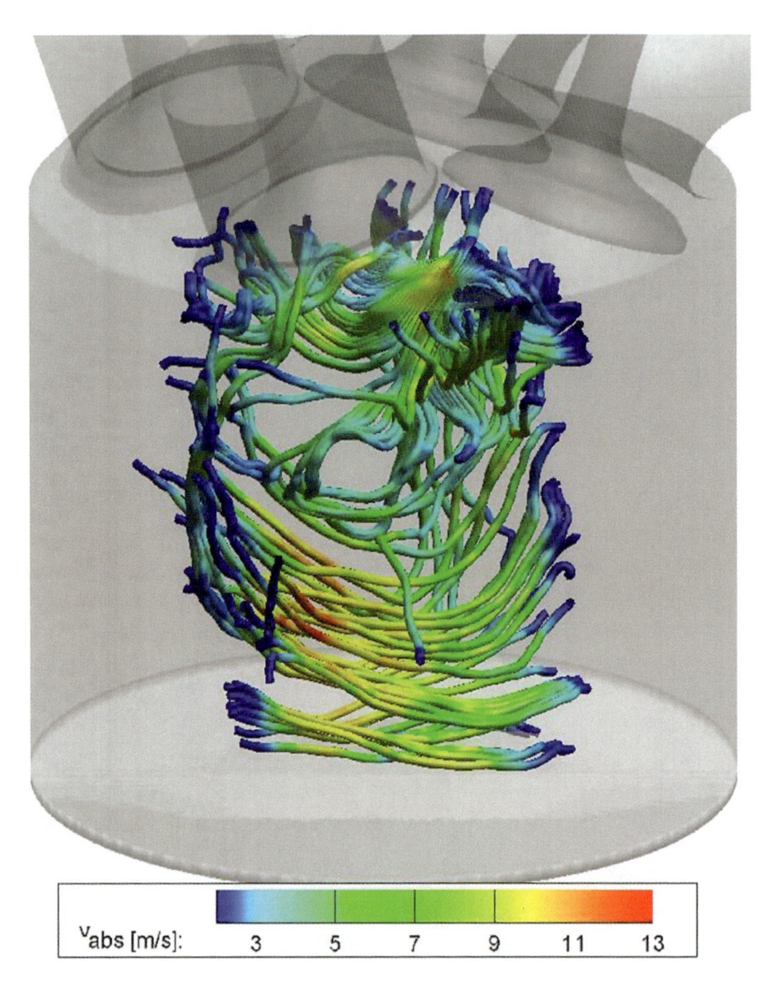

Fig. 28 Three dimensional view of the inflow below the inlet valves. The streamtubes are *color coded* with the absolute velocity

absolute velocity can be located. The first is between the inlet valves, whereas the second can be identified in the lower part of the tumble. The flow enters the combustion chamber between the two inlet valves, and rolls up beneath the valves. This forms a helical flow structure. However, the main part of the flow is directed in a downward direction and forms the tumble flow, which was already explained before. Figure 29 shows the three-dimensional view of the intake flow below the inlet valves in more detail. The stream rods show the flow direction and they are color coded with the absolute velocity. The three-dimensional structure of the two counter-rotating vortices, which were already shown before, can clearly be seen.

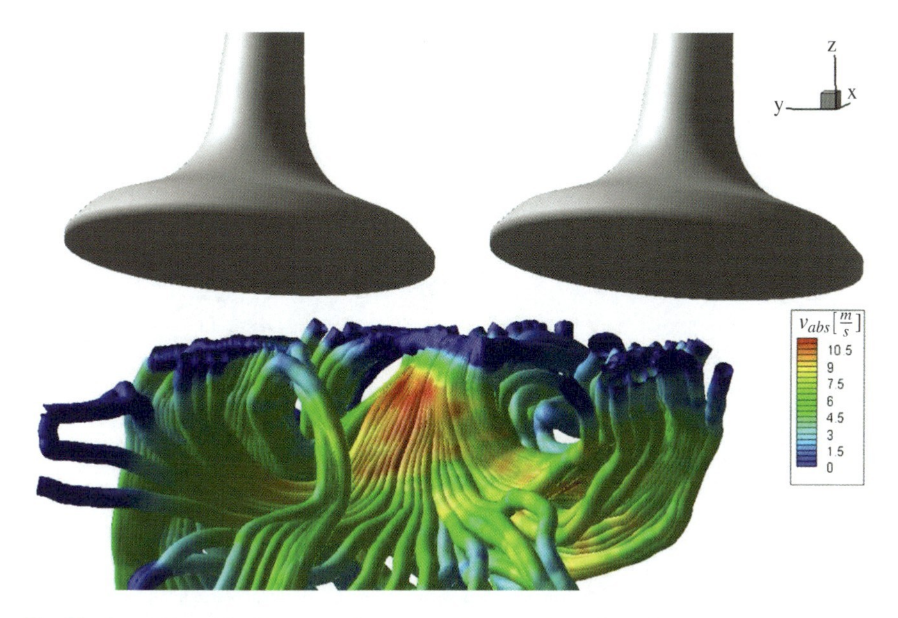

Fig. 29 Three dimensional view of the inflow below the inlet valves. The streamtubes are *color coded* with the absolute velocity

4 Conclusions

The PIV measurement technique has been explained in detail and a general overview of the technique has been given. Furthermore, the requirements concerning tracer particles, as well as the generation of theses are explained. Furthermore, different light sources, as well as the generation of light sheets, have been explained. This is complemented by an explanation of several post-processing methods.

2D2C-PIV measurements in several planes, stereoscopic PIV measurements in a set of planes, and volumetric holographic PIV measurements were presented and the flow inside the cylinder of two four-valve piston engines at a crank angle of 160° atdc has been analyzed. To visualize the in-cylinder flow, mean velocity fields for the 2D2C-PIV measurements, as well as for the stereoscopic PIV measurements, have been determined by 150 instantaneous vector fields for the 2D2C-PIV measurements and 70 instantaneous vector fields for the stereoscopic PIV measurements, respectively. For the 2D2C-PIV measurements, the turbulent kinetic energy has been calculated. Furthermore, for the stereoscopic PIV measurements, vortices where detected by means of the Γ_1 criterion. The three measurement techniques show the feasibility of PIV measurements for the analysis of the flow inside a piston engine. Especially the 3D measurement techniques, which are stereoscopic PIV and holographic PIV, are promising tools for the analysis of the highly three-dimensional flow inside a piston engine.

Acknowledgments This research is part of the the Cluster of Excellence "'Tailor-Made Fuels from Biomass"', the NRW Forschungsschule "'Brennstoffgewinnung aus nachwachsenden Rohstoffen (BrenaRo)"' and the collaborative research center SFB686 which is funded by the German Research Association (Deutsche Forschungsgemeinschaft, DFG). The support of the DFG is gratefully acknowledged.

References

1. Arroyo, M.P., Greated, C.A.: Stereoscopic particle image velocimetry. Meas. Sci. Technol. **2**, 1181–1186 (1991)
2. Bücker, I., Karhoff, D., Dannemann, J., Pielhop, K., Klaas, M., Schröder, W.: Comparison of piv measured flow structures in two four-valve piston engines. In: Stab2010 (2010)
3. Born, M.: Optik. Springer, Berlin (1981)
4. Born, M., Wolf, E.: Principles of Optics. Cambridge University Press, London (1999)
5. Coudert, S., Schon, J.-P.: Back-projection algorithm with misalignment corrections for 2D3C stereoscopic PIV. Meas. Sci. Technol. **12**, 1371–1381 (2001)
6. Dannemann, J., Klaas, M., Schröder, W.: Three dimensional flow field within a four-valve combustion engine measured by particle-image velocimetry. In: ISDV 14 (2010)
7. Dannemann, J., Pielhop, K., Klaas, M., Schröder, W.: Cycle-resolved multi-planar particle-image velocimetry measurements of the in-cylinder flow of a four-valve combustion engine. In: 8th International Symposium on Particle Image Velocimetry—PIV 09 (2009)
8. Dannemann, J., Pielhop, K., Klaas, M., Schröder, W.: Cycle resolved multi-planar flow measurements in a four-valve combustion engine. Exp. Fluids **50**, 961–976 (2010)
9. Eichler, J., Eichler, H.J.: Laser, Bauformen, Strahlführung. Springer, Anwendungen (1998)
10. Elsinga, G.E., van Oudheusden, B.W., Scarano, F.: Experimental assessment of tomographic-PIV accuracy. In: 13th International Symposium on Applications of Laser Techniques to Fluid Mechanics (2006)
11. Estevadeordal, J., Goss, L.: PIV with LED: particle shadow velocimetry (PSV). In: 43rd AIAA Aerospace Sciences Meeting and Exhibit, Meeting Papers (2005)
12. Gabor, D.: A new microscopic principle. Nature **161**, 777–778 (1949)
13. Gaydon, C., Raffel, M., Willert, C., Rosengarten, J., Kompenhans, J.: Hybrid stereoscopic particle image velocimetry. Exp. Fluids **23**, 331–334 (1997)
14. Graftieaux, L., Michard, M., Grosjean, N.: Combining PIV, POD and vortex identification algorithms for the study of turbulent swirling flows. Meas. Sci. Technol. **12**, 1422–1429 (2001)
15. Große, S., Schröder, W., Klaas, M., Klöckner, A., Roggenkamp, J.: Time resolved analysis of steady and oscillating flow in the upper human airways. Exp. Fluids **42**, 955–970 (2007)
16. Grout, I.A.: Integrated Circuit Test Engineering. Springer, London (2006)
17. Guibert, P., Murat, M., Hauet, B., Keribin, P.: Particle image velocimetry measurements: application to in-cylinder flow for a two stroke engine. SAE Paper 932647 (1993)
18. Heflinger, L.O., Stewart, G.L., Booth, C.R.: Holographic motion pictures of microscopic plankton. Appl. Opt. **17**, 951–954 (1978)
19. Herrmann, S.F.: Three-dimensional optical flow measurements with short coherence holography. Ph.D thesis, Carl von Ossietzky Universität Oldenburg, July 2006
20. Heywood, J.B.: Internal Combustion Engine Fundamentals. McGraw-Hill, New York (1988)
21. Huang, R.F., Huang, C.W., Chang, S.B., Yang, H.S., Lin, T.W., Hsu, W.Y.: Topological flow evolutions in cylinder of a motored engine during intake and compression strokes. J. Fluids Struct. **20**, 105–107 (2005)
22. Janesick, J.R.: Scientific charge-coupled devices. In: SPIE—The International Society for Optical Engineering (2001)

23. Konrath, R.: Strömungsanalyse im Zylinder eines 4-Ventil-Motors mit der holographischen Particle-Image Velocimetry. Ph.D thesis, Rheinisch Westfälische Technische Hochschule Aachen (2003)
24. Konrath, R., Schröder, W., Limberg, W.: Holographic particle image velocimetry applied to the flow within the cylinder of a four-valve internal combustion engine. Exp. Fluids **33**, 781–793 (2002)
25. Kreis, T.: Handbook of Holographic Interferometry. Wiley, Weinheim (2005)
26. Leith, E.N., Upatnieks, J.: Reconstructed wavefronts and communication theory. J. Opt. Soc. Am. **52**, 1123–1130 (1962)
27. Li, Y., Zhao, H., Peng, Z., Ladommatos, N.: Particle image velocimetry measurements of in-cylinder flow in internal combustion engines—experiment and flow structure analysis. Proc. Institut. Mech. Eng. **216**(Part D), 65–81 (2002)
28. Milonni, P.W., Eberly, J.H.: Lasers. Wiley, New York (1988)
29. Prasad, A.K., Adrian, R.: Stereoscopic particle image velocimetry applied to liquid flows. Exp. Fluids **15**, 49–60 (1993)
30. Prasad, A.K., Jensen, K.: Scheimpflug stereocamera for particle image velocimetry in liquid flows. Appl. Opt. **34**, 7092–7099 (1995)
31. Pu, Y., Meng, H.: An advanced off-axis holographic particle image velocimetry (HPIV) system. Exp. Fluids **29**, 184–197 (2000)
32. Raffel, M., Willert, C.E., Wereley, St.T., Kompenhans, J.: Particle Image Velocimetry. Springer, Berlin (2007)
33. Reuss, D.L., Rosalik, M.: PIV measurements during combustion in a reciprocating internal combustion engine. SAE Technical Paper (1998)
34. Ruck, B.: Einfluß der Tracerteilchengröße auf die Signalinformation in der Laser-Doppler-Anemometry. Technisches Messen - tm **57**, 284–295 (1990)
35. Schröder, F., Klaas, M., Schröder, W.: Multiplane-stereo piv measurements for steady flow in the first two bifurcations of the upper human airways during exhalation. In: Proceedings of the ECCOMAS Thematic International Conference on 043 Simulation and Modeling of Biological Flows (SIMBIO 2011) (2011)
36. Siegmann, A.E.: Lasers. University Science Books, Sausalito (1986)
37. Soloff, S.M., Adrian, R., Liu, Z.-C.: Distortion compensation for generalized stereoscopic particle image velocimetry. Meas. Sci. Technol. **8**, 1441–1454 (1997)
38. Soodt, T., Schröder, F., Klaas, M., Schröder, W., van Overbrüggen, T.: Experimental investigation into the transitional bronchial velocity distribution using stereoscanning PIV. Exp. Fluids **52**(3), 709–718 (2011)
39. Stansfield, P., Wigley, G., Justham, T., Catto, J., Pitcher, G.: PIV analysis of in-cylinder flow stuctures over a range of realistic engine speeds. Exp. Fluids **43**, 135–146 (2007)
40. Svelto, O.: Principles of Lasers. Springer, New York (2009)
41. Tennekes, H., Lumley, J.L.: A First Course in Turbulence. MIT Press, Cambridge (1972)
42. Thyagarajan, K., Ghatak, A.: Lasers: Fundamentals and Applications. Springer, Boston (2010)
43. van de Hulst, H.C.: Light Scattering by Small Particles. Wiley, New York (1957)
44. van Doorne, C.W.H., Westerweel, J.: Measurement of laminar, transitional and turbulent pipe flow using Stereoscopic-PIV. Exp. Fluids **42**, 259–279 (2007)
45. Westerweel, J.: Efficient detection of spurious vectors in particle image velocimetry data. Exp. Fluids **16**, 236–247 (1993)
46. Westerweel, J.: Fundamentals of digital particle image velocimetry. Meas. Sci. Technol. **8**, 1379–1392 (1997)
47. Willert, C., Stasicki, B., Moessner, S., Klinner, J.: Pulsed operation of high power light emitting diodes for flow diagnostics. Technical Report, Institute of Propulsion Technology, German Aerospace Center (DLR) (2010)

Towards Model-Based Design of Tailor-Made Fuels from Biomass

J.J. Victoria Villeda, M. Dahmen, M. Hechinger, A. Voll and W. Marquardt

Abstract In face of the continuous depletion of fossil carbon resources alternative liquid energy carriers have to be identified to guarantee sustainable future mobile propulsion. In this context, the Cluster of Excellence (CoE) "Tailor-Made Fuels from Biomass" (TMFB) at RWTH Aachen University aims at identifying sustainable fossil fuel surrogates from biomass by means of a holistic approach from biomass supply to engine combustion. As the fuel identification process requires the screening of a tremendous number of possible fuel candidates, solely experimental methodologies cannot be applied. To this end, a research team at AVT.PT contributes to a model-based fuel design (MBFD) methodology which is based on an integrated product and process design approach, considering aspects of both fuel combustion and fuel production. It aims at identifying possible fossil fuel surrogates from a database of rigorously generated molecular structures. These fuel surrogates have to comply with a set of pre-defined constraints, which has been elaborated by interdisciplinary collaboration within the CoE. The present contribution illustrates the status quo and future perspectives of model-based fuel design and its integration into the research context of the TMFB cluster.

1 Introduction

Constantly rising global energy demands, fast growing markets and increasing greenhouse gas emissions force society to explore novel concepts for mobility. Predictions of fossil carbon reserves and the potential for their economic recovery have always been and still are inconclusive [1, 2]. In recent years, rapidly increasing oil prices and a growing awareness of sustainability have induced a larger interest in renewable carbon resources. In face of electric mobile propulsion being heavily

J.J. Victoria Villeda · M. Dahmen · M. Hechinger · A. Voll · W. Marquardt (✉)
Aachener Verfahrenstechnik—Process Systems Engineering (AVT.PT), RWTH Aachen University, Turmstraße 46, 52064 Aachen, Germany
e-mail: wolfgang.marquardt@avt.rwth-aachen.de

M. Klaas et al. (eds.), *Fuels From Biomass: An Interdisciplinary Approach*,
Notes on Numerical Fluid Mechanics and Multidisciplinary Design 129,
DOI 10.1007/978-3-662-45425-1_12

discussed nowadays, carbon-based liquid energy carriers remain fundamental since they offer high volumetric energy contents and enable leveraging the results gained in decades of extensive research in combustion science. Until recently—due to seemingly unlimited availability of crude oil at low production cost—industry and academia have so far concentrated on optimizing the chemical value-chain targeting fossil (long-chained) hydrocarbon mixtures as raw materials [3]. Today, in the context of fuels as their corresponding products, the inevitable transition to renewable raw materials constitutes a major challenge for industry and society being reliant on further increasing needs for transportation.

The research cluster "Tailor-Made Fuels from Biomass" (TMFB) at RWTH Aachen University gears towards a holistic approach for the utilization of entire plants to derive novel tailor-made fuels characterized by low pollutant emissions (particles and NO_x) and high-efficiency in low-temperature combustion [4]. Rather than breaking the macromolecules native to biomass into C_1 building blocks followed by their reassembling into target molecules, the research strategy in the TMFB cluster aims at preserving the natural molecular structures in the raw materials in order to exploit the synthesis power of nature to the extent possible [3, 5]. Consequently, highly functionalized molecular structures have to be gently depolymerized and catalytically re-functionalized to achieve higher carbon efficiency and a lower entropic loss compared to existing thermochemical processes like gasification [6]. In contrast to petroleum-based chemical engineering, low-temperature liquid phase reactions and separation technologies have to be employed. Therefore, establishing integrated bio-refineries presents a true challenge in chemical engineering enquiring into re-thinking the entire value-chain [6].

As a partner in the TMFB cluster, AVT.PT follows an integrated approach to model-based design of biofuels and their production pathways [3]. The molecular search space under consideration contains an enormous number of yet unexplored chemical compounds rendering the identification of fuel candidates entirely based on experiments impossible. To this end, AVT.PT aims at the provision of methodologies for model-based fuel design by applying and tailoring Computer Aided Molecular Design (CAMD) techniques to the biofuel design problem. Research partners from within the cluster define desired product features, i.e., constraints for combustion relevant properties. Predicting the thermodynamic and combustion kinetic properties of a fuel solely from its molecular structure is achieved by means of data-driven analysis incorporating physical knowledge to constitute semi-empirical property models [3]. The models, checked for stability, robustness and predictive capability, are used for a rapid evaluation of a large number of molecular structures. Facing insufficient availability of training data for kinetic property prediction, an iterative model-assisted experimental design strategy is conceived to reduce experimental effort needed for establishing such property models. At a first stage of process design, Reaction Network Flux Analysis (RNFA) provides the classification of the variety of potential synthesis pathways [3]. Starting from the major fractions of native biomass, a reaction network is constructed towards the target molecule based on reaction steps obtained from literature or explored in the interdisciplinary context of the cluster. Application of optimization techniques enables an early screening for efficient reaction routes and an early assessment of synthesis opportunities [7, 8].

This article will first review requirements on fossil fuel surrogates. Examining operation ranges of today's Diesel and Gasoline engines defines criteria used in virtual screening for potential bio-renewable substitutes. Thereafter, the methodological framework of model-based fuel design at AVT.PT will be presented in detail. At first, product and process design methods will be introduced and discussed separately, followed by concepts of their integration. Each part will be split into a status quo report and an outlook section stating current and future research efforts.

2 Product and Process Specifications

One of the most important working hypotheses chosen in the research context of TMFB cluster is to consider biofuels not solely as carriers of chemically bound energy but rather as a conglomerate of different properties. This offers the possibility to tailor the fuel to the specific requirements of existing combustion processes or even allows the development of new combustion engine concepts [9].

Furthermore, several assumptions are made which result in additional constraints in the search for novel biofuels regarding product and process specifications, with the most general claiming that biofuels are organic molecules only containing carbon, hydrogen and oxygen, which are the three major constituents of biomass in general.

A very strong and important design parameter is the oxygen content of a fuel molecule. In particular, an oxygen content of less than 5 wt% significantly reduces the combustion engine performance and increases the level of soot formation [10]. Furthermore removing oxygen is usually hydrogen-intensive thus reducing the economic efficiency of the corresponding production process. A higher fuel oxygen content avoids these effects, but leads to a decrease in the heat of combustion, since energy-rich carbon-hydrogen bonds are replaced by carbon-oxygen bonds containing significantly less chemically bound energy.

Selected thermodynamic and combustion kinetic properties of the fuel are crucial for the proper function of modern combustion engines. The most important thermodynamic properties are liquid density and normal boiling point. In current engines, with a fixed maximum fuel injection volume, a high fuel density results in less fuel consumption but in an increase in soot and particle emissions, while low density fuels behave the other way round. The boiling point strongly influences the vaporization kinetics of the fuel. If it is too high, combustion is incomplete and will consequently result in higher emissions of unconverted fuel. Accordingly, a high enthalpy of evaporation causes a large amount of heat to evaporate the fuel resulting in a strong decrease in the temperature inside the combustion chamber. Surface tension has to be considered as drop formation during injection has to fulfill certain requirements. A high value of the surface tension leads to the formation of large droplets in the spray which can coalesce and produce a film on the wall of the combustion chamber, leading to incomplete combustion [11].

Combustion-kinetic properties such as the laminar burning velocity or the ignition delay time are important for the determination of the ignition angle during engine operation. Efficient combustion and extended durability of certain parts of the engine (such as the crank shaft) are strongly influenced by a proper setting of the ignition angle [11].

On the processing side, direct re-functionalization of the biomass is desired in terms of maximizing the energy efficiency of the production process. Hence, the molecular structure shall be conserved to the extent possible while simultaneously matching the given product requirements [3].

These and other specifications of biofuel properties drive the formulation of the model-based fuel design problem: Model-based fuel design inevitably has to address requirements from both, the product and the process, if a sustainable biofuel production process and an energy-efficient, low-emission combustion should be achieved. Product design has to be based on property estimates related to molecular structure, because the set of possible fuel candidates is very large and measured property data are typically lacking. In a first step, structure-property-relationships have to be established to allow the prediction of key properties in a fast and accurate manner to identify fossil fuel surrogates. Process design requires the multi-criteria evaluation of a variety of possible reaction pathways to assess key performance indicators of fuel production. A tight integration of product and process design should be envisioned at least as a long term goal, because of the strong interdependence of the respective decisions.

3 Product Design: Identifying Promising Fuel Candidates

Various methodologies have been proposed in the literature to solve product design problems, i.e., to link desired macroscopic properties of the product (or, more precisely, the target properties of the fuel) to its molecular structure. These methodologies are widely associated with the term Computer-Aided Molecular Design [12]. The two major tasks of model-based product (or fuel) design can thus be summarized as follows:

1. Develop a quantitative relation between macroscopic target properties and molecular structure. This relation should not only describe the behavior of known molecules, but it also has to reliably predict the properties of novel molecular structures beyond the experimentally investigated range.
2. Based on such structure-property relations, molecular structures should be identified which meet the desired macroscopic target properties best.

While the first task aims at deriving a predictive property model, the second task exploits the information reflected by this model to identify promising molecular structures of the fuel.

The procedure for designing a future fuel is sketched in Fig. 1. The fuel design process is initiated by a rigorous generation of organic molecular structures. To this end, commercially available structure generators such as the Molgen software package [13] are employed to create feasible molecular structures only considering valence rules but without assessing thermodynamic stability or any other thermodynamic criteria of the generated structures. Structure generation is followed by several analysis steps which aim at reducing the number of potential fuel candidates. The properties of the generated structures are predicted by some property model; molecular structures outside a desired property window are excluded from further consideration. The complexity of the properties and the property models employed are incrementally increased during the refinement process. In particular, few and simpler and thus computationally less demanding models are applied to the very large number of compounds at the beginning of the screening, whereas more complex structure-property relations are only used for a comparably small number of promising fuel candidates. Thus, while model complexity is incrementally increased, the number of potential structures is reduced accordingly. This ensures a targeted and computationally feasible process for the identification of possible fuel candidates.

In order to realize the product design strategy displayed in Fig. 1, predictive property models for the fuel relevant refinement criteria are required. As shown in Fig. 1, the type of model employed depends on the property considered. The different property modeling approaches employed in the fuel design process are presented in the following. We distinguished between models which directly relate macroscopic properties to molecular structure and so called semi-empirical models, which integrate physically motivated sub-models in a data-driven modeling strategy.

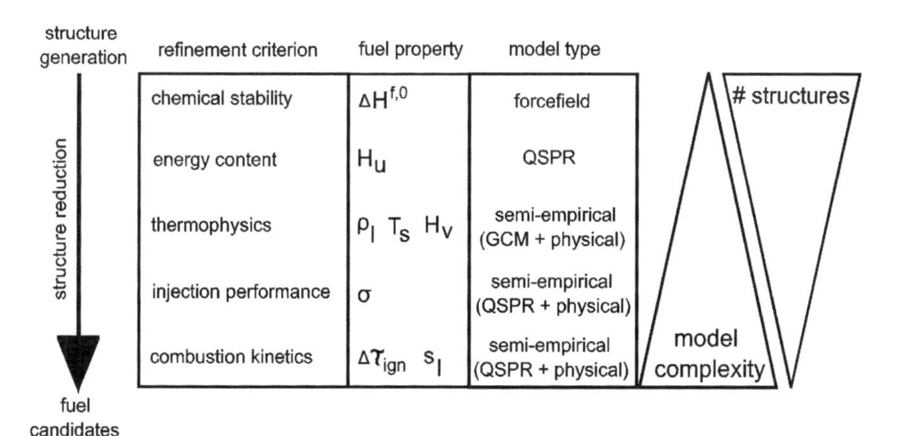

Fig. 1 Identification of promising fuel candidates

3.1 Relating Fuel Properties to Molecular Structure

Three types of modeling methods are included in the fuel design process (Fig. 1), which directly link molecular structure to the macroscopic fuel properties of interest, namely force field methods (FFM), group contribution methods (GCM) as well as quantitative structure-property relations (QSPR).

Thermodynamic stability of the generated structures is assessed by predicting the enthalpy of formation $\Delta H^{f,0}$ at standard state. The standard enthalpy of formation of a compound is defined as the change of enthalpy involved when 1 mol of a substance is formed from its constituent elements at standard state. The higher this enthalpy, i.e., the energy level of the compound, the less stable is the molecular structure. Molecular mechanics [14] can be used to determine the enthalpy of formation of a given molecular structure. Classical molecular mechanics assume a molecule to constitute of a system of point masses (the atoms) and connecting springs (the bonds). This mass-spring system is parameterized by some force field (or potential function) which defines the attractive and repulsive forces between the participating atoms. Force field parameters are adjusted to match measured thermophysical properties of selected training compounds. The force field is typically minimized to determine the molecules conformation with minimum energy.

The basic force field concept can be extended towards the treatment of reactive molecular mixtures [15]. Rather than regarding molecules as mass-spring systems, the so-called reactive force field (ReaxFF) considers atoms as charged entities which are able to form and break bonds depending on their distance. This formalism allows the simulation of the course of a reaction, and hence a dynamic rearrangement of atoms between molecules. Thereby, a compound's enthalpy of formation can be derived from reaction enthalpies considering the elements involved. Assessment of molecular stability is enabled at the moderate computational effort of molecular mechanics approaches. ReaxFF is thus a reasonable tool to screen the initially large number of molecular structures for thermodynamic stability and to remove unstable molecules from further consideration.

Besides molecular mechanics motivated force-fields, at least two other types of molecular property models are applied in product design, i.e., group contribution methods (GCM) and quantitative structure-property relations (QSPR). A first GCM has been proposed by Lydersen [16]. Later refinements were introduced by many researchers, including Joback and Reid [17] and Marrero and Gani [18]. The GC approach decomposes the molecular structures into substructures, so-called groups. A numerical contribution to a macroscopic molecular property is assigned to each group such that the addition of these group contributions results in the measured value of some property of the molecule. Once the group contributions have been identified via regression, novel molecules can be assembled from those groups and the properties of the new structures can be estimated. Moreover, groups can be assembled such that a desired property value is met. Thus, various product design tasks have already been approached by GCM, including solvent selection for biodiesel [19] and bio-butanol [20] processes as well as the design of ionic liquids [21].

While GCMs have been successful in addressing particular product design problems, they fall short in predicting more complex properties such as those related to combustion kinetics of a fuel. Despite first successes in predicting the kinetics of particular types of single reactions such as hydrogen abstraction by GCM [22, 23], the kinetics of more general reaction mechanisms have not yet been reliably predicted from group contributions. In addition, only few measured datasets on reaction (or combustion) kinetics are available. Therefore, the GCM has been shown not to be suitable for predicting the laminar burning velocity [24], because of the large number of parameters contained in any property model stemming from GCM.

Accordingly, QSPRs have been chosen as an alternative modeling strategy, because it results in a lower number of parameters in the property models and therefore facilitates an application to combustion kinetic property modeling as part of fuel design. Katritzky et al. [25] provide a comprehensive review on the theory and application of QSPR for prediction of various types of properties. Similar to GCM, QSPR have been widely applied to successfully predict thermophysical properties [26, 27]. However, contrary to GCM, QSPR relate macroscopic molecular properties to molecular descriptors. While the groups of GCM can be interpreted as a dedicated set of molecular descriptors, the descriptors used in QSPR incorporate a large variety of different molecular attributes which can be derived computationally from its molecular structure. These descriptors are deduced from existing theories such as quantum mechanics and information or graph theory. The descriptors cover a wide range of types including simple shape descriptors such as the asphericity of a molecule or complex quantum-mechanical descriptors as for instance the lowest unoccupied molecular orbital energy [28]. The generality of descriptors in QSPR reduces the number of adjustable parameters in a typically linear regression model for a desired property considerably and therefore renders the method attractive for property estimation, if a restricted set of measurements is available (such as it is the case in combustion kinetics). Based on the rich information gathered in a large number of descriptors to reflect molecular properties, tailored regression algorithms determine those descriptors which have the strongest effect on the macroscopic property of interest. These descriptors then enter a linear property model of the form

$$\boldsymbol{y} = \beta_0 + \beta_1 \cdot \boldsymbol{x}_1 + \beta_2 \cdot \boldsymbol{x}_2 + \cdots + \beta_m \cdot \boldsymbol{x}_m. \tag{1}$$

In this equation, $\boldsymbol{y}$ denotes the property vector of the database molecules used to establish the model, $\boldsymbol{x}_\mathrm{i}$ denotes the ith molecular descriptor and β_i are adjustable model parameters. Model descriptors $\boldsymbol{x}_\mathrm{i}$ are selected such that they reflect the variance in the macroscopic property $\boldsymbol{y}$ within known molecules (training set data). For meaningful extrapolation towards novel fuel structures, careful selection of the training set data and employment of validation techniques are crucial. Details on model identification using QSPR as well as improved methodologies such as the targeted QSPR method can be found elsewhere [29, 30].

As shown in Fig. 1, QSPRs are employed to predict the fuels' energy content by means of the lower heating value H_u. The high accuracy of this modeling approach has been demonstrated elsewhere [3]. However, in other cases, where the complexity of the physical phenomena determining the target properties of interest is higher, direct correlations between molecular structure and target property result in higher uncertainty when applied to the wide range of molecules generated by a molecular generator. Hence, to overcome this problem, the empirical, direct correlation of a property and molecular structure has to be replaced by a physically motivated structured model, where the sub-models could be represented by a QSPR. Such models are called hybrid, grey or semi-empirical.

3.2 Semi-empirical Property Models

The extrapolative capabilities of a model increase as more physically founded, rigorous descriptions of the underlying phenomena are incorporated in a structured model. However, the actual physical relations between molecular structure and macroscopic properties are often too complex or not entirely understood such that a rigorous property model can often not be formulated. Semi-empirical modeling constitutes a promising alternative to both, rigorous and empirical property modeling [31]. These types of models combine a physically motivated, "rigorous" model structure with empirical sub-models. The model structure is typically derived from first-principle models postulating reasonable simplifying assumptions and mathematical approximations. The level of rigor retained in the semi-empirical model should be as high as possible and is determined by the required level of accuracy for fuel selection. As shown in Fig. 1, semi-empirical models are employed to predict thermophysical fuel properties, properties related to injection as well as combustion kinetics of the fuel.

In order to improve the prediction of vapor-liquid equilibrium, GCM are combined with a physically motivated structured model, namely a cubic equation of state. Equations of state were first introduced by van der Waals [32] and have been steadily improved since then, with the two most important contributions being the Soave-Redlich-Kwong (SRK-EoS) [33] and the Peng-Robinson (PR-EoS) equation of state [34]. Generally, these equations take into account an attractive and a repulsive term to describe molecular interactions. They therefore relate the three intensive variables pressure p, molar volume V_m and temperature T by

$$p = \frac{RT}{V_m - b} - \frac{a\alpha}{V_m(V_m + b)} \tag{2}$$

with

$$a = f(T_c, p_c) \tag{3}$$

$$b = f(T_c, p_c) \tag{4}$$

$$\alpha = f(\omega). \tag{5}$$

The two terms of the equation of state (2) depend on parameters a and b which themselves are functions of the critical temperature T_c, the critical pressure p_c and the acentric factor ω. These three properties can be accurately predicted by means of GCM and afterwards applied to the equation of state under consideration to calculate liquid and vapor densities. To improve the quality of liquid density prediction, a volume correction term can be introduced into the equation [35]. As all cubic equations of state are unable to describe the discontinuous behavior of the isotherms in the two-phase-regime, they can not be used directly to calculate further equilibrium properties. To overcome this drawback, the Maxwell criterion is applied to correct the equation leading to accurate predictions at phase equilibrium [36]. This strategy allows the calculation of the normal boiling point T_b, which is one of the key fuel properties, as well as the vapor pressure curve of the considered molecule. Furthermore, by introducing the Clausius-Clapeyron equation (i.e. [37]), one can retrieve the enthalpy of vaporization in a physically motivated manner.

In addition to the properties at phase equilibrium, it is possible to calculate the surface tension σ based on the Sugden equation [38]

$$\sigma = [P(\rho_L - \rho_V)]^4, \tag{6}$$

where P denotes the so-called parachor, a temperature-independent parameter, which—together with the density difference—relates the surface-tension of a compound to its molecular structure. This parameter can be calculated by means of a QSPR [39], while the density of the vapor phase ρ_V and the liquid phase ρ_L are retrieved from an equation of state. As the estimation of the surface tension is strongly influenced by the density difference, high quality density estimates are essential to obtain reliable results.

Finally, combustion kinetic properties are predicted by a physically motivated model reflecting the combustion kinetic phenomena, where the sub-models are represented by QSPR. In particular, the so-called laminar burning velocity s_l, which is a telling indicator for assessing the combustion behavior of gasoline fuels, is approximately expressed by the physically sound expression [40]

$$s_l = A\left(T^0\right) \cdot Y_{F,u}^m \cdot \frac{T^u}{T^0} \cdot \left(\frac{T^b - T^0}{T^b - T^u}\right)^n \tag{7}$$

with

$$T^0 = -\frac{E}{ln(p/B)}. \tag{8}$$

Here, $Y_{F,u}$ denotes the mass fraction of the fuel in the unburned gas, and T^u and T^b are the temperatures in the unburned and burned gas, respectively. T^0 is the so-called crossover temperature within the flame, p is the pressure and B, E, F, G, m and n are fuel-specific, empirical parameters. Again, as in case of thermophysical property prediction outlined above, QSPR are employed to relate the molecular structure of the fuel to fuel-specific parameters in a physically sound model for the burning velocity rather than to the burning velocity itself. The resulting semi-empirical model has been shown to reliably predict the burning velocity of ethanol within measurement accuracy [24].

Likewise, the ignition delay time $\Delta\tau_{ign}$ is a fundamental quantity to characterize combustion of Diesel fuels. It describes the time between fuel injection and auto-ignition. While too short ignition delay causes engine knock, too long ignition delay considerably lowers the engine efficiency. In contrast to the burning velocity, which is also affected by transport effects in the engine, such as diffusion and convection, the "true" (chemical) ignition delay is entirely controlled by the reaction kinetics. Hence, an appropriate structure of an ignition model has to be identified and the occurring parameters have to be related to molecular structure. Unique values for the model parameters constitute a necessary requirement for meaningful correlation to a fuel's molecular structure. Therefore, mathematical identifiability analysis [41] should be carried out to select an appropriate model structure. Experiments can be conducted in a so-called rapid compression machine. A semi-empirical model has to account for heat loss over the cylinder walls increasing the measured ignition delay compared to a truly adiabatic constant-volume scenario. Model-based optimal experimental design techniques help to minimize experimental effort in screening a fuel [42].

The combination of property prediction methods and physically motivated modeling approaches leads to hybrid models with a sufficiently large range of applicability to predict fuel-relevant properties for a broad variety of potential organic fuel molecules. In the context of the fuel design approach presented above, predictive property models allow for a targeted reduction of the rigorously spanned molecular search space to finally identify the most promising structures complying with fuel requirements. The remaining molecular structures can be subsequently assessed with respect to the performance of a possible production process as described in the process design section.

3.3 Outlook

Despite the thoroughly chosen property model types and their educated combination in fuel design, there is still significant room for improvement for both, property modeling and candidate fuels screening.

A large uncertainty is related to the use of proper 3D descriptors in the QSPR framework. 3D descriptors are derived from the three-dimensional molecular structure, which has to be computed by some appropriate methodology before descriptor evaluation. Possible methods for geometry optimization range from simple force field methods as described above such as AM1 [43] to rigorous methods based on density functional theory like B3LYP [44]. While the computational effort increases considerably with the modeling detail, the level of accuracy required for a proper descriptor calculation is still unknown. However, complex fuel-relevant properties such as those related to fuel injection or combustion are likely to be insufficiently described by the molecular topology only. Hence, reliable 3D descriptors seem to be crucial for fuel identification in order to be able to span the entire range of fuel relevant properties.

Another related unresolved issue concerns the treatment of the conformational arrangement of molecular structures. Since molecules are not present in a single spatial configuration but in a temperature-dependent distribution of different conformers, molecular descriptors have to be evaluated for each conformer and averaged to properly reflect the physical properties of the compound. While this methodology is well known in pharmaceutical drug design [45], it is hardly accounted for in chemical engineering applications of QSPR.

The QSPR identification strategy itself also leaves room for further improvement. Although the descriptors are physically sound, their correlation to a property of the molecule under consideration is empirical in nature. Therefore, a careful examination of the derived models with respect to robustness, stability and predictive capability is obligatory before applying them to evaluate large sets of diverse molecular structures. This issue got significant attention in drug design, where specific guidelines to derive predictive models have been stressed recently [46] and where commonly used validation measures were shown to be insufficient to assess the predictive quality of a QSPR model [47]. Especially in case of a low number of training compounds complemented by multi-collinearity among the large number of descriptors, QSPR modeling might result in useless chance correlations. Moreover, instead of relying on a single best model, a set of rival models could be considered. If, for example, noisy training data prohibit agreeing on a single best model, averaging predictions of equally possible rival models might improve the prediction quality. In particular, the consensus hits approach has successfully been applied along these lines [48].

Moreover, while targeted QSPR can achieve high accuracy by focusing on a group of similar molecules [26], it may fall short for the rapid evaluation of a large number of compounds, since an assessment of a tailored similarity group and the subsequent model building have to be repeated for each compound to be predicted. Since some QSPRs showed good performance over a broad molecular range [3], global models might be better suited for the generate-and-test methodology pursued here. Additionally, nonlinear variable selection strategies such as the k-nearest-neighbor approach [48] might also prove to constitute a favorable complement to property modeling based on QSPR.

Besides the relatively simple thermodynamic fuel properties, combustion kinetic properties like ignition delay, laminar burning velocity or soot formation rely on temperature, pressure, air-to-fuel-ratio and dilution. These four degrees of freedom not only complicate the collection of consistent and comparable measurement data from literature, they also significantly increase the effort for obtaining new measurements. Despite the fast experiments performed in specialized devices such as rapid compression machines, flow reactors or high repetition rate shock tubes, the large variety in both, fuel candidates and experimental conditions, results in an enormous experimental effort. This effort can be reduced if guided experiments are carried out at conditions which maximize the information gain with respect to property model development [49]. As these optimal design of experiment techniques not only minimize the number of experiments required for a single fuel, these methods can likewise be employed to identify the ideal sequence of investigated fuels to obtain reliable property models with large applicability ranges at minimum experimental effort. Moreover, if the focus is extended from pure fuels towards mixtures which are likely to be superior over single compound energy carriers, the number of degrees of freedom within the fuel identification process increase even further, hence rendering a targeted model-based fuel design even more inevitable.

Finally, an alternative to the proposed generate and test-methodology could be pursued to solve the fuel design problem: the property models are inverted to compute a molecular structure which matches the values of a set of target fuel properties. While this method was demonstrated to be feasible if GCM are used [50] models based on QSPR can typically not easily be inverted due to their inherent complexity.

While the entire fuel design strategy thus still leaves a lot of space for research along various lines, the proposed approach represents a promising framework for a guided identification of novel transportation fuels.

4 Process Design: Identifying Promising Reaction Pathways

Most chemicals and biofuel components can be produced in various ways, in particular, a range of raw materials can be converted into the target product by different reaction pathways applying a variety of reaction and separation steps. Overall, this diversity of reaction pathways gives rise to a broad range of process alternatives. An early farsighted comparison of these alternatives requires a systematic evaluation which enables an efficient guidance of future research in chemistry and chemical engineering to finally come up with the most sustainable process.

To this end, Reaction Network Flux Analysis (RNFA) is presented as an optimization-based screening method for the evaluation of reaction pathways in an early stage of the process design procedure. The systematic analysis of mass and energy balances is complemented with meaningful thermodynamic, ecological and economic evaluation criteria, which are adapted to the requirements of an early stage of process design. RNFA provides a first insight in the performance of a future production process and reveals the challenges to be addressed in the next design steps. Based on the results of RNFA, a ranking of the different reaction pathways is possible including the identification of potentials for improvement or of unavoidable bottlenecks [7, 8].

The RNFA concept is based on metabolic flux analysis, which is an established method in systems biology [51]. In RNFA, material flows are traced in a network, which comprises of reactions connecting raw materials, intermediates and desired products. Usually, these reaction networks can be constructed according to the open literature (e.g. [52–54]) or according to preliminary results of catalysis research (as carried out, for example, in the TMFB cluster [55]). The substances and reactions can be represented in terms of a graph as nodes and arcs, respectively. The reaction network model is formulated by a set of material balances for all the substances. In general, any substance can be formed by several reactions such that the resulting system of equations is typically underdetermined. Thus, mathematical optimization techniques can favorably be utilized to detect those reaction pathways, which are composed of the optimal set of reaction steps required for the conversion of a starting material into a target compound. The optimization problem can be written as

$$\begin{aligned} &\min_{f,b} \phi \\ s.t.\, &\boldsymbol{A} \cdot \boldsymbol{f} = \boldsymbol{b}, \\ &\boldsymbol{f}, \boldsymbol{b} \geq 0. \end{aligned} \tag{9}$$

The matrix $\boldsymbol{A}$ comprises the stoichiometric coefficients of all reactions, the vector $\boldsymbol{f}$ stands for the reactive flux through the network, while the vector $\boldsymbol{b}$ indicates the products and by-products formed. The objective function ϕ can be selected to reflect the requirements of the considered application. For example, the product yield can be maximized. It is also possible to choose a multi-criteria objective function to discuss, e.g., the trade-off between ecological and economic benefit. In addition, yield constraints can be implemented to model the limited conversion of a reaction. As the yield coefficients of the single reaction steps are subject to ongoing research, they are varied from ideal conversion (of 100 %) to experimentally demonstrated yields. After solving problem (9) and thus identifying all possible reaction pathways, the solutions can be additionally evaluated according to previously defined performance criteria such as by-product formation, hydrogen need, energy or carbon efficiency, environmental impact or production cost among others. In some cases, the calculation of the performance criteria requires physical property data (e.g. combustion enthalpies), which have not yet been measured for all the involved substances. Thus, property prediction models of any kind are applied, including empirical and semi-empirical models based on GCM or QSPR [3, 16–18] as described above.

4.1 Outlook

As the results of RNFA are sensitive to structural changes in the reaction network, as many reactions as possible should be included from the beginning. At the same time, a detailed literature review is very time-consuming, even though web-based search engines and subject-specific databases support the search for patents and journals. To facilitate the general workflow and to assure the achieved results, an integrated database is currently set-up containing the relevant organic reactions reported in literature. In addition, the yield coefficients and the property data of the considered substances will be stored. The database will also provide the possibility to select reactions of interest in a comfortable way. Furthermore, it will be complemented by an algorithm for the automatic construction of reaction networks providing the input information for the network analysis (e.g. the matrix of stoichiometric coefficients and the property data) in the required format.

Besides, RNFA will be linked more tightly to the subsequent process design task. Therefore, the RNFA methodology will be expanded to take separation and purification steps into account. The overall goal is to estimate the feasibility and the effort of certain separation steps and to predict the consequences for the performance of the corresponding reaction pathway. The feasibility of a separation step can be tested by physically motivated indicators [56], which can directly be integrated in the network evaluation. By defining the composition of the mixture to be separated, RNFA provides essential information for subsequent conceptual design, for example, for the set-up of integrated reaction-separation sequences as part of the design framework under development at AVT.PT [57]. Once the starting and target material mixtures and their composition are known, process flowsheet alternatives are generated (manually based on engineering intuition and experience) at different levels of detail, they are classified by short-cut models and finally optimized with rigorous models to enable efficient variant generation and evaluation of the process for the reaction step at hand. This way, entire separation sequences, which are required to reach the desired specifications, can be designed in addition to the reactor system. Typically, the recycle structure for solvents and unconverted reactants is included. Based on these results, the energy demand as well as the operating and investment cost can be estimated. Thus, this refinement enables an evaluation of candidate processes by more detailed performance criteria, eventually resulting in a complete life-cycle assessment during the design process.

5 Integrating Product and Process Design

To sustainably produce fossil fuel surrogates from biomass, product and process design need to be combined to exhibit all constraints resulting from product application but also to identify and evaluate possible synthesis routes of the considered molecules. Although a certain molecular structure may be suitable as a fuel in terms of desired thermodynamic or combustion-kinetic properties, it can be infeasible or very expensive to synthesize this molecule due to structural features, required reactants or an excessive production of certain by-products.

The combination of product and process design can be applied in two different ways [3]. The first alternative is based on a reaction network comprising all considered molecules and their corresponding reaction routes starting from the platform chemicals. Once calculated, predicted thermodynamic properties of the molecules can be used as additional criteria in the evaluation of the various reaction routes. This first procedure for integrated product-process design can be applied, if target molecular structures for certain applications have already been identified in previous steps and a comparison of alternative production routes shall be carried out.

Within the presented model-based fuel design approach, promising fuel molecules have to be selected first from a database of rigorously generated molecular structures, which have typically not been characterized experimentally before. Therefore, the presented approach to product-process design integration is not feasible. Thus, a second possible combination of product and process design is introduced, where fuel candidates are identified first through comparison of calculated properties with property constraints, experimentally derived in an application test bench. The molecules matching these constraints are then passed to the process design task for the generation and evaluation of possible production pathways. The goal is to analyze process feasibility (in terms of specific criteria such economics, safety, health and environment) for the production of promising molecules.

5.1 Outlook

The main focus of future work will be the tight integration of product and process design leading to an efficient screening procedure for an enormous number of molecules which have to match the requirements of a given product applications. As the generation of the reaction network by means of literature search is one of the most time consuming parts, improvements will largely focus on supporting and facilitating this step. For this purpose, the identification of suitable platform chemicals based on molecular complexity measures [58, 59] and substructure similarity [60] is a first essential step for network generation. With additional information on the molecular structure of possible starting materials, literature research for recorded reactions can be intensified. If no suitable platform chemical is available, no reaction network should be generated. Hence, this strategy emphasizes molecules which are structurally similar to given starting materials and are therefore expected to be more promising for efficient synthesis. In order to hint to unknown reactions, the second line of action refers to the application of rigorous reaction generators using transition tables [61] which codify certain transformations of functional groups or single atoms, thus leading to a variety of possible molecules which can be derived from the considered starting molecule. This methodology can help chemists to find ways to avoid known bottlenecks by indicating alternative pathways with the same target molecule. Based on the molecular structure of platform chemicals and target molecules a desirable last step would be the definition of a numerical measure which expresses the ease of product synthesis, similar to the synthetic accessibility score used in the pharmaceutical industry [62]. Taking all these methods and improvements into account, the long-term goal is a software package providing a user-friendly framework for the efficient identification of application-specific biofuels.

6 Conclusions

This contribution presents the status quo of and an outlook on future work in model-based fuel design carried out within the Cluster of Excellence "Tailor-made Fuels from Biomass" at AVT.PT. By now, model-based fuel design is already an important part of the search for next-generation tailor-made biofuels. Its importance will grow further, because we intend to consider the whole range of possible molecular structures which is prohibitive for a solely experimental-based fuel design. In face of the ideas presented in the outlook sections of this paper, it becomes clear that still a lot of effort is needed to develop and to implement a sound model-based fuel design process. However, the methodological capabilities for the characterization of promising substances on the basis of their molecular structure will continuously advance. This will not only improve the prediction quality and thus will reduce the uncertainty in the screening process. It will also significantly decrease the time for the whole identification and evaluation process.

Of course, research on model-based fuel design will not be self-sufficient to successfully solve the fuel design problem. It will substantially benefit from progress in many fields of research carried out in the TMFB cluster such as fuel property and combustion kinetics prediction, ignition engine modeling and design, conceptual process synthesis and bio- and chemo-catalysis and of course experimental validation. Model-based fuel design is expected to provide a unifying methodological framework for the efficient and targeted search for sustainable fuels derived from biomass.

Acknowledgments One of the authors, Juan José Victoria Villeda, wants to thank the state government of North Rhine-Westphalia for granting a PhD scholarship within the Research School BrenaRo "Brennstoffgewinnung aus nachwachsenden Rohstoffen". All other authors appreciate the financial support off the Cluster of Excellence "Tailor-Made Fuels from Biomass", which is funded by the Excellence Initiative of the German federal and state governments to promote science and research at German universities.

References

1. Fayazz, S., Frenzel, P., Köster, M., Kollmeier, B., McIntyre, J., Meier, K., Müller, M., Schmidt, J., Schmidt, P., Somoza, I., Weber, N., Weinert, T., Ayesteran, J., Kopriwa, N., Pfennig, A.: Wie können wir zukünftig ausreichend Energie nachhaltig bereitstellen? CLB **60**, 32–39 (2009)
2. Gorelick, S.M.: Oil Panic and the Global Crisis. Wiley-Blackwell, Oxford (2010)
3. Hechinger, M., Voll, A., Marquardt, W.: Towards an integrated design of biofuels and their production pathways. Comput. Chem. Eng. **34**, 1909–1918 (2010)
4. RWTH Aachen University: Tailor-Made Fuels from Biomass—Excellence RWTH Aachen University establishes a Fuel Design Center. NatureJobs 502 (2007)
5. Sanders, J., Scott, E., Weusthuis, R., Mooibroek, H.: Bio-refinery as the bio-inspired process to bulk chemicals. Macromol. Biosci. **7**, 105–117 (2007)
6. Marquardt, W., Harwardt, A., Hechinger, M., Kraemer, K., Viell, J., Voll, A.: The biorenewables opportunity—toward next generation process and product systems. AIChE J. **56**, 2228–2235 (2010)
7. Voll, A., Marquardt, W.: Reaction network flux analysis: optimization-based evaluation of reaction pathways for biorenewables processing. AIChE J. (2011) (accepted)
8. Voll, A., Marquardt, W.: Multi-objective screening of biorefining processes in the early design stage by reaction network flux analysis. ECOS 2011, Novi Sad, Serbia
9. Pischinger, S., Müther, M.: Potentials and Challenges of Tailor-Made Fuels from Biomass, RWTH Aachen University, Aachen
10. Janssen, A., Jakob, M., Müther, M., Pischinger, S.: Massgeschneiderte Kraftstoffe aus Biomasse—Potenzial biogener Kraftstoffe zu Emissionsreduktion. MTZ **12**, 922–928 (2010)
11. Basshuysen, R., Schäfer, F.: Lexikon Motorentechnik. Vieweg, Wiesbaden (2006)
12. Gani, R.: Chemical product design: challenges and opportunities. Comput. Chem. Eng. **28**, 2441 (2004)
13. Gugisch, R., Kerber, A., Kohnert, A., Laue, R., Meringer, M., Rücker, C., Wassermann, A.: Molgen 5.0, http://molgen.de (2010)
14. Jensen, F.: Introduction to Computational Chemistry, 2nd edn. John Wiley and Sons, Chichester (2006)
15. van Duin, A.C.T., Dasgupta, S., Lorant, F., Goddard, W.A.: ReaxFF: a reactive force field for hydrocarbons. J. Phys. Chem. A **105**, 9396–9409 (2001)
16. Lydersen, A.: Estimation of Critical Properties of Organic Compounds. Eng Exp Stn Rep, Madison, Wisconsin (1955)
17. Joback, K.G., Reid, R.C.: Estimation of pure-component properties from group-contributions. Chem. Eng. Commun. **57**, 233–243 (1987)
18. Marrero, J., Gani, R.: Group-contribution based estimation of pure component properties. Fluid Phase Equilib. **183**, 183–208 (2001)
19. Karunanithi, A., Gani, R., Achenie, L.E.K.: Biodiesel process design through a computer aided molecular design approach. In: Proceedings of FOCAPD 2009, CRC Press, Boca Raton (2009)
20. Kraemer, K., Harwardt, A., Bronneberg, R., Marquardt, W.: Separation of butanol from acetone–butanol–ethanol fermentation by a hybrid distillation–extraction process combining extraction and distillation. In: Proceedings of Escape 20, Elsevier, Amsterdam (2010)
21. McLeese, S., Eslick, J., Hoffmann, N., Scurto, A., Camarda, K.: Design of ionic liquids via computational molecular design. In: Proceedings for FOCAPD 2009, CRC Press, Boca Raton (2009)
22. Sabbe, M.K., Reyniers, M.F., Waroquier, M., Marin, G.B.: Hydrogen radical additions to unsaturated hydrocarbons and the reverse b-scission reactions: modeling of activation energies and pre-exponential factors. ChemPhysChem **11**, 195–210 (2010)

23. Saeys, M., Reyniers, M.F., Speybroek, V.V., Waroquier, M., Marin, G.B.: Ab initio group contribution method for activation energies of hydrogen abstraction reactions. ChemPhysChem **7**, 188–199 (2006)
24. Hechinger, M., Marquardt, W.: Targeted QSPR for the prediction of the laminar burning velocity of biofuels. Comput. Chem. Eng. **34**, 1507–1514 (2010)
25. Katritzky, A., Kuanar, M., Slavov, S., Hall, C.: Quantitative correlation of physical and chemical properties with chemical structure: utility for prediction. Chem. Rev. **110**, 5714–5789 (2010)
26. Shacham, M., Brauner, N., Cholakov, G., Stateva, R.: Property prediction by correlations based on similarity of molecular structures. AIChE J. **50**, 2481–2492 (2004)
27. Wakeham, W., Cholakov, G.S., Stateva, R.: Liquid density and critical properties of hydrocarbons estimated from molecular structure. J. Chem. Eng. Data **47**, 559–570 (2002)
28. Todeschini, R., Consonni, V.: Molecular Descriptors for Chemoinformatics. Wiley, Weinheim (2009)
29. Shacham, M., Brauner, N.: The SROV program for data analysis and regression model identification. Comput. Chem. Eng. **27**, 701–714 (2003)
30. Kahrs, O., Brauner, N., Cholakov, G.S., Stateva, R.P., Marquardt, W., Shacham, M.: Analysis and refinement of the targeted QSPR method. Comput. Chem. Eng. **32**, 1397–1410 (2008)
31. Kahrs, O., Marquardt, W.: The validity domain of hybrid models and its application in process optimization. Chem. Eng. Process. **46**, 1054–1066 (2007)
32. van der Waals, J.D.: Over de Continuiteit van den gas- en vloeistoftoestand. Dissertation. University of Leiden (1873)
33. Soave, G.: Equilibrium constants from a modified Redlich-Kwong equation of state. Chem. Eng. Sci. **27**, 1197–1203 (1972)
34. Peng, D., Robinson, D.B.: New 2-constant equation of state. Ind. Eng. Chem. Fund. **15**, 59–64 (1976)
35. Peneloux, A., Rauzy, E.: A consistent correction for Redlich-Kwong-Soave volumes. Fluid Phase Equilib. **8**, 7–23 (1982)
36. Stein, W.: Die Anwendbarkeit des Maxwell-Kriteriums zum Berechnen der thermodynamischen Zustandsgrößen im Zweiphasengebiet. Forsch. Ingenieurwes. **34**, 193–195 (1968)
37. Callen, H.B.: Thermodynamics and an Introduction to Thermostatistics. Wiley, New York (1985)
38. Sugden, S.: The variation of surface tension with temperature and some related functions. Chem. Soc. **125**, 32–41 (1924)
39. Gharagheizi, F., Eslamimanesh, A., Mohammadi, A., Richon, D.: QSPR approach for determination of parachor of non-electrolyte organic compounds. Chem. Eng. Sci. **66**, 2959–2967 (2011)
40. Müller, U.C., Bollig, M., Peters, N.: Approximations for burning velocities and Markstein numbers for lean hydrocarbon and methanol flames. Combust. Flame **108**, 349–356 (1997)
41. Quaiser, T., Mönnigmann, M.: Systematic identifiability testing for unambiguous mechanistic modeling—application to JAK-STAT, MAP kinase, and NF-kB signaling pathway models. BMC Systems Biology (2006)
42. Walter, E., Pronzato, L.: Identification of Parametric Models from Experimental Data. Springer, Paris (2006)
43. Dewar, M., Zoebisch, E., Healy, F., Stewart, J.: AM1: A new general purpose quantum mechanical molecular model. J. Am. Chem. Soc. **107**, 3902–3909 (1984)
44. Stephens, P., Devlin, F., Chabalowski, C., Frisch, M.: Ab initio calculation of vibrational absorption and circular dichroism using density functional force fields. J. Phys. Chem. **98**, 11623–11627 (1994)
45. Pissurlenkar, R.R.S., Khedkar, V.M., Iyer, R.P., Coutinho, E.C.: Ensemble QSAR: a method based on conformational ensembles and metric descriptors. J. Comput. Chem. **32**, 2204–2218 (2011)

46. Golbraikh, A., Shen, M., Xiao, Z., Xiao, Y., Lee, K., Tropscha, A.: Rational selection of training and test sets for the development of validated QSAR models. J. Comput. Aid Mol. Des. **17**, 241–253 (2003)
47. Golbraikh, A., Tropscha, A.: Beware of q^2! J. Mol. Graph. Model. **20**, 269–276 (2002)
48. Shen, M., LeTiran, A., Xiao, Y., Golbraikh, A., Kohn, H., Tropscha, A.: Quantitative structure-activity relationship analysis of functionalized amino acid anticonvulsant agents using k nearast neighbor and simulated annealing PLS methods. J. Med. Chem. **45**, 2811–2823 (2002)
49. Bard, Y.: Nonlinear Parameter Estimation. Academic Press, New York and London (1974)
50. Gani, R., Achenie, L.E.K., Venkatsubramanian, V.: Introduction to CAMD. In: Gani, R., Achenie, L.E.K., Venkatsubramanian, V. (eds.) Computer-Aided Molecular Design: Theory and Practice. Elsevier, Amsterdam (2003)
51. Varma, A., Palsson, B.: metabolic flux balancing—basic concepts, scientific and practical use. Nat. Biotechnol. **12**, 994–998 (1994)
52. Corma, A., Iborra, S., Velty, A.: Chemical routes for the transformation of biomass into chemicals. Chem. Rev. **107**, 2411–2502 (2007)
53. Huber, G., Iborra, S., Corma, A.: Synthesis of transportation fuels from biomass: Chemistry, catalysis, and engineering. Chem. Rev. **106**, 4044–4098 (2006)
54. Kamm, B., Gruber, P., Kamm, M.: Biorefineries—Industrial Processes and Products, vol. 1, 2. Wiley-VCH, Weinheim (2005)
55. Geilen, F., Engendahl, B., Harwardt, A., Marquardt, W., Klankermayer, J., Leitner, W.: Selective and flexible transformation of biomass-derived platform chemicals by a multifunctional catalytic system. Angew. Chem. Int. Edit. **49**, 5510–5514 (2010)
56. Jaksland, C., Gani, R., Lien, K.: Separation process design and synthesis based on thermodynamic insights. Chem. Eng. Sci. **50**, 511–530 (1995)
57. Kossack, S., Refinius, A., Brüggemann, S., Marquardt, W.: Konzeptioneller Entwurf von Reaktions-Destillations-Prozessen mit Näherungsverfahren. Chem. Ing. Tech. **79**(10), 1601–1612 (2007)
58. Bertz, S.: The first general index of molecular complexity. J. Am. Chem. Soc. **103**, 3599–3601 (1981)
59. Barone, R., Chanon, M.: A new simple approach to chemical complexity. Application to the synthesis of natural products. J. Chem. Inf. Comp. Sci. **41**, 269–272 (2001)
60. Jochum, C., Gasteiger, J., Ugi, I.: The principle of minimum chemical distance (PMCD). Angew. Chem. Int. Edit. **19**, 495–505 (1980)
61. Fontain, E., Reitsam, K.: The Generation of reaction networks with RAIN. 1. The reaction generator. J. Chem. Inf. Comp. Sci. **31**, 96–101 (1991)
62. Gasteiger, J.: De novo design and synthetic accessibility. J. Comput. Aid Mol. Des. **21**, 307–309 (2007)

Biofuels for Combustion Engines

Aspects of Modern Engine Development in the Context of Future Biofuels

Johannes Richenhagen, Florian Kremer, Carsten Küpper, Tobias Spilker, Om Parkash Bhardwaj and Martin Nijs

Abstract The requirements on the development of combustion engines have dramatically changed in the past decade. This includes strict emission laws, CO_2 emission reduction, different propulsion concepts including powertrain electrification and a reduced time to market with an increased number of engine variants. One alternative to mitigate both the need for fossil burnings and the CO_2 emission reduction is the use of alternative fuels from biomass. Thus, different legislation authorities aim for higher proportions of alternative fuels on the market. However, this strategy involves changes on different development domains for combustion engines. This paper presents ongoing research taking place within the interdisciplinary activities at the Institute for Combustion Engines. The effects on the control system as one enabler of further investigations are presented from the perspective of variant management and complexity handling. Proceedings of the research on innovative control algorithms for fuel adaption are outlined. At third, we discuss the impact of direct injection of alternative fuels on liner wetting and piston ring development. At last, the combustion of fuels from biomass with regards to the emissions formation is investigated from two points of view: for gasoline combustion methods, the characteristics of gaseous emission are presented. For Diesel combustion, we show the different formation of particles by applying diverse measurement methods.

Abbreviations

AUTOSAR	Automotive Open System Architecture
CA	Crank Angle
CAN	Controller Area Network
CO	Carbon monoxide
CO_2	Carbon dioxide

J. Richenhagen · F. Kremer (✉) · C. Küpper · T. Spilker · O.P. Bhardwaj · M. Nijs
Institute for Combustion Engines (VKA), RWTH Aachen University, Aachen, Germany
e-mail: kremer@vka.rwth-aachen.de

M. Klaas et al. (eds.), *Fuels From Biomass: An Interdisciplinary Approach*,
Notes on Numerical Fluid Mechanics and Multidisciplinary Design 129,
DOI 10.1007/978-3-662-45425-1_13

CLCC	Closed Loop Combustion Control
CN	Cetane Number
COC	Center of Combustion
CSL	Combustion Sound Level
DBE	Di-n-butylether
DiCoRS	Digital Combustion Rate Shaping
DOC	Diesel Oxidation Catalyst
DPF	Diesel Particulate Filter
EC	Elemental Carbon
ECU	Electronic Control Unit
EGR	Exhaust Gas Recirculation
EU	European Union
FSN	Filter Smoke Number
HC	Hydrocarbon
HECS	High Efficiency Combustion System
HC	Unburned Hydrocarbons
ILC	Iterative Learning Control
IMEP	Indicated Mean Effective Pressure
ISF	Insoluble Fraction
ISNOx	Indicated Specific Nitrogen Oxides
ISO	International Standardization Organization
MTHF	Methyltetrahydrofurane
NEDC	New European Driving Cycle
NO_x	Nitrogen Oxides
OC	Organic Carbon
OEM	Original Engine Manufacturer
PD	Proportional Differential
PM	Particulate Matter
RCP	Rapid Control Prototyping
RTE	Runtime Environment
SDD	Software Design Description
SOF	Soluble Organic Fraction
SPICE	Software Process Improvement and Capability Determination
US	United States of America
TGA	Thermo-Gravimetric Analysis
ULSD	Ultra Low Sulphur Diesel
PCU	Piston Cylinder Unit
MBS	Multi Body System
MSC ADAMS	Commercial Multi Body System Software
DI	Direct Injection Technology for Internal Combustion Engines
FORTRAN	General-Purpose Programming Language
IL	Control Variable for the Liner Circumference
JL	Control Variable for the Liner Height
n_{IL}	Number of Surface Sections in Circumferential Direction

n_{JL}	Number of Surface Sections in Height Direction
φ	Circumferential Angle of the Liner
Z	Height Axis of the Liner
h_{liner}	Liner Height
h_{ring}	Piston Ring Height
$n_{segments}$	Number of Segments of a Piston Ring
m_i	Mass Flow of a Component 'i' in the Lubricant
g_i	Mass Transfer Coefficient of a Component 'i' in the Lubricant
c_i	Concentration of a Component 'i' in the Lubricant
c_{gas}	Concentration of the Lubricant in the Gas
α_i	Heat Transfer Coefficient of a Component 'i' in the Lubricant
ρ_i	Density of a Component 'i' in the Lubricant
c_p	Heat Capacity of the Gas
Le_i	Lewis number of a Component 'i' in the Lubricant
λ_i	Thermal Conductivity of the Gas
$D_{i,gas}$	Diffusion Coefficient of a Component 'i' in the Lubricant
p	Pressure in the combustion chamber
d_{liner}	Diameter of the Liner
T_{gas}	Temperature of the Gas
C_I	Variable #1 to Describe the Combustion Process
c_m	Mean Velocity of the Piston
V_h	Cylinder Displacement Volume
T_I	Temperature at the Beginning of the Compression
p_I	Pressure at the Beginning of the Compression
V_I	Cylinder Volume at the Beginning of the Compression
p_0	Pressure without Compression
dt	Time Increment
h_{start}	Local Oil Film Height before Ring Slides over the Surface Cell
$h_{min}(t)$	Local Minimum Distance between Ring and Liner Surface
$h(t)$	Distance between Ring and Liner Surface
u	Injector Actuation Voltage
w	Setpoint Trace
u_φ	Circumferential Length of a Liner Cell
v_{ring}	Velocity of The Ring
$c_{off,bead}$	Constant to Cover the Oil Release from the Ring Flank
LCS	Load Capacity Scope
$R_{rms(100\ \%LCS)}$	RMS Roughness outside the Load Capacity Scope
$V_{reservoir}$	Reservoir Volume
$h_{res,max}$	Maximum Oil Height in the Reservoir
$dh_{reservoir}$	Change of Oil Height in the Reservoir
$R_{tb};R_{bb}$	Radius of top and bottom Bead
c_{flow}	Flow Coefficient
CFD	Computer Fluid Dynamics
E0	Ethanol fuel blend with volumetric ethanol content of 0 %

E20	Ethanol fuel blend with volumetric ethanol content of 20 %
E85	Ethanol fuel blend with volumetric ethanol content of 85 %
TWC	Three-way catalyst

1 Introduction

The requirements in the development process of modern combustion engines can directly be derived from automotive needs. Technical, economical and emotional aspects are to be taken into account. Nowadays, modern powertrains must minimize fuel consumption and exhaust gas emissions at low costs for diverse international markets, needs and fuel qualities. New propulsion technology trends such as an increasing electrification lead to higher quality and safety standards. Besides the extension of the driving comfort, non-technical requirements (brand building and "fun to drive") have to be met [1].

In the past decade, the reduction of greenhouse gasses has been one of the main targets, mainly driven by legislation. By 2020, the fleet CO_2 production and hence fuel consumption in the European Union shall not exceed 95 g CO_2/km which is equivalent to a reduction of nearly 40 % compared to 2009 [2]. In the United States (US), greenhouse gases are to be reduced by higher fuel efficiencies which, converted to CO_2 emissions, follow the same tendency by reducing fuel consumption by 12 % [3]. Figure 1 shows both legislation tendencies as an overview.

In order to achieve a reduction of both greenhouse gasses and fossil fuel dependency, there exist several legislation approaches to achieve the efficiency targets by applying renewable fuels from biomass to the automotive market. Both US and EU legislation authorities assume a significant carbon dioxide emission reduction potential by applying fuels from biomass to the automotive sector. Hence, reductions of up to about 80 % for Diesel and 70 % for Gasoline engines are expected (Fig. 2).

In the European Union, 10 % of the energy content consumed by passenger vehicles in 2020 shall be gained from renewable sources [4]. The US aim for prescribing the total amount of renewable fuels to be sold from 14 billions of gallons per year in 2011 up to 36 billions of gallons in 2022 [3]. In relation to the total US transportation energy consumption in 2009, this directive equals a consumed energy fraction of 15 % in 2022 [3, 5] (Fig. 3).

The outlined legislation targets lead to the conclusion, that the efficient use of alternative fuels from biomass in combustion engines is a core development requirement for both research institutions and the automotive industry (Fig. 4).

In the following sections, these requirements are presented in detail from the perspective of diverse engine development domains.

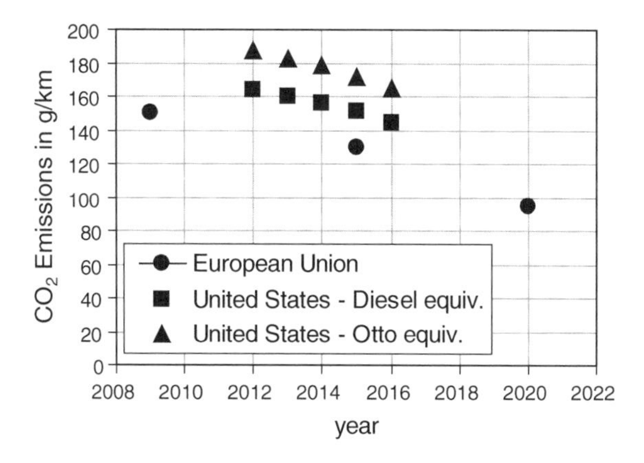

Fig. 1 CO_2 reduction targets of European Union and United States according to US and EU legislation [2, 3]

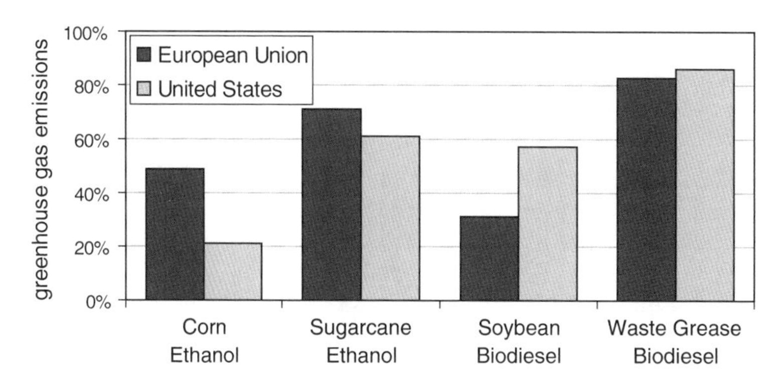

Fig. 2 Greenhouse gas emission reduction potential estimations for exemplary renewable fuels according to US and EU [4, 60]

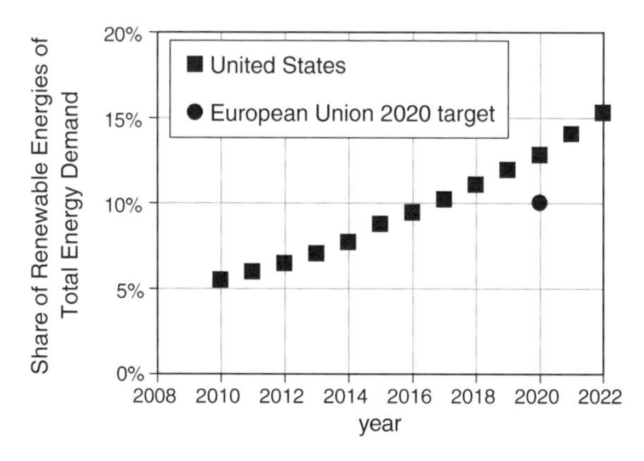

Fig. 3 Target energetic market share of energy from biomass according to US and EU legislation [3–5]

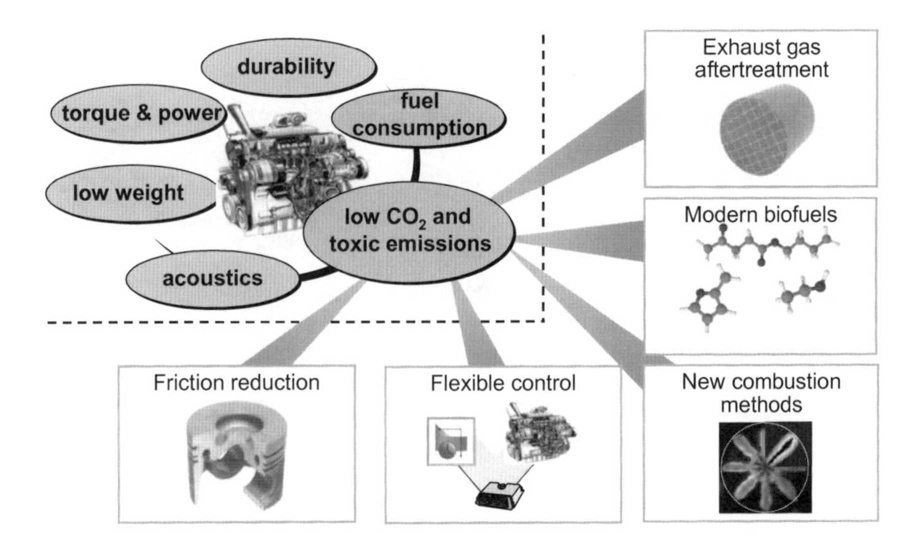

Fig. 4 Global engine requirements for higher fuel efficiency with fuels from biomass

Besides friction reduction and combustion improvement effective control systems are needed to enable further engine research. The variety of fuels from biomass increases the number of control software variants dramatically. Moreover, affected safety-relevant functions cause a functionality extension. Additionally, the fuel injection control needs to adapt quickly to the different fuels that are expected to occur on the market. One of the major differences encountered with alternative fuels is the significantly different evaporation behaviour. In combination with state-of-the-art direct fuel injection, the cylinder wall-wetting affects the construction of sealing components, especially the piston rings. Moreover, the emission formation changes for both gaseous and solid emissions. At the end of this chapter, an overview of first research results and future development trends is given.

2 Agile Control Software Development

In this first section, the impact of alternative fuels on the control software development is outlined. On the control system level, the fuel change comes along with modifications on different components. Especially in the case of gaseous fuels, additional effort must be spent to control the fuel system. The low pressure supply system is equipped with an additional tank pressure sensor whereas on high pressure injection level the pressure increase and proper charge mixture must be controlled [6, 7]. For bivalent operation with gaseous and liquid fuels, the mode switch and fuel quality detection is added to the control [8–10]. If liquefied gas is applied, the injection strategy changes dramatically due to the different evaporation behaviour and fuel characteristics. For a cold engine start, additional effort must be spent, if Ethanol is used [10]. Additionally due to change in exhaust gas characteristics and temperature profiles of exhaust gas, advanced control structures are desired to ensure the efficient operation of complex after-treatment systems in modern engines.

These system changes affect the control software of the engine management system in various ways. The number of software variants increases since different fuels from renewable sources are dominant in international markets. Aside the functional changes and extensions that come along with the new components mentioned above, additional safety and diagnosis functions must be added for the fuel detection and tank pressure sensor to avoid hazardous combustion behaviour [7, 10].

For the software development process, the additional flexibility and safety requirements lead to a merge of functions from different control domains. Hence, the configuration of software components to one control becomes a development task. For gas vehicle engine control, the integration of gasoline, diesel, commercial vehicle and gas-specific functions into a gas vehicle control software may be necessary [11]. Since the additional control requirements caused by renewable fuels are in line with additional functionalities caused by other technology trends such as powertrain electrification, the software development industry encounters a continuously increasing extent of the software resource consumption and complexity at a shrinking time-to-market (Fig. 5).

With the outlined challenges, the introduction of renewable fuels puts additional requirements from three areas of control software development:

- Complexity increase due to additional functionality
- Higher number of software variants to be maintained at the same time
- Process effort for software configuration and safety-related functions

The following sections show, how process, complexity and variant handling can be dominated by a proper function development process and a flexible control software architecture.

2.1 Function Development Requirements

The application of renewable fuels involves special requirements on the software development. In the following sections, the focus is put on two aspects of software development: architecture design and configuration management and integration. The requirements that arise in these research areas are outlined in more detail.

According to the V cycle reference model, software architecture design starts directly after the definition of software requirements. Hence it is the base for any further development steps. The detailed software architecture development requirements can be derived from reference models such as the Automotive SPICE standard defined by the association of German car manufacturers (Fig. 6). The purpose of architecture development is to provide a design that implements and

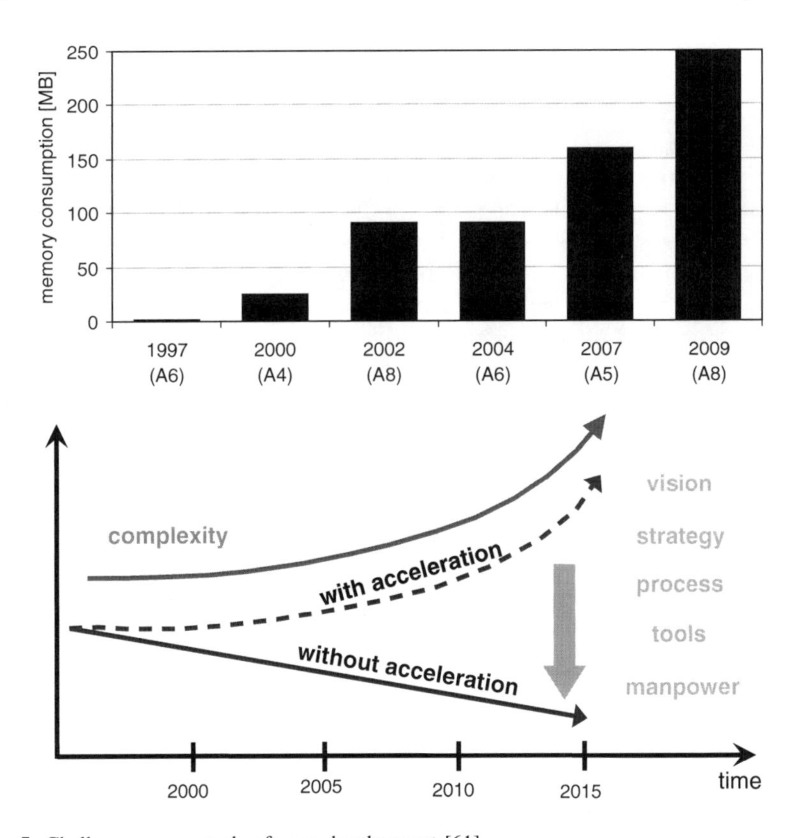

Fig. 5 Challenges on control software development [61]

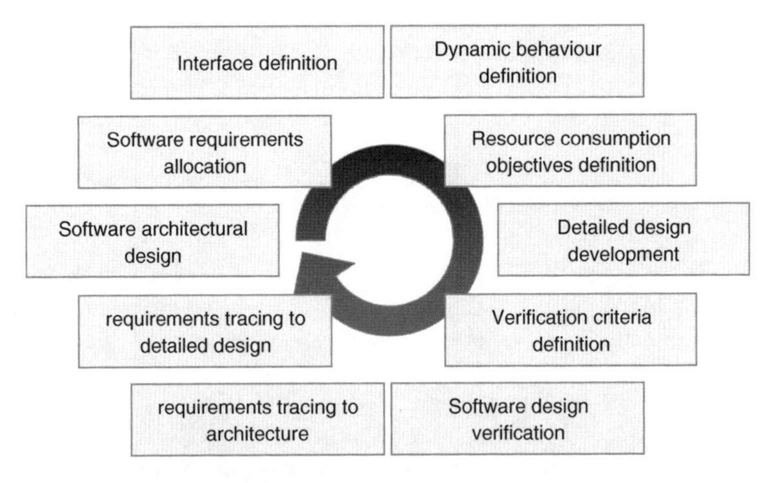

Fig. 6 SW architecture development process according to automotive SPICE ©

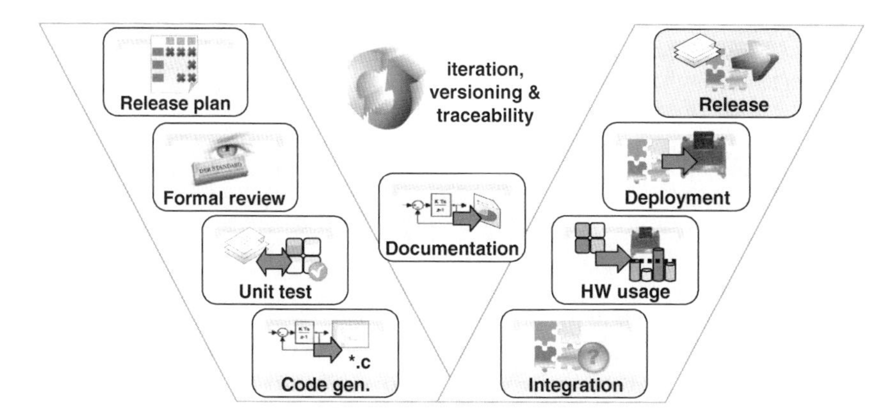

Fig. 7 Development process requirements according to ISO 26262 [19]

can be verified against the software requirements [12]. At the beginning, the software architecture is defined on the component level. This includes components related to application as well as supplied components. All needed software functions are split up and distributed to components.

In order to achieve an effective handling of fuel and strategy variants, additional criteria must be considered. These so called quality criteria are specified and detailed in ISO/IEC 25000 [13]. According to these criteria, the requirements on the software architecture for renewable fuels are the following (Fig. 7):

- Functionality
  - Suitability: functional requirements for renewable fuels
  - Interoperability: proper interaction with control hardware
- Reliability
  - Maturity: decreased error probability with each application
  - Fault tolerance: faulty code must not cause a software crash
- Usability
  - Understandability: software must be documented
  - Operability: configuration for algorithm research
- Efficiency
  - Time Behaviour: timely and correctly ordered execution of tasks
  - Resource Utilization: no exceed of memory and processor limits
- Maintainability
  - Analyzability: automation enabling through consistent naming
  - Changeability: introduction and maintenance of variants
  - Testability: manageable function sizes and interfaces

- Portability
  - Adaptability: quick response to control system changes
  - Installability: transparent installation package of needed data
  - Replaceability: seamless software and hardware change

Important criteria in terms of the application of fuels from biomass are the modifiability of the software for different control functions and strategies and the ability to be integrated into existing control software frameworks providing vehicle integration functions.

After the architectural design, the requirements are allocated to the components and the interfaces (used and produced signals of each component) are defined. To specify the dynamic behaviour such as execution order, function sampling rate and operating modes, a strong interaction with the following step is necessary: hardware resource and time constraints must be considered. Since the actual extend of the final software and the implementation is unknown yet, the resource consumption must be estimated by appropriate methods [14]. When the detailed design is conducted, the function size must be kept testable, e.g. through manageable interface size and functionality extend. The remaining architecture design steps deal with the assignment of requirements to the software units to allow for proper testing after function implementation.

The quality criteria mentioned above do not apply for engines running with alternative fuels only but form a general requirement portfolio. Moreover, the automotive industry has some specific market conditions with a strong influence on the software architecture. Typically, the development of the control hardware and the corresponding hardware assessment through software drivers is conducted by first-tier hardware suppliers. The original engine manufacturer (OEM) develops the application specific function algorithms. Depending on the project work split, the interface between the supplier and OEM software changes without technical reasons. As a result, there is a substantial risk, that the application software is hardware-specific depending on the agreement between the involved parties [15].

In order to achieve less hardware dependency, relevant stakeholders of the automotive industry are developing the standard AUTomotive Open System ARchitecture (AUTOSAR) that intends the standardization of software components introducing a layered software architecture (Fig. 8).

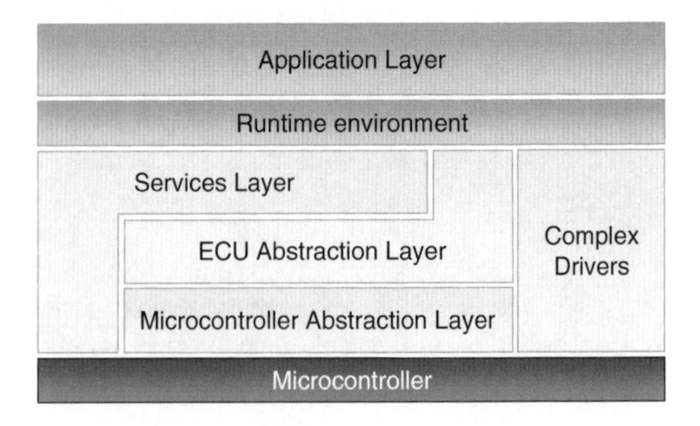

Fig. 8 AUTOSAR layered architecture [62]

Basic design principle is the separation of the software into layers that are interfaces for their upper layers. The microcontroller hardware is assessed by three layer types. Sensors and actuators are controlled via the Microcontroller Abstraction Layer which offers a hardware-independent interface to the ECU abstraction layer. The services layer offers diagnostic and safety functions. These functions can be used by the application software components for e.g. event notification or data provision. The ECU (Electronic Control Unit) abstraction layer provides diverse types of input and output signals being received and transmitted. The control of some exceptional sensors and actuators is highly hardware-dependent. Thus, AUTOSAR allows for the definition of so called "Complex Drivers" that integrate all layer functionalities and directly interface application software and the microcontroller. The Runtime Environment (RTE) manages the data flow in between the layers and within the application software. The latter is organized in components that represent the function entities. These components are defined during the application level software architecture design. The layers below the RTE, also called basic software, are usually provided by the first-tier.

Aside the increased hardware independency, this structure comes along with further advantages [15]:

- more stable requirements for basic software components and application software interfaces
- avoidance of late requirement change risks
- reduction of complexity and increase of reliability
- function-oriented view onto the software
- component reuse
- quicker software integration
- increased product quality.

While the basic software components are defined precisely with functional requirements, interfaces and timing and calibration specification, the application software components for powertrain applications are not fully standardized. A functional design principle is indicated for engine control units, but further notice is not given. However, when developing a control software architecture, the design principles of the AUTOSAR framework should be regarded since it is the only existing architecture standard under definition.

The configuration management and integration requirements are derived from the standard for the development of safety-critical systems since it affects all necessary aspects of function development. This involves precise documentation of all development activities, management of requirements and test case management and execution on different levels throughout the entire development [16].

The recurrent and iterative execution of tests, reports and documentation is currently considered to be the main software development cost driver [17, 18]. As a result, many process steps are conducted for few milestones only, errors and problematic issues are detected lately.

For suitable countermeasures, new quality assurance approaches are necessary. The possible verification and validation must be integrated within a testing strategy in an agile development framework that allows for minimized manual testing effort, continuous software inspection and traceability of test results over the entire project to re-enable frontloading of software error mitigation. So far, there exists no framework that satisfies this need. Existing tools cover either single inspection steps such as unit test, guideline check and code generation or do not fulfill all needs of an integrated quality assessment from the requirement to executable code on the target [19].

The requirements on the framework are detailed and split up into three categories in [19]:

- Model-based function development principles
  - Architecture conformance tests
  - Unit design and integration tests
  - Configuration and documentation of software releases
- Software integration principles
  - Automation of all tests
  - Meaningful software quality metrics
- Safety-compliant process principles
  - Code generation and compilation including deployment
  - Creation of a delivery package

Summarizing the requirements on an improved function development method, the challenges caused by renewable fuels can be solved successfully by defining a flexible software architecture as the base for efficient control implementation and testing. A design along the AUTOSAR standard is recommendable. Secondly, the software architecture and the implementation of control algorithms must be continuously assessed and where possible improved by an automated integration process that is scalable to the project needs. State-of-the-art quality guidelines for a proper software assessment must be regarded.

2.2 Flexible Propulsion Control Software Architecture and Development Process

The requirements on function development for renewable fuels require a control software architecture that both allows both for the integration of non-fuel-specific functions as well as the development of new algorithms to fulfill the specific needs. Additionally, up-to-date design principles tailored for automotive needs such as the AUTOSAR standard must be regarded. Moreover, the compliance with this standard and the quality requirements given by ISO 25,000 must be tracked during the

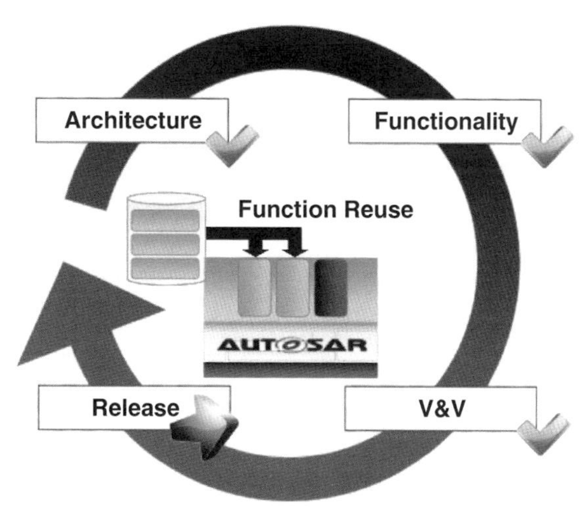

Fig. 9 Strategy for an agile software development process

software development process to avoid big integration and error detection efforts at the end of the product lifecycle. Therefore, the architecture itself but also the inspection framework must be developed (Fig. 9).

Being the enabler for a control software with reusable components, the inspection framework was designed first. Similar to continuous integration frameworks for text-based software development for PC applications, the system architecture of the integration framework is split up in different entities. Distributed developers using personal computers, one central data repository managed by a version-control system and an integration server that pulls continuously the latest software status to perform the integration operations are connected (Fig. 10). The integration script is triggered via a command-line interface by the task scheduler of the operating system. Since the scripts are executed on a daily basis when developers are absent (i.e. at night), the framework is called "nightly build".

In contrast to existing frameworks in various application domains, the integration operations are conducted by scripts of the domain-specific language for automotive applications. As an example, the tool Matlab © was applied since it offers model-based development tools (e.g. Simulink ©) as well as a scripting framework. This technology was selected for different purposes.

On the one hand, software developer use Matlab tools to perform model development steps, e.g. for model parameterization, guideline checking, data dictionary management or code generation. For compilation and code generation from the model, Simulink has a Matlab scripting interface. Using the same technical base, the integration server is able to use identical tools as the developer. In the reverse direction, developers can perform build operations to check, if their models conform to the implementation guidelines. Thus, redundant tool implementation is avoided.

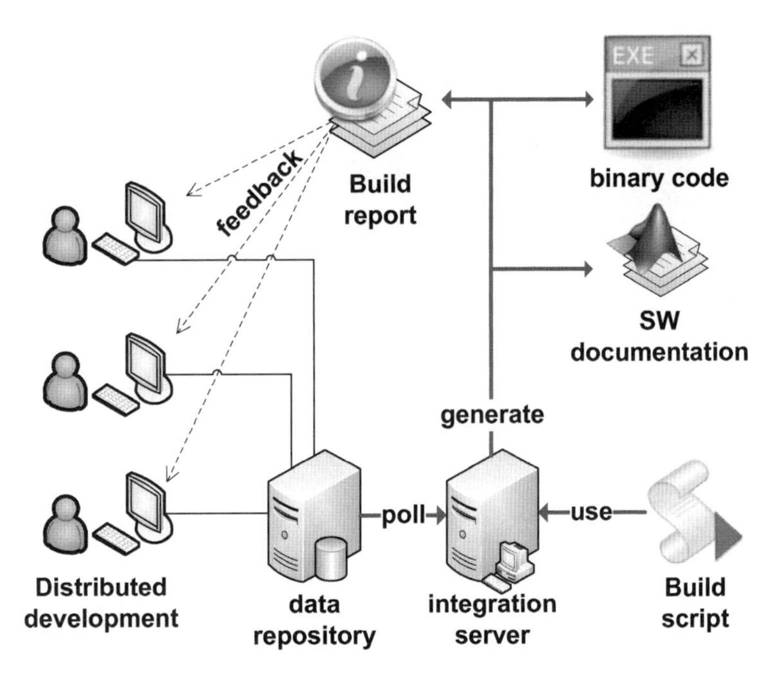

Fig. 10 Nightly build system architecture

On the other hand, besides various scripting functionalities, Matlab offers a command-line and COM/.NET interface for a comfortable integration of third-party software such as requirements management, bugtracking or version management tools. At last, the integration process by itself can be called via command line e.g. by task scheduling programs. Hence, all requirements deducted from safety-compliant process principles can be fulfilled. With this decision, some drawbacks are accepted. In comparison to mere text-based model checking, the execution time and hardware resources increase and have to be mitigated by parallel computing and a high performance build server. Additionally, the integration framework is built up in a tool-specific environment.

Concerning integration functions, the focus is being put on syntax check automation and metric development.

In order to achieve a high flexibility of the integration configuration for various projects, customers, implementation guidelines and tools, the integration software architecture is structured on a modular base. After a definition of all (project-specific) integration parameters, the process can be set-up using tasks that communicate over common interfaces. The first task is the interpretation of the release plan. A project information interface and an integration log are initialized. Afterwards, different integration tasks can be compiled using a dedicated data store for logging and result information. Each task produces results such as a status overview (e.g. as matrix output in Excel), software documentation, a detailed check report

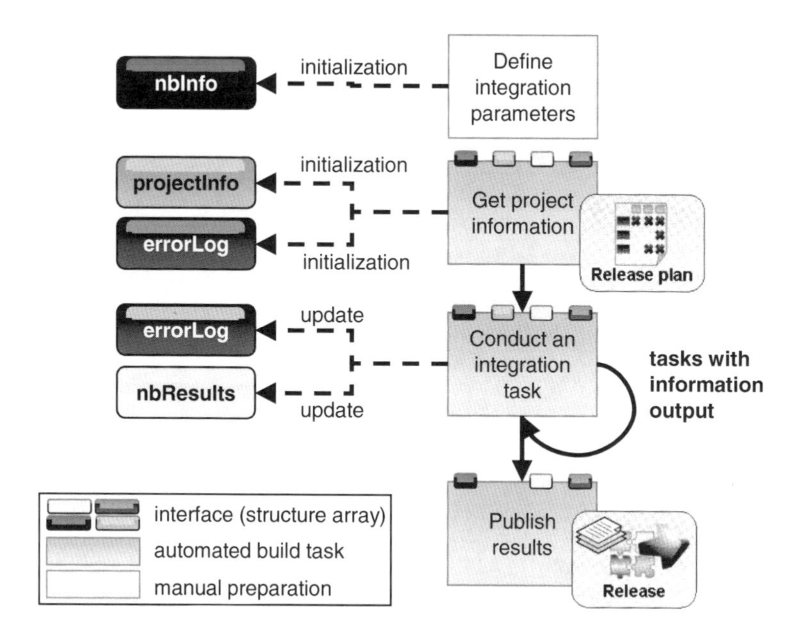

Fig. 11 Script architecture for quality check automation

and a command-line log for detailed debugging. At the end, integration results are published via upload on a server and to the version management system (Fig. 11).

So far, with the given scripting framework, the quality inspection requirements (see Fig. 7) have been implemented by about 50 % for the left branch of the software V cycle. This includes the set-up of a release plan that defines the software models to be assessed at certain milestones. Based on an automated documentation tool, printed software documentation (Software Design Description—SDD) and application guidelines can be produced. Moreover, all software units can be checked against standard style guides for proper model implementation. Functional unit tests including testing coverage and pass/fail evaluation can be conducted on a daily base. Finally, the code of all software models and its compilation for the target can be executed continuously.

The existing features were applied in a series software development project to assess the feasibility and applicability under project conditions. In line with project planning activities such as the definition of system requirements and a project plan, corresponding development activities are executed. Software models are implemented along functional and formal requirements and daily assessed according to a release plan which is derived from project milestones. As a result of the nightly build, an overview report of all software models and all conducted checks is created and serves both as status information for project stakeholders and a quality report for the software developer. This includes the execution of 150 style checks and 600 functional test cases, the creation of 2,500 pages software documentation and 1.1 mio lines of code and the logging of all errors and events during the build process

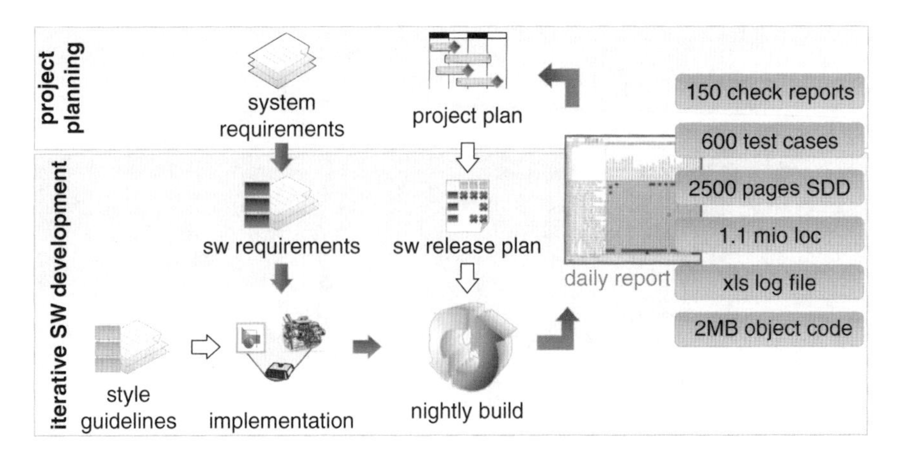

Fig. 12 Application of the nightly build in a series software project

(Fig. 12). Workload to the amount of estimated 20 men working days is executed daily allowing for early detection of integration and quality errors.

Aside the design of the continuous integration framework, the software architecture of the control system must be defined. In order to assess the feasibility of the software concept, the software is applied to a two-stage turbocharged gasoline engine in a passenger car being controlled by a Rapid Control Prototyping (RCP) system which allows for both quick integration of new functions with the hardware and control under real-time conditions.

As proposed by AUTOSAR, the software is split up into three layers (Fig. 13). On the lowest layer, hardware-specific functions are allocated. These include mainly drivers for the different sensors and actuators as well as for the communication network. Interrupt triggers for ignition timing and knocking control are allocated here also.

In the basic software layer, three functionality types are implemented. All control tasks are called by a scheduler which in series control units would be the operating system. On an electrical level, basic diagnostics on the sensors and actuators are conducted. For RCP systems, these functionalities can be rather small since the required error tolerance level is lower as for series controls. Hence, checks of the electric range for the indication of shortcuts to the ground or the battery are conducted. Moreover, all electrical signals are provided with an error status as signals hiding the hardware layout below. Checks for proper communication with other control units (e.g. via Controller Area Network—CAN) such as time outs or bus-off detections are allocated here also. The third basic functionality is the processing and prevention of errors. Any errors detected with the electrical checks are debounced to gain secure error information. Moreover, critical errors can be logged for the entire power cycle and can be read or written from/to the persistent memory. Hence, diagnostic functions such as cycle recognition being necessary for the fulfillment of onboard diagnosis requirements can be implemented.

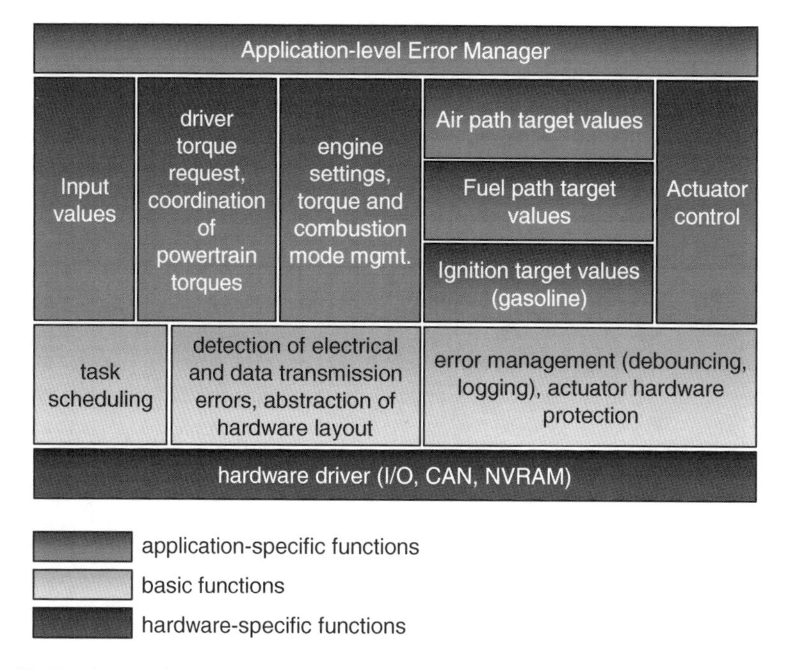

Fig. 13 Top-level software architecture for a flexible engine management system

The application layer top-level architecture is split up in component compositions that reflect the physical structure of the powertrain [20]. All input values delivered by the basic software are preprocessed for further use. This includes the conversion to physical values, setting of replacement values in the case of an error, manual override or modeling of missing sensors. Hence, a universal interface for application functions is provided that allows for the required development of control algorithms with minimized integration effort. For vehicle integration, the driver's torque request and the coordination of the different powertrain torque elements (engines, motors, brakes) is implemented. For mere combustion engine propulsion, the latter contains rather simple functionality whereas it becomes more complex with hybrid powertrains. The determination of the engine combustion mode and the target values for air, fuel and ignition path is done within the torque and mode management. Then, the engine torque request is realized via slow (i.e. air) and fast paths (i.e. fuel, ignition for gasoline engines). Finally, dedicated component contains controllers for the positioning of actuators.

The functional requirements on the software components have been defined by functional experts. The next development steps will be the detailing of the top-level architecture and its breakdown to atomic entities which are implemented and tested according to assigned requirements. In order to maintain the software and avoid architecture erosion, any further development of functions takes place within the nightly build framework.

2.3 Achievements and Further Research

Analyzing the requirements of the use of renewable fuels on combustion engines in automotive applications, a need for flexible control system architectures, especially affecting the control software, has been detected. As a second step, an analysis of the requirements on this control system was conducted. In line with series software development requirements it became clear, that a flexible architecture is to be maintained continuously. This includes regressive testing, checking for formal compatibility, function improvement and function extension. Since there exists no tool that satisfies these needs, a continuous quality inspection framework was developed, that conducts all required operations directly on model level.

The framework was applied in a series development project saving a significant amount of work by check automation. Hence, the continuous integration methodology will be applied and further developed within the design of the control software architecture. The latter is set-up at the architecture top level according to the AUTOSAR software standard.

Further development steps will include on the one hand the finalization of the integration framework. Further research on effective integration check metrics and the estimation of hardware resource metrics is needed to extract more information of the daily software checks. On the other hand, the software architecture of the control system has to be finalized assigning all functional requirements to corresponding components and functions. Moreover, control algorithms tailored for the requirements of renewable fuels must be developed and implemented. One control algorithm example will be outlined in the following section.

3 Pressure-Based Injection Control with Fuel Adaption

The achievements in the field of structural software design as described in the previous chapter open up the way and set the boundary conditions for the development of algorithms to control the combustion in the internal combustion engine. In this field it is not limited to the application of biofuels but a general trend that the percentage of closed-loop control structures in the combustion engine application is continuously increasing [21]. For that, classical open-loop structures such as the injection timing in diesel engines or the spark timing in gasoline engines are being transferred into closed-loop control circuits [22]. Especially when applied to the modern passenger car diesel engine, the closed-loop combustion control (CLCC) offers a variety of advantages in the application process of the engine: First of all, the stability of the combustion event can significantly be improved by CLCC. Variances and abnormalities such as different cylinder filling in multi cylinder engines or differing injector opening behaviour can be detected and compensated for. Also, ageing effect of all combustion relevant components can be balanced out over engine lifetime. In general, any difference in the boundary conditions of the combustion event can be identified and compensated for. This especially includes a possible variance in fuel quantity, which may range from slight differences between winter and summer fuels to completely different fuel structures like they are to be expected for newly developed, next generation biofuels.

There is one essential prerequisite for the realization of a CLCC structure: An on-line cylinder pressure signal is required as input for the control algorithm. Therefore, the engine needs to be equipped with a pressure transducer that supplies the ECU with the required pressure data. As most research engines are equipped with pressure sensors, fundamental investigations for CLCC can be carried out without difficulty. In addition to that, pressure transducers have been developed that can be integrated into series production engines so that CLCC has already made its way into serial application [23].

This section of the paper deals with the possibilities that advanced CLCC algorithms offer, especially in the context of future biofuel application in passenger car diesel engines. Newly developed biofuels are characterized by a wide range of fuel properties such as boiling and evaporation tendency or self-ignition behaviour. When being used in yet existing combustion engines, these variances cause drastic alterations in the course of the combustion with direct effects on the engine's performance and emission production. In order to compensate for these changes and allow a stable and comfortable engine operation independently of the fuel, an advanced control algorithm is presented that allows for the realization of pre-defined pressure traces in a modern diesel engine.

3.1 Closed-Loop Combustion Control Strategies

In general, different approaches to control the combustion in a closed-loop structure can be separated and grouped by their dimensionality, e.g. the number of variables that are being controlled (Fig. 14). The most obvious approach is a 1-dimensional control of the center of combustion (COC). The COC describes the point in time at which 50 % of the fuel mass brought into the cylinder have been converted and can be described as one of the major characteristics of the diesel combustion process. The COC can be controlled by the timing of the main injection which has been in the focus of a number of fundamental investigations [21, 24]. In addition to that, first serial applications have been reported [23, 25].

The COC can be extended by an additional maximum heat release control, increasing the dimensionality of the concept to 2. The maximum heat release rate is one of the main drivers in the generation of combustion noise and therefore important to be considered during the development process of modern diesel engines. It can be controlled by adjusting the fuel rail pressure, where a lower rail pressure will result in lower rates of maximum heat release as the amount of fuel brought into the cylinder during the ignition delay is smaller [26].

The optimum solution and the aim of the investigations presented here is the control of the complete combustion trace. This ideal approach can be described as

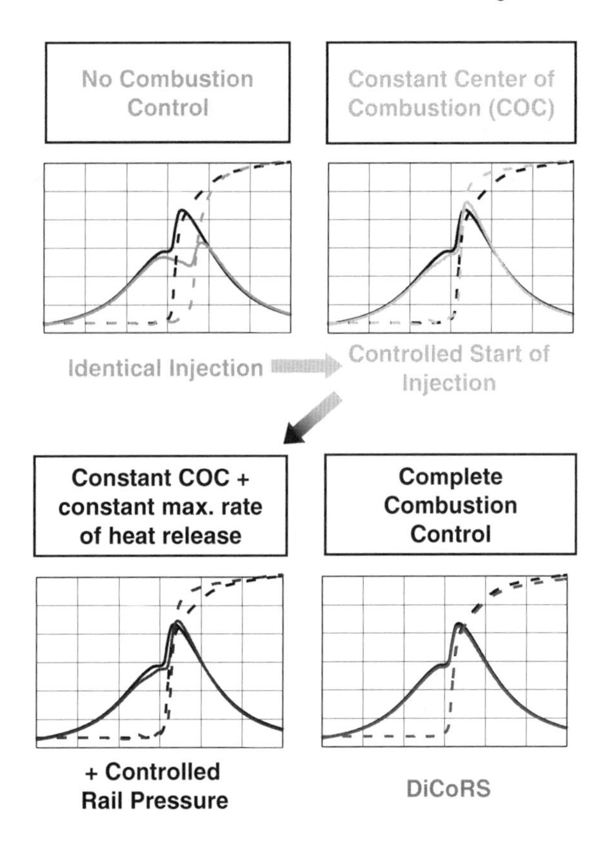

Fig. 14 Possible steps of closed-loop combustion control, comparison of standard diesel (*black*) and low-cetane fuel (*grey*)

n-dimensional as every step of the pressure trace is being controlled. In Fig. 14, the evolution of the dimensionality, peaking in the n-dimensional complete combustion control, can be explained. In this figure, a combustion comparison between standard diesel fuel and a low-Cetane number diesel-like fuel is depicted. The combustion is represented by its pressure trace as well as the corresponding heat release rate. As the lower Cetane number indicates the second fuel's lower self-ignition tendency, the combustion is retarded notably when no control at all is applied, left top case. When COC is applied, both traces converge since now the center of combustion is equal for both combustions. The addition of a maximum heat release control structure (3rd case) results in a further approximation of both traces as now the maximum gradient of the heat release curve is equal for both fuels. Still, even with this 2-dimensional case there are notable differences in the course of the combustion. Therefore, the 4th, n-dimensional case aims to compensate all fuel differences and realize an identical combustion for both fuels. In order to do so, more complex control structures need to be applied. One possible approach is Digital Combustion Rate-Shaping which is presented in the next chapter.

3.2 Advanced Closed-Loop Combustion Control

The algorithm used to realize a complete combustion control that allows holistic fuel compensation was developed based on iterative learning control theory and has evolved in different development stages. Iterative Learning Control (ILC) emerged from fundamental robotic research and describes a step-by-step minimization of the control error towards a final error of zero. The first approach utilized a prototype Rate-Shaping injector with which the rate of injected fuel was controlled in a way that the targeted cylinder pressure trace was met [27, 28]. The results of these investigations demonstrated the feasibility of the application of iterative learning theory to combustion control. Different pre-set processes were analyzed, resulting in the choice of a combustion process with a constant pressure rise starting in the Top-Dead Center position, called the alpha-process. By controlling the injection in a way that the alpha-process was run under varying boundary conditions (load, engine speed, egr-rate), benefits in pollutant emissions and especially combustion noise could be identified.

Based on the findings of these investigations, the control structure was refined so that the same functionality could be utilized without having to rely on rate-shaping injection equipment [29, 30]. The iterative nature of the algorithm remained unchanged but the central injection strategy calculation was modified as the pressure control was aimed to be realized with standard injection equipment based on multiple injection events. Because of the digital nature of the injection actuation (active or non-active) this approach is called "Digital Combustion Rate Shaping" (DiCoRS).

DiCoRS combines the theory ILC with a 2-point control methodology. Since with standard Common-Rail equipment the only possible variation of the injector actuation over time is the activation or deactivation of the injector, a 2-point control structure needs to be implemented that allows a decent control of the pressure trace. Therefore, the desired pressure trace is compared to the actual, measured cylinder pressure trace at every node of the sampled data (that is 0.2 °CA). With a self-adjusting delay model, each node of the pressure trace is related to a node of the injector actuation signal.

In Fig. 15, the general structure of an ILC is depicted. The setpoint vector w (desired pressure trace) is compared to the actual pressure trace y and processed

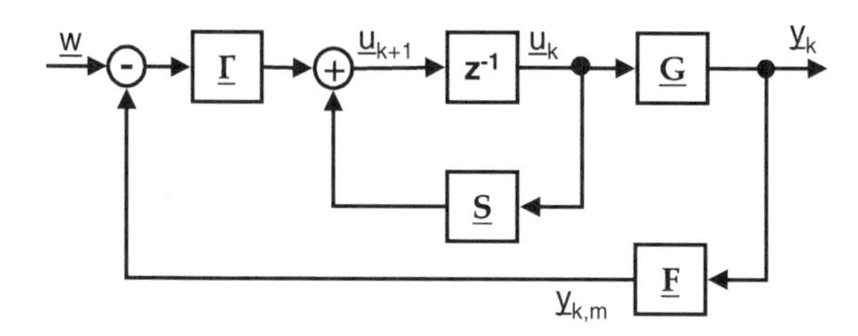

Fig. 15 Iterative learning control

by the so called learning operator Γ. The decisive part for the iterative nature is the memory element z^{-1}: That element returns the actual injector actuation voltage u_k to the engine G, while, based on the learning operator Γ, the voltage for the next cycle u_{k+1} has already been calculated. The learning law for this ILC structure can then be formulated as follows:

$$u_{k+1}(i) = Su_k(i) + \Gamma e_k(i) \tag{1}$$

This learning low shows how the effect of the injector actuation in time step k directly influences the voltage of the next cycle u_{k+1}, while also the actual control error is being considered for the next cycle. This combination ideally leads to a step-wise iteration towards a final control error of zero.

For DiCoRS, the ILC law is applied to each node of the pressure trace separately. At each node, the desired and the actual pressure trace are compared and then processed in a Proportional/Differential manner (PD) by the learning operator Γ. This means that both the direct control error as well as the gradient of the control error are being considered by the learning operator. In order to prevent the controller from pulsating, an additional hysteresis rule is integrated.

In Fig. 16, the desired pressure trace, an exemplary measured pressure trace as well as the hysteresis criteria and the injector energizing curve is shown. When the actual pressure trace the indicated node leaves the boundaries set by the hysteresis curves around the pressure set point and the injection at time step k is active at the corresponding node of the injector energizing curve, the injection is deactivated at that node for the next time step k + 1. For an overshooting this criteria applies correspondingly. By that, the energizing function is adjusted iteratively so that the final injection signal will trigger the combustion as defined by the desired pressure trace. Thereby, the number of possible injection events is not limited. For reasons of material durability only the minimum length of one injection and the injection pauses are limited so that the injector cannot be activated for an unphysically short amount of time.

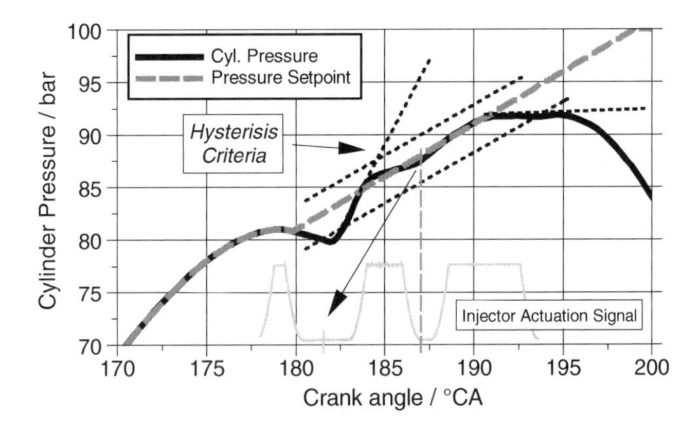

Fig. 16 DiCoRS pressure trace analysis

Based on the ILC 2 point controller it is possible to control different pressure traces with standard Common-Rail injection equipment, using multiple injections. With these boundary conditions being set, experimental investigations can be carried out to analyze the effect of advanced closed-loop combustion control on the thermodynamics of the diesel combustion. First investigations of the general functionality of the method and deeper analysis of different controlled pressure traces have been executed [29, 30]. It was shown that for part-load operation an alpha-process with a pressure gradient of 3 bar/°CA establishes an optimal compromise between pollutant emissions, fuel efficiency and combustion noise emissions. Those analyses did not take alternative fuels into account, yet. In the investigations presented here, DiCoRS is being applied to alternative fuels in order to find out about the potential it offers with regards to fuel compensation. Therefore, experimental testing with a one-cylinder research diesel engine was carried out.

3.3 Methodology

The one cylinder engine used for the experimental investigations (engine 1) was developed in order to assess the potential of future biofuels in a modern, state-of-the-art combustion system. The engine has a displacement volume of 0.4l and combines different elements of downsizing concepts like high boost pressures and elevated rail pressures. These hardware specifications as well as the boundary conditions chosen for the investigated load points assume a modern EURO 6 combustion system. Table 1 displays the engine specifications. The rail pressure, the boost and exhaust back pressures as well as the temperature of the intake air can be chosen individually as function of the investigated load point.

The engine is integrated into a test cell which supplies all necessary surroundings for a stable engine operation. Air, fuel and oil can be conditioned externally and independently of the engine mode to simulate different operating conditions. All static temperatures and pressures in the air path, the fuel supply system and the oil gallery are measured and supervised continuously. The in-cylinder pressure is

Table 1 Specifications engine 1

Stroke/mm	88.2
Bore/mm	75.0
Displacement Volume/cm3	390.0
Compression Ratio/−	15.0
Valves/−	2 Intake/2 Outtake
Maximum cylinder pressure/bar	220.0
Injection system	Common Rail, max. Rail Pressure 2000 bar
Boosting system	3 external Roots compressors, max. boosting pressure 3.4 bar

measured and recorded at a frequency of 0.1 °CA so that the trace of combustion can be followed and investigated at a high accuracy. The exhaust gases are analyzed with modern gas sampling equipment. With that, nitrogen oxides (NOx), hydrocarbon (HC) and carbon monoxide (CO) emissions as well as particulate matter (PM) in the exhaust gas can be detected. The fuel consumed is determined with a volumetric measurement system. The desired rate of recirculated exhaust gas (EGR-rate) is controlled via an externally driven egr-valve, while the rail pressure is controlled with the engine control unit (ECU). In normal operation without CLCC, also the injection is defined with the ECU.

The load points that are typically investigated with engine 1 are shown in Table 2. These 4 part load points were chosen to represent the load spectrum that a typical medium class car would face in the European certification cycle NEDC [31]. The first two points represent the very low-load, urban part of the cycle, while load point 3 and 4 take the higher part load operation in the later phase of the cycle into account.

For the DiCoRS investigations the test cell is extended by a Rapid Control Prototyping (RCP) system. The integration of that system is displayed in Fig. 17. The control structure as explained above is being executed on the RCP system which, based on the calculated injector actuation curves, controls the injection. The calculated injection profile is amplified by an output stage and directly applied to the injector. The in-cylinder pressure signal is fed to the RCP system where the explained comparison and recalculation of the injection signal takes place. During operation, the functionality of the control algorithm can be supervised by the Control Desk interface. In order to prevent non-logical or possibly harming injector actuation traces from being applied to the engine, every new injection profile has to be confirmed by the test bench technician. Until the confirmation for the new profile k + 1 has been sent, the old profile of time step k is being applied to the engine.

In order to assess the potential of DiCoRS to compensate for different fuel properties, the combustion of a standard Diesel fuel is compared to a newly developed biofuel of the latest generation. Led by the Chair for Combustion Engines (VKA) at RWTH Aachen University, a Cluster of Excellence named "Tailor-Made Fuels from Biomass" was installed in 2007 that aims to develop new biofuels in an integrated process, combining chemistry, process engineering and combustion science. In this cluster, new processes have been developed that allow

Table 2 Operating points engine 1

Operating point	1	2	3	4
Indicated mean effective pressure (IMEP)/bar	4.3	6.8	9.4	14.8
Engine speed/min^{-1}	1,500	1,500	2,280	2,400
Boost pressure/bar	1.07	1.5	2.29	2.6
Exhaust gas back pressure bar	1.13	1.6	2.39	2.8
Rail pressure/bar	720	900	1,400	1,800
Charged air temperature/°C	25	30	35	45

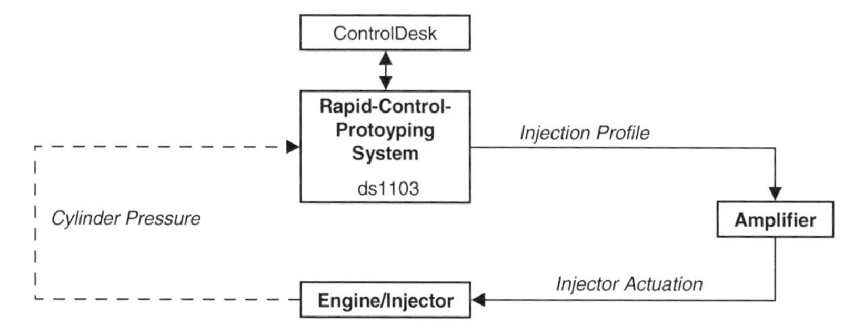

Fig. 17 Interaction between engine and control system

the production of biofuels that can directly be derived from glucose by new catalysts [32]. One of the fuels developed in this Cluster is 2-methyltetrahydrofurane (2-MTHF). This molecule forms a ring structure and contains oxygen that directly originates from the glucose the fuel was derived from. Due to its very low self-ignitability (Cetane Number 15) it was blended with 30 vol% of dibuthyl ether (DBE, Cetane Number 100) to improve its ignition behaviour. Table 3 depicts an overview of the most important fuel properties of the 2-MTHF/dibuthyl ether in comparison to Diesel fuel.

From the presented table it can be seen that the self-ignition properties of 2-MTHF/DBE are completely different than those of Diesel fuel. 2-MTHF/DBE contains more than 18 % of oxygen which contributes to the lower heating value of only 34.9 MJ/kg in comparison to 42.9 MJ/kg for Diesel. Both the oxygen content as well as the lower aromatic content reduce soot production during the combustion which enables the use of high EGR-rates in order to reduce nitrogen oxide emissions. It was shown that this partially homogenized combustion (cold combustion) allows a soot free diesel combustion [32].

Besides the positive effects of the lower self-ignitability of 2-MTHF/DBE (longer mixing, therefore better homogenization), there are also disadvantages of the long ignition delays that are caused by the low Cetane Number. One major disadvantage is the notably higher combustion noise [32]. Due to the long mixing period a large amount of fuel is being prepared during the ignition delay which burns rapidly after the ignition and causes a steep pressure rise at the start of the

Table 3 Fuel properties

	Diesel EN590	70 % 2-MTHF/30 % DBE
Cetane number/−	56.5	~30
Oxagen content/m−%	0.14	16.8
Heating value/MJ/kg	42.9	34.86
Boiling range/°C	180–350	80–142
Aromatic content/m−%	~25	0

combustion. This steep pressure rise is responsible for unacceptable high noise emissions.

The aim of these investigations is to avoid these differences in the course of the combustion and realize an equal energy conversion throughout the whole combustion event. On the one hand, this will enable a dedicated analysis of the effects of different fuel properties on the emission behaviour as now there are no detectable differences in the event of energy conversion over time. On the other hand, it will further increase the usability of tailor-made fuels from biomass as obviously higher noise emissions could be avoided.

3.4 Main Results

The assessment of the potential of advanced CLCC for the compensation of fuel properties is analyzed in two steps. First, the combustion behavior of both investigated fuels is characterized without DiCoRS. Therefore, the fuels are analyzed with one main injection that is set to meet an identical Center of Combustion, see Fig. 15. For both fuels the rate of recirculated exhaust gas is chosen in a way that an identical indicated specific NOx-emission level of 0.25 g/kWh is reached. This level represents a modern, Euro 6 conform combustion system. Both the equal COC as well as the identical NOx-level guarantee, from an emission standpoint of view, constant boundary conditions that allow a basic analysis of the different behavior of both fuels.

In the next step both fuels are investigated with the DiCoRS method. Therefore, again the egr-rate is chosen in a way that equal NOx-emissions are achieved with both fuels. Now, the injection is controlled by the DiCoRS algorithm implemented onto the RCP system. For each operating point, the step-by-step iteration towards the desired pressure trace is executed until the set point course of combustion is realized with a satisfying accuracy. Both fuels are investigated with the alpha-process with a pressure gradient of 3 bar/°CA as this is the process that previously had been identified as optimal choice [28, 29]. For the investigations presented here the third load point of Table 2 was chosen.

In Fig. 18, the burn function, heat release rates and cylinder pressure traces of all 4 combustion processes are depicted.

In the course of the cylinder pressure traces the effect of the DiCoRS can easily be seen. Both controlled combustions start exactly at Top Dead Center (Crank Angle 0°) and carry out a constant pressure rise of 3 bar/°CA. The earlier start of the combustion for the controlled processes can also be identified in the burn function and the heat release rates. According to the pressure traces, the energy conversion starts earlier.

As the combustion advances, the difference between both fuels becomes evident. Although the injection strategy between the non-controlled and the DiCoRS processes is essential, the later fuel conversion of both processes with 2-MTHF/DME is notably quicker than with Diesel as can be seen in the evolution of the burn

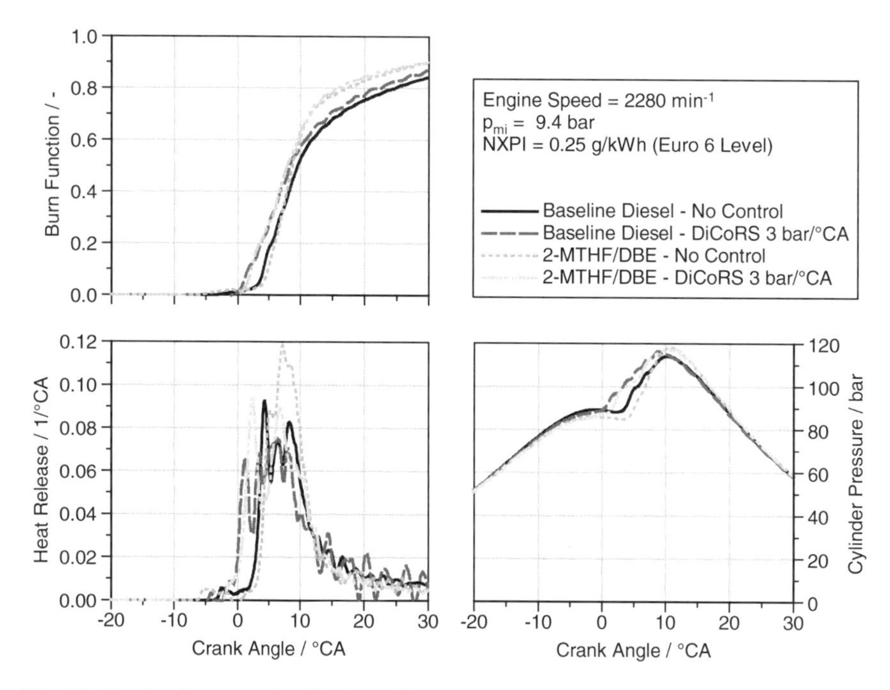

Fig. 18 Combustion traces for diesel and 2-MTHF/DBE

function. This is related to the longer ignition delay and therefore better mixing as well as the lower boiling curve of the biofuel that enhances the oxidation also in the later phase of the combustion.

In Fig. 19, the efficiency, the pollutant emissions and the combustion noise are displayed at a constant indicated NOx-level of 0.25 g/kWh. Hereby, the efficiency and the emissions are related to the uncontrolled Diesel process which is considered to be the base calibration. It can be seen that, due to nearly equal centers of combustion, there is no big difference in the efficiency of the four processes. The positive effect of the oxygenated fuel its longer ignition delay becomes visible in the particulate emissions. Both the uncontrolled as well as the controlled process with 2-MTHF/DBE burn nearly soot-free.

The positive effect of the biofuel's long ignition delay on the mixing process negatively effects the CO and HC emissions. During the ignition delay, badly mixed zones form close to the cylinder walls which are the origin of the elevated HC and CO emissions. For 2-MTHF/DBE, no big influence of the DiCoRS controlled process can be identified with regard to HC, CO and PM emissions.

With regard to the produced combustion noise which is depicted in the model-based value Combustion Sound Level (CSL), the positive effect of DiCoRS becomes clear. For both fuels, the combustion noise can be reduced significantly with a maximum reduction of more than 3 dB for the 2-MTHF/DBE mixture in comparison to the uncontrolled process. These 3 dB correspond to a division into half of the sound pressure level.

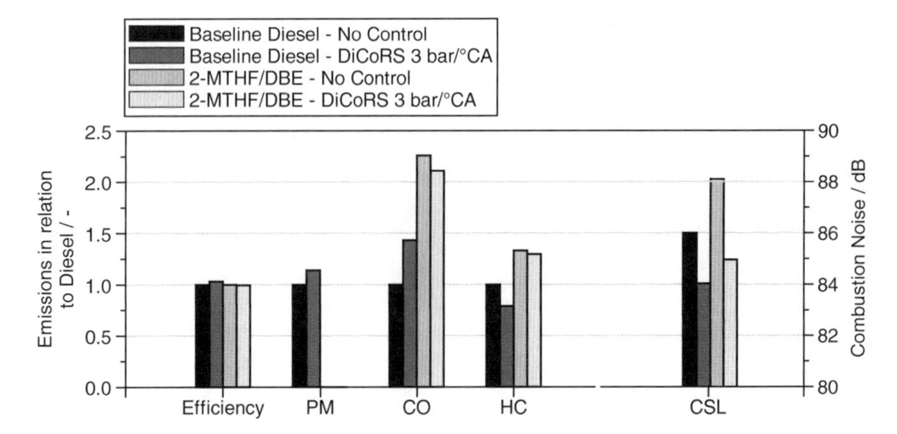

Fig. 19 Efficiency, emissions and combustion noise for diesel and 2-MTHF/DBE, NOx-level 0.25 g/kWh

The combustion noise for both DiCoRS controlled processes differs less than 1 dB. Theoretically, it would be completely equal, but the long ignition delay produces a little steeper first pressure gradient for the 2-MTHF/DBE which results in a slightly higher sound level of 85 dB. Nevertheless, in comparison to the level of the uncontrolled biofuel combustion of more than 88 dB it was reduced significantly.

The presented example indicates how advanced control structures enable an engine operation with biofuels that on the one hand allow the utilization of the emission benefits of these fuels but at the same time reduce the negative side-effects like higher combustion noise. With advanced control algorithms, the way is opened up for the use of alternative fuels in yet existing combustion systems. In this context, many different aspects have to be investigated in order to qualify the impact of the fuel property on the whole engine. Among these, combustion aspects as well as mechanical issues play a role. As an example of both, in the following the effect Ethanol on the piston ring lubrication as well as the cold-start behavior in a Gasoline engine is investigated.

4 An Advanced Oil Distribution Model for 3D Piston Ring Dynamics

4.1 Introduction—State of the Art

Increasing fuel prices, environmental changes and legislative regulations have been the driving forces since several years now. Hence, in the recent years the development effort for combustion engines has been focused on the fuel consumption, emission improvement and alternative fuels. The contribution of piston rings to theses topics are:

- Increase of the degree of efficiency by sealing the combustion chamber, minimizing the blow-by gas and lower the friction in mixed lubrication regimes
- Regulation of the oil distribution on the liner, since oil evaporation directly effects the emission quality
- Alternative fuels as ethanol show a significantly different boiling behavior. In combination with the direct injection technology, this leads to oil dilution under low temperature conditions. A piston ring wear affecting lubrication regime of oil and fuel is the result. In diesel combustion this liner wetting occurs due to exhaust after treatment actions/measures.

Additionally to these technical challenges, the boundary conditions for the engineering process have been changed by globalization. The shortened development cycles in the engine development require powerful and reliable tools to reduce the testing effort.

For the piston ring development, this means that an advanced simulation tool has to consider more than 100 parameters of geometrical, lubricant and gas properties. Besides the amount of parameters the objectives as friction, wear, blow-by and oil consumption have to be covered with a three dimensional model.

Due to the complexity, a lean and fast platform is required to build up such a comprehensive tool. In [33] 70 % of the oil in the piston cylinder unit (PCU) is evaporated, 20 % are lost by throw of from the top land and the last 10 % are lost by reverse blow-by. Therefore, most effort is put in a detailed description of the evaporation approach.

4.2 Multibody System Piston Ring Module

The presented requirements regarding to the complexity and calculation duration are fulfilled by the multibody system (MBS) formulation. Based on the platform of MSC ADAMS the user is enabled to generate a three dimensional model of the cranktrain and its piston group while keeping the calculation effort at a minimum. The piston rings are created as segments which are connected by 6×6—stiffness matrices. The mechanical functionality has been presented in [34].

To implement a suitable the oil consumption of the PCU-MBS model a large number of physical effects has to be considered. One of these is the piston secondary motion which directly influences the piston ring dynamics. Additionally the gas dynamics from the combustion chamber to the crankcase—the path of the blow by gas—must be included. The resulting dynamic intermediate gas pressures together with the ring pretensions determine the contact pressure between the ring and the liner. The flexibility and the dynamic behavior of each ring as well as the contacts to the piston groove must be included to cover the phenomena like ring twist, bulge and ring flutter.

To implement these functionalities into a MBS model various interactions have been formulated as modules of interaction. In the past fully coupled modules for ring-liner interaction, ring-groove interaction and ring-gas interaction have been developed. The relevant effects like full three dimensional resolution, hydrodynamic-mixed lubrication, gas flow through the grooves and damped contact in the grooves by oil squeezing are considered.

To extend the piston ring calculation with an oil consumption model and wear indication, an oil module is developed and linked to the piston ring architecture of the MBS model. This module also has the capability to include the effects of regular and alternative fuels by adopting the properties of the lubrication regime. The particular influence of biofuels on the lubricant is discussed in the following section.

4.2.1 Oil Dilution by Biofuels

The relevance the oil dilution has risen since the direct injection (DI) technique is state of the art in diesel engines, respectively it will be state of the art in a few years in gasoline engines as well. By means of the emission regulations post injection has become necessary for diesel engines to regenerate the diesel particle filter or the NOx-catalyst. The post injection timing and the position of the piston permit a fuel entrancement into the oil film. The resulting oil dilution can have a severe impact on the wear behavior of the piston group.

Since ethanol blends are introduced for DI gasoline engines the oil dilution has become an important object of investigation. The hydrophilicity of ethanol in combination with short distance operation and cold start conditions lead to serious oil dilution. Beginning with the lower evaporation point of the ethanol, blends with a high amount of ethanol require an early pre-injection which enables the injector to blow the ethanol along the piston to the liner. The result is a reduced viscosity of the lubrication regime. On the one hand by means of the fuels and on the other hand due to water molecules. The lubrication regime for the piston group becomes even worse when then the engine is operated for a longer distance, the lubricant becomes hot and the water evaporates. As a consequence a large mixed lubrication regime and heavy wear can occur.

But even for biofuels that have different properties regarding the hydrophilicity, they have to be investigated regarding the wear behavior in case of an oil dilution. Since the biofuel viscosity is significantly lower than the oil viscosity severe changes regarding the hydrodynamic film build up take place.

To cover this, the presented oil model contains a multi component approach for the lubricant and offers different balance scopes to describe the oil flow on the liner and piston ring flanks.

4.3 Oil Household Module

To create a comprehensive oil model for the PCU of an internal combustion engine, several mechanisms have to be considered. The basic equation imposed is the conservation of mass. The most important oil covered surface of the system is the liner. The liner is modeled as being coated with a thin oil film, effectively described by the local film height. The liner surface is partitioned into rectangular cells with constant surface areas. The amount of oil that resides on each surface area is captured into only one parameter: the average surface oil film height. Thus the mass conservation applied to the liner surface must include:

- evaporation
- liner oil reservoir
- oil beads at ring flanks and throw off mechanism
- reverse blow by

Finally the oil supply by the oil mist in the crankcase can be considered to avoid oil starvation on the liner.

These mechanisms are implemented in a FORTRAN subroutine. To provide parameters which are necessary for the governing equations, a common data pool for all subroutines, including those for the contact interaction, is created. This enables the oil routine to request any required information. Moreover, this data pool interface ensures a simultaneous interaction between the subroutines.

In the following the framework for embedding the mechanisms and the fundamental loop architecture in the governing subroutine is presented.

4.3.1 Framework for Transport Approaches

In this section the discretization architecture of the ring and the liner as well as the classification of the balance scopes for the oil household model is presented.

The liner is divided into n_{IL} circumferential cuts and n_{IJ} height cuts (see Fig. 20). The result is a rectangular grid with $n_{IL} \times n_{IJ}$ fields. The minimum size of the fields is determined by the ring geometries and its circumferential discretization. To ensure that in axial direction a maximum of two liner cells simultaneously get into contact with a ring, the minimum axial distance of the height cuts is limited to the factor $1.2 \times h_{ring}$. The circumferential discretization is related to the segmentation cuts of the piston ring. It is assumed that all rings have the same number of segments. Hence, the liner is divided into $c_{IL} \times n_{Segments}$. Here, c_{IL} is the factor which is used to increase the liner resolution by a multiple of the ring segments. It must be borne in mind that the ring subroutine architecture already provides a higher resolution on each segment by means of the contact models to the liner and to the piston groove. This enables a detailed balancing for the oil flux between the liner and the ring segment flanks. Each cell of the liner grid contains the relevant information like the oil properties and the corresponding oil height.

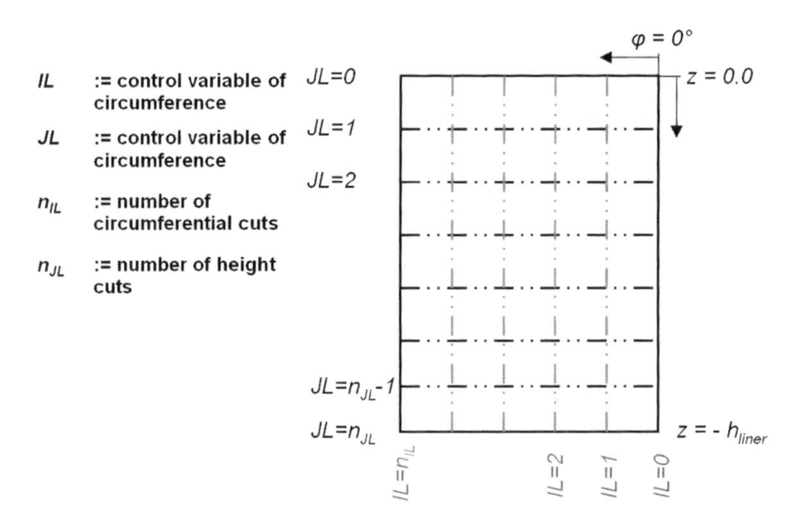

Fig. 20 Discretization of the cylinder in the FORTRAN subroutine

To create the oil household model, several mechanisms have to be included. In this approach the ring flanks are able to pick up the oil in a bead that is scraped from the liner surface if the oil film height in front of the ring is higher than the minimum distance to the liner at that point of the circumference. A release of the oil from the bead on the flank onto the liner occurs in two cases. Firstly, if the bead is on the back flank with respect to the motion direction, the oil deposit will directly flow on the surface cell, where this case happens. The second case happens when oil film height of a cell is smaller than the local piston ring to liner distance. The latter case will rarely occur, as the piston rings will most of the time scrape off the oil from the liner surface and release it when the direction of motion changes.

The second included mechanism describes the oil transfer between the oil film and the honing grooves of the liner. The honing grooves serve the purpose to 'fill up' areas on the liner that suffer from oil starvation. The nearby feeding of the oil film from the honing grooves is understood as being physically caused by film creep. This implies a finite and limited to small region of oil film height the oil creep occurs.

Furthermore, a mechanism of oil evaporation—based on the heat transfer analogy—is implemented. Because the oil is partitioned into its particular components the oil dilution by fuel entrainment can be incorporated as well.

Due to the different transport mechanisms the region surrounding the ring top flank and the ring bottom flank is divided into five regions for each ring to detect the relevant mechanisms for each liner cell (see Fig. 21). Using the example of the first ring the balance scopes are explained in the following.

The first region starts from the upper edge of the cylinder and ends at *border #1* where the upper bead starts (Fig. 22). In this area the mechanisms of oil evaporation and oil transport between the reservoir and the oil film can occur. The evaporation

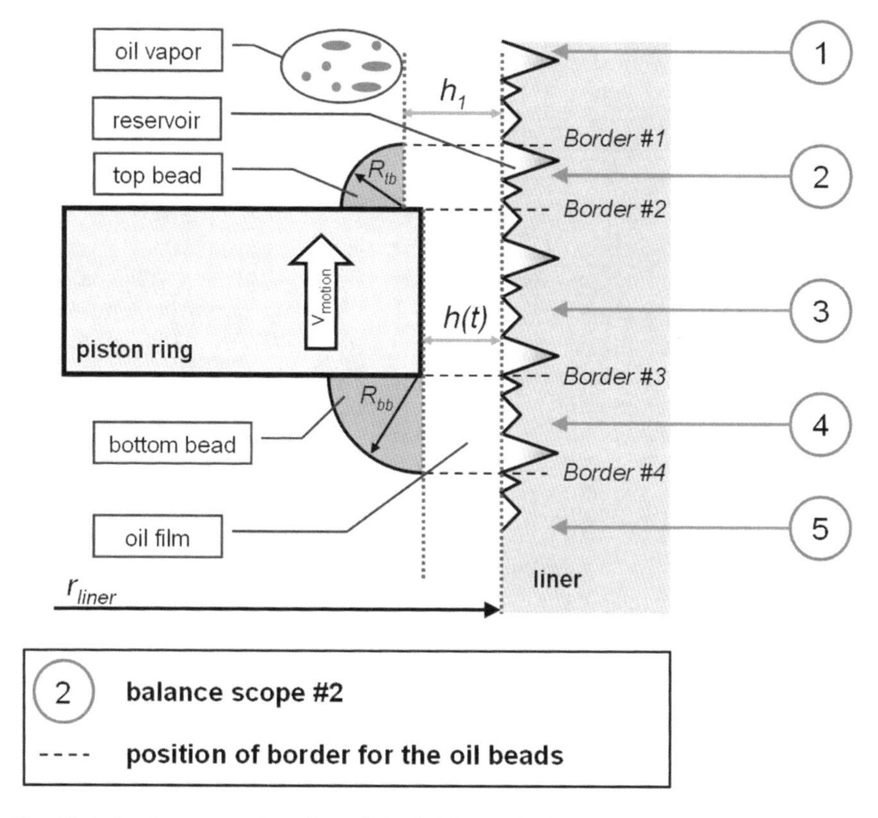

Fig. 21 Balancing scopes along the cylinder height considering the piston ring position

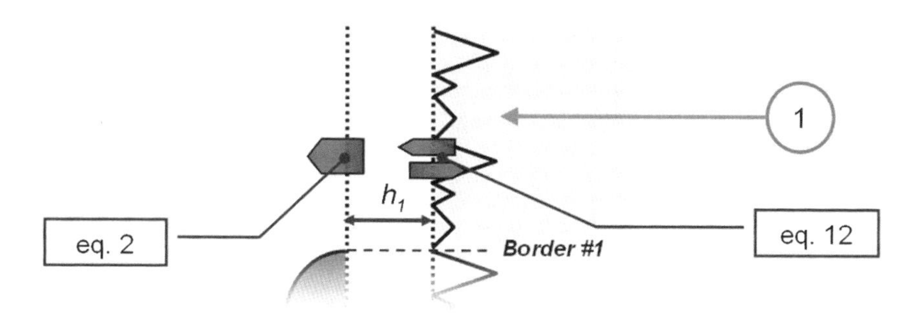

Fig. 22 First balancing scope of the considered piston ring

takes place if the oil film height is greater than zero. The oil supply by the reservoir is active if its oil volume is higher than the specific oil reservoir height. This allows oil to creep from the honing grooves to the starved liner (oil film height is zero). The reservoir is filled with oil, if the oil film height is non zero and the reservoir oil height below its maximum value.

The second balance scope contains the exchange between bead and oil film as well as between oil film and reservoir. By the fact that the elements of the balance scope are the same as in the fourth section the following statements are valid for both (see Fig. 23). Here, the sliding direction of the piston ring in relation to the position of the bead is decisive for the flow direction of the oil. On the one hand the ring flank can scrape the oil, which is then transferred to the bead or on the other hand the bead can release the accumulated oil. The reservoir is filled by a surplus amount of oil of the prevailing oil film.

In the third balance scope (see Fig. 24) an exchange between oil film and reservoir can only occur under particular conditions. The oil exchange modes are covered by part (a) and (b) of Fig. 28.

The fifth balance scope (see Fig. 25) is very similar to the first one, whereas here an oil source is integrated to supply the reservoir in case of a critical small oil volume. This can be explained by the fact that oil starvation usually does not take place. The oil mist in the crankcase supplies the liner with oil. Nonetheless, the parameter for the oil supply has to be adjusted carefully.

In the following section the specific transfer elements are presented in detail.

4.3.2 Oil Evaporation

Most of the oil evaporation which contributes to the oil consumption occurs above the first and second piston ring.

The total evaporation mass rate per time unit results from the sum of the evaporation rate of each lubricant component $\dot{m}_i$.

The evaporated mass flow rate per component is determined by the concentration difference of the component between the liner surface and the gas in the combustion chamber. Hence it is:

$$\dot{m}_i = g_i \cdot \left(c_{i,oil\ component} - c_{gas,\infty}\right) \tag{2}$$

$$\dot{m} = \sum_{i=1}^{nc} \dot{m}_i \tag{3}$$

Due to the high gas exchange rate compared to evaporation mass flow, the assumption of a zero-oil-concentration ($c_{gas,\infty} = 0$) in the gas is valid. The evaporation of the components is also useful to extrapolate a tendency regarding composite properties after a defined time slot. The new concentration $c_{i,\ new}$ of each component in the lubricant is assumed to be:

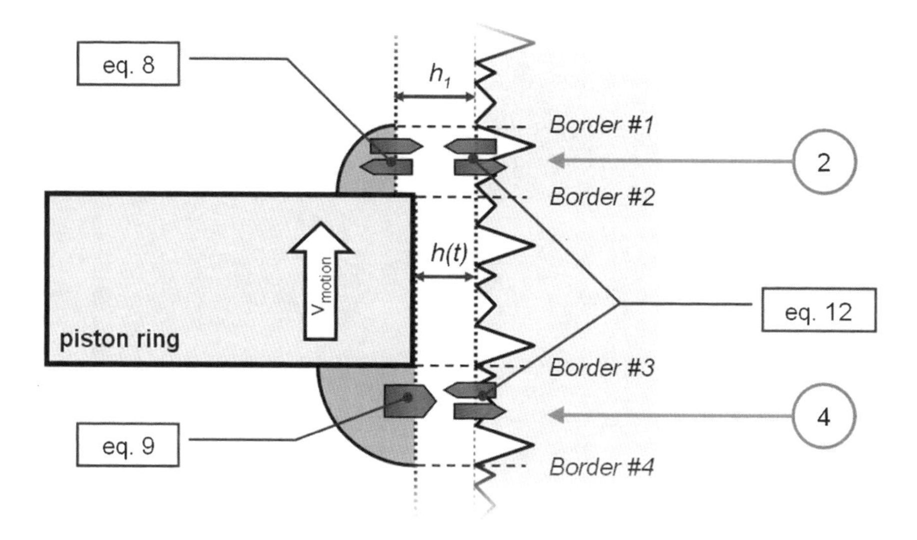

Fig. 23 Second and fourth balancing scope of the considered piston ring

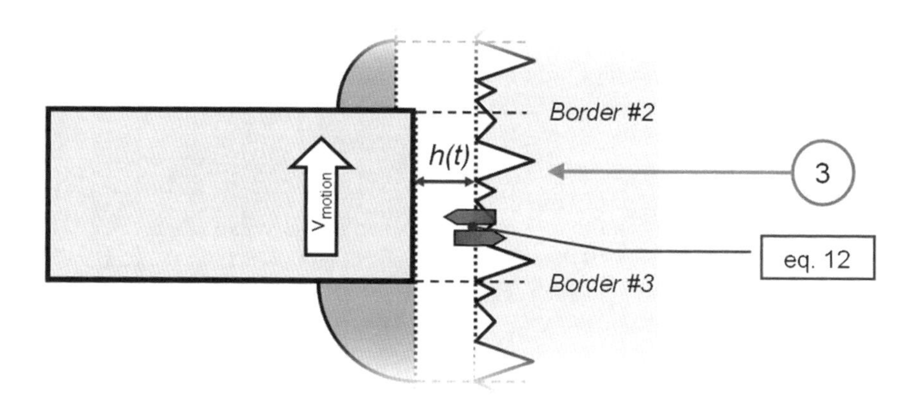

Fig. 24 Third balancing scope of the considered piston ring

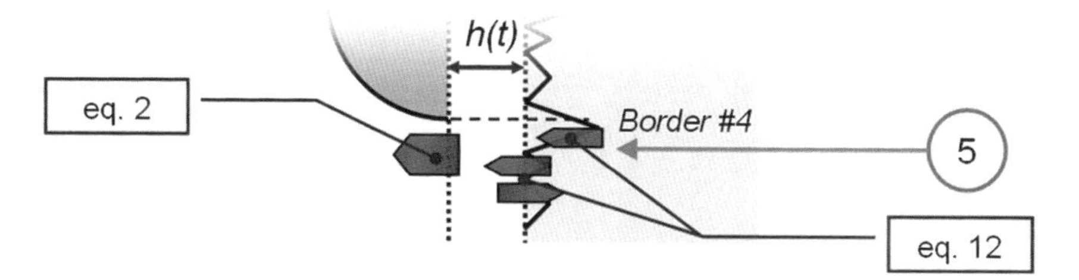

Fig. 25 Fifth balancing scope of the considered piston ring

$$c_{i,new}(\Delta t) = \frac{m_i - \dot{m}_i \cdot dt}{m_{oil}} \tag{4}$$

The Eq. (2) needs the input of the starting concentration and the mass transfer coefficient. The calculation of this coefficient is presented in the following.

By the analogy of Chilton and Colburn [35] the mass transfer g_i coefficient is:

$$g_i = \frac{\alpha}{\rho_{gas} \cdot c_{p,gas} \cdot Le_i^{2/3}} \tag{5}$$

Here, the heat transfer coefficient α is calculated by the Woschni Correlation, which is:

$$\alpha = 0.013 \cdot \frac{p^{0.8}}{d_{liner}^{0.2} \cdot T_{gas}^{0.53}} \cdot \left[C_1 \cdot c_m + C_2 \cdot \frac{V_h \cdot T_1}{p_1 \cdot V_1} \cdot (p - p_0) \right]^{0.8} \tag{6}$$

Depending on the injection technology and combustion process the coefficients C_1 and C_2 are adopted to provide adequate 'heat transfer analogy'.

The Lewis number Le_i is provided by the relation of inertia and viscosity properties which can be expressed by the Schmidt number and the Prandtl number. Reduced to key parameters [36]:

$$Le_i = \frac{\lambda_i}{\rho_i \cdot c_p \cdot D_{i,gas}} \tag{7}$$

The diffusion coefficient $D_{i \to gas}$ can be obtained by the Wilke-Lee relation [36]. This includes approaches of the molecular behavior of the relevant components taking into account the prevailing temperature.

Within these formulas (Eqs. 2–7) a detailed approximation of the mass transfer rate of each component is provided.

4.3.3 Oil Beads at Ring Flanks

Two different modes have to be considered for the oil beads. These are scraping and drain off.

In the scraping mode (see Fig. 26) the difference between the ring to liner distance and the oil supply height in front of the ring is made. This amount of oil is accumulated per time step and added to the oil volume in the bead. The Eq. (8) describes the change of oil volume in the bead:

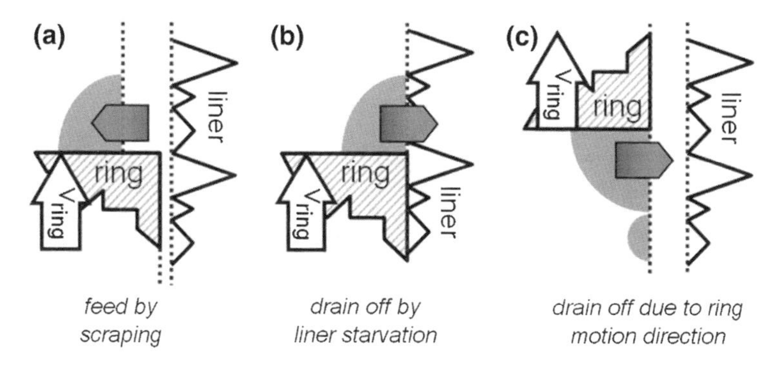

Fig. 26 Transport mechanisms of the accumulated oil at the ring flanks

$$dV_{bead}(t) = (h_{start} - h_{\min}(t)) \cdot u_{\varphi} \cdot |v_{ring}| \cdot dt \tag{8}$$

In case of oil starvation on the liner h_{start} would be zero and dV_{bead} could be negative. This implies an oil mass flow rate from the bead to the liner surface.

The case (c) of Fig. 26 occurs if the ring moves away from the bead. This leads to an immediate drain off of the bead. The drain off is depends on the ring velocity and can be adjusted by the parameter c_{drainoff}. The drain off volume $dV_{bead,drainoff}$ is characterized by the following equation (see Eq. 9):

$$dV_{bead,drainoff}(t) = c_{off.bead} \cdot u_{\varphi} \cdot v_{ring} \cdot dt \tag{9}$$

The balancing of the oil beads on the first compression ring top flank allows embedding an equation which indicates the oil throw off. Depending on the surface tension of the lubricant in the oil bead and the distribution of the lubricant on the upper ring flank, the formation of droplets is determined. In further steps of the model development more detailed work will be done with respect to the aspect of droplet formation.

4.3.4 Liner Oil Reservoir Approach

Since quite a long time liner surfaces are particularly machined to provide a defined —by experience developed—amount of oil for the along sliding piston and its rings. Because detailed models with interacting surfaces require detailed micro-structures, statistical approaches are used to keep the information density as small as possible.

Using the Abbot Firestone curve [37], the statistical surface structure can be divided in two sections. These sections are determined by the contact description of the mixed friction mode. The first section, the load capacity scope (LCS) describes the range of contact surface which occurs in the typical elastic body contact (Fig. 27). Below this scope, neither an elastic penetration, nor a relevant amount of

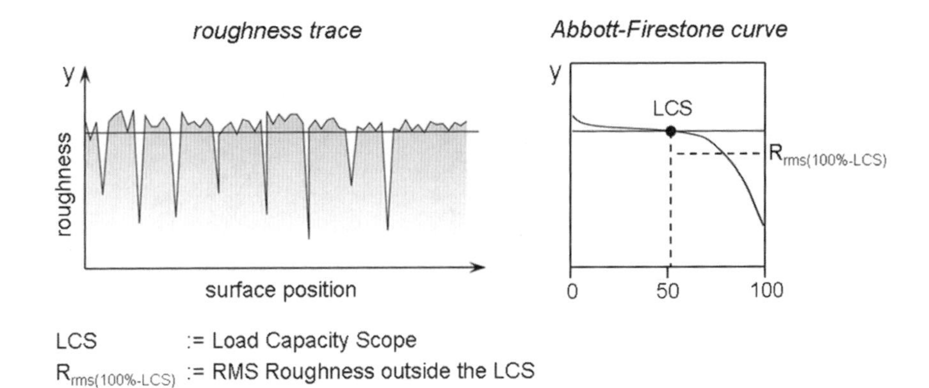

Fig. 27 Dependency of the liner surface roughness and the LCS

wear occurs. From this it follows that the roughness volume capacity below the LCS represents the reservoir volume $V_{reservoir,cell}$. With the representing section area of the liner this can be formulated as:

$$V_{reservoir,cell} = \left(R_{LCS} - R_{rms(100\ \%-LCS)}\right) \cdot \frac{\pi \cdot d_{liner} \cdot h_{liner}}{n_{IL} \cdot n_{JL}} \tag{10}$$

An assumption is made that the oil volume is distributed in the honing groove similar to a sphere:

$$h_{res.\max} = \sqrt[3]{\frac{3 \cdot V_{reservoir,cell}}{4\pi}} \tag{11}$$

Basically the groove form will be different. But due to the lack of information, the approach of the sphere leads to the most practical simplification.

To balance the oil flow, there are three modes which change the oil reservoir height by the variable $dh_{reservoir}$. The first one describes the emptying mechanism (Fig. 28a). By the oil creeping effect the oil drains off the reservoir in case of it is filled by at least 90 % and the oil film height is near zero.

$$dh_{reservoir} = c_{flow} \cdot h_{\max.\mathrm{res}} \tag{12}$$

Otherwise the reservoir can be fed by the oil film until it reaches its maximum oil height (Fig. 28b). The last mode occurs in case of starvation when the oil height in the reservoir is below a critical value which activates an oil supply in reservoir (Fig. 28c). This mechanism is installed to apply the oil support from the crankcase. By the complexity of the MBS model, the attached subroutines, the 3D resolution and the requirements regarding calculation time, this simplification is made to neglect the oil transport at the piston and the bridging effect which guides the oil back to the liner.

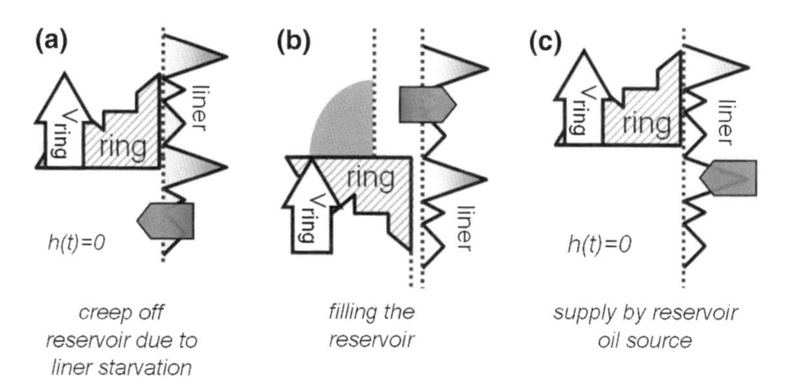

Fig. 28 Transport mechanisms liner reservoir model

4.3.5 Discharging of Lubricant Dissolved in Blow-by Gas

Due to the pressure differences between the combustion chamber and the crankcase effects like blow-by and reverse blow-by occur. To minimize these gas flows a high effort is spent by optimizing the piston ring properties. Nonetheless, during the combustion phase the pressure gradient between the combustion chamber and the first land of the piston leads to a gas flow through the ring gap of the first compression ring. The pressure level in the first land increases up to 20 bars and more [38]. After that, the exhaust valves open for the gas exchange phase. By means of the low pressure in the combustion chamber and the remaining blow-by gas volume in the first land, the first compression ring does a contact alteration. This creates an additional gap to the ring gap. Through this gap the pressure gradient is released between the first land and the compression chamber. This reverse blow-by gas contains fractions of the lubricant, which has been dissolved. The lubricant fractions—discharged through the exhaust manifold—become part of the oil consumption.

This process is basically an interdisciplinary task of piston ring dynamics and the fluid dynamics. A detailed description on the one hand needs the dynamic behavior of the piston and piston ring related to the liner and on the other hand the gas flow and oil distribution information by means of a highly resolved three dimensional CFD calculation is required. This effort is not applicable to the current MBS model. Therefore, an empirical approach based on measurements [39] is used. The measurements show an amount of discharged oil in the exhaust gases of 0.05–0.1 % oil mass fraction per reverse blow-by mass fraction.

To evaluate the gas dynamics a gas model subroutine is linked to the MBS model. Thus, concerning the reverse blow-by the corresponding discharged oil amount can be determined. It is assumed that the dissolved lubricant, which is not discharged, is fed back to the oil sump.

4.4 Simulation Strategy

Besides the architecture and the embedding of the subroutine, it is important to give the user the choice between the most accurate simulation and the fastest simulation. By means of the high data streams for the book keeping of state variables and for the data output generation, an option is implemented to decide the frequency of the subroutine call. If the oil household routine is used in every time step of the MBS model the state variable are calculated simultaneously to the ring dynamics. A kind of a cosimulation approach is applied if the subroutine is called only in user defined intervals. This can reduce the calculation duration significantly, but leads to less precise results.

Nonetheless, it will be a challenging task possible to set up calculations with about 20 cycles, while having the results at the same day. Therefore, long term (in the scale of complex simulation models) oil transport is not the prior objective of the simulation strategy.

4.5 Conclusion and Outlook

A model has been developed to integrate the oil transport and evaporation at the cylinder of a combustion engine within a FORTRAN subroutine into a commercial multibody system software. The model is able to provide a simultaneous interaction not only with the piston ring dynamics of the MBS model but also with the corresponding contact subroutines for hydrodynamic and mixed lubrication interaction.

The presented model contains the oil transport aspects as evaporation, accumulation of oil at the ring flanks by scraping, oil throw of from the first compression ring, oil consumption by reverse blow by and oil accumulation and release in the honing structure of the cylinder. The influence of changed lubrication properties by fuel entrainment are considered by means of the implementation of an oil component model. This allows the user to take into account the oil dilution by pure ethanol as well as ethanol blends or any arbitrary biofuel.

In spite of the three dimensional modeling, the estimated calculation duration will be significantly below those of comparable explicit Finite Element models. Optionally, the calculation effort can be further reduced by adjusting the frequency of the subroutine call.

Further aspects like the oil transport on the piston lands, through the piston grooves and the bridging from the piston to the liner has been investigated by experiment [40] and partly modeled for two dimensional systems [40]. These investigations showed, inter alia, that the oil transport on the piston lands took in several configurations between 25 and 50 cycles. Bearing in mind the complexity of the presented three dimensional models, for the present, the balancing scopes for the oil model will be kept focused on the cylinder.

Nonetheless, further extensions can be made to integrate local fuel entrainment with the data which is provided by CFD calculations [41]. The resolution and structure of the cylinder data grid allows this enhancement without significant changes in the FORTRAN code.

As presented, the knowledge of the oil properties is important to describe the oil distribution on the liner. Additionally, the oil dilution can have a severe impact on the wear behavior of the piston group. Besides this, one further important objective for biofuels is the emission quality. In the following section the impact of ethanol blends on the exhaust emissions is discussed.

5 Effects of Ethanol Fuel Blends on Gasoline DI Engines

The different composition of ethanol fuel blends and the differences in the combustion process of these fuels lead to an unequal composition of the exhaust gas and require adapted strategies for pollutant reduction. This chapter describes the impact of the usage of three fuel blends on the exhaust gas emissions in a dynamic test cycle.

5.1 Introduction to Ethanol Fuel Blends

Two different ethanol fuel blends and one basic gasoline fuel are investigated. The base fuel is a gasoline fuel without any ethanol content and with a research octane number of 95 (E0). The other two fuel blends consist of 20 % (E20) and 85 % (E85) ethanol by volume. Detailed information about the research fuels is given in Table 4.

The non-linear behavior of ethanol fuel blends with regard to the resulting fuel properties (e.g. vapor pressure of E20) results from the formation of azeotropes between ethanol and certain hydrocarbons of the gasoline fuel. This azeotropic

Table 4 Fuel properties

	E0	E20	E85
Research octane number, RON (−)	95.3	102.2	106.1
Density at 15 °C (kg/m^3)	738.5	748.4	785.5
Ethanol volume content (%)	<0.1	19.6	85.8
Oxygen mass content (%)	0.0	7.2	29.9
Stoichiometric air requirement (−)	14.6	13.4	9.8
Vapor pressure DVPE (kPa)	59.2	64.6	38.3
Lower heating value (kJ/kg)	43,433	40,042	29,157

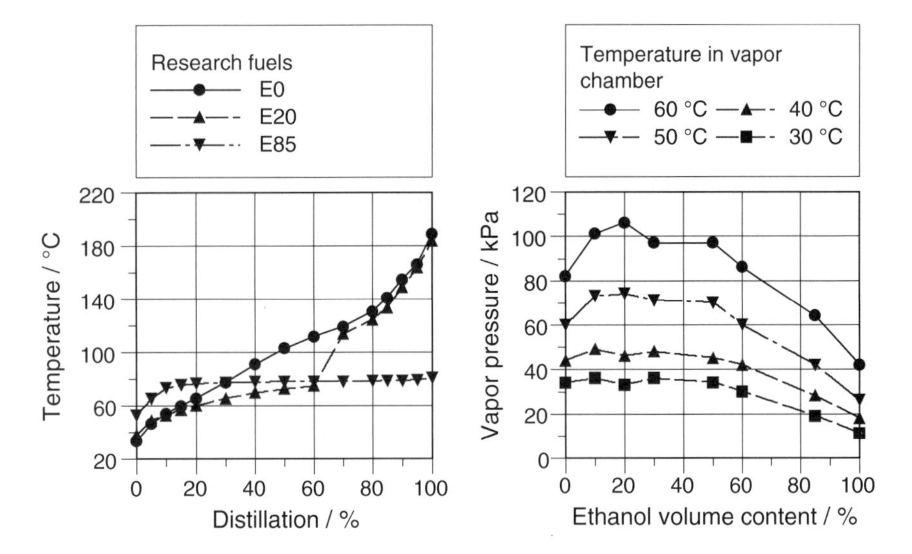

Fig. 29 Distillation curve of ethanol fuel blends and vapor pressure curves [63, 64]

behavior results in the vapor pressure behavior of ethanol fuel blends shown in Fig. 29.

The described fuel properties have significant influences on the operation of gasoline engines and have to be considered during the engine design phase and the engine calibration process. Especially the low vapor pressure, the high enthalpy of vaporization and the high oxygen content of fuels with high ethanol content need to be considered.

5.2 Research Engine and Investigation Methods

Investigations are based on a 1.8 l gasoline engine with direct injection and turbo charging, Fig. 30. Fuel is dosed via high pressure pump with three lobe actuation and piezo-electronic controlled injectors (Bosch HDEV4).

The injector nozzle opens to the outside and forms a fuel spray in the shape of a hollow cone with a cone angle of 85°. Injector and spark plug are located between intake and exhaust valves and their relative position is shown in Fig. 30. Table 5 gives an engine data overview.

The dynamic test cycle is performed on a dynamic engine test bench with exhaust gas analysis upstream and downstream three-way catalyst (TWC) whereas engine load and engine speed are derived from a demonstrator vehicle equipped with the research engine. This research vehicle is based on a Ford Focus ST (model year 2005) in which its original gear box and exhaust gas system are combined with the research engine.

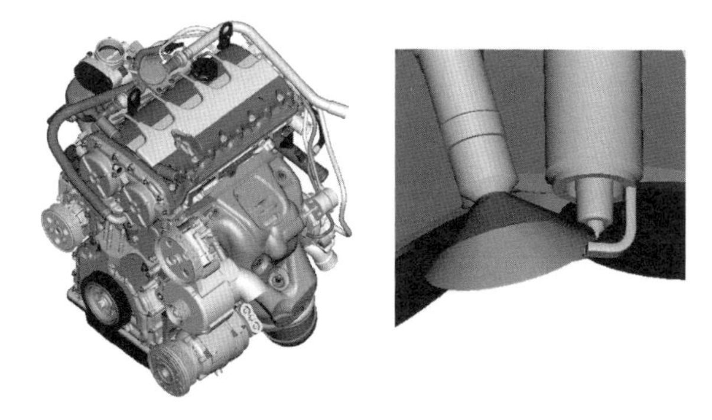

Fig. 30 Research engine #2 and injector spray layout [65]

Table 5 Engine data—engine #2

Bore/stroke (mm)	81/87
Valves per cylinder (−)	4
Max. mean effective pressure (bar)	22.4
Compression ratio (−)	11.5
Dimensions of TWC (inch × inch)	4.66 × 5
Cell density of TWC (cpsi)	400

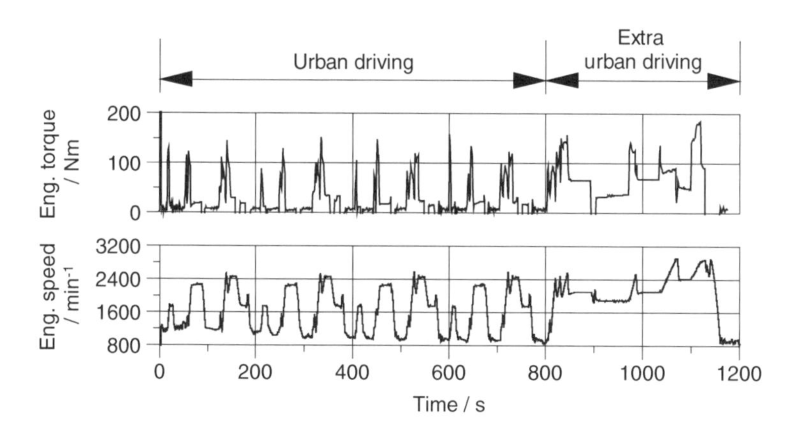

Fig. 31 Engine speed and torque in the dynamic test cycle

The resulting engine speed and engine torque during the dynamic cycle is displayed in Fig. 31. In the first part from, 0 s till 800 s, urban driving is simulated, while the second part refers to extra-urban operating conditions. The dynamic cycle is performed at an ambient temperature of 20 °C.

The calibration of the prototype engine control unit (ECU) is as realistic as possible within the given functionalities. Functions as catalyst heating, lambda control and dependencies of engine temperature, e.g. engine idle speed and ignition timing, have been integrated. But the performance of a rapid prototyping ECU is still not on the same level as a series production ECU, leading to some inadequateness in mixture precontrol.

5.3 Fuel Depended Engine Calibration

Due to the different fuel properties of the research fuels the engine calibration has to be individually adapted for each fuel. Especially properties like oxygen content, octane number and enthalpy of evaporation have to be considered. They result, for example, in different cold start behavior, exhaust gas temperatures and knock resistance.

Because of the high oxygen content of ethanol fuel blends the stoichiometric air requirement and the heating value are reduced. The combination of decreased air requirement and different fuel densities require an adaptation of the injected fuel mass for E20 +7.8 % and for E85 +32.9 %.

5.3.1 Engine Cold Start with Ethanol Fuel Blends

One major topic in dynamic engine investigation is the engine cold start, especially when it comes to exhaust gas emissions. At this operation point the exhaust aftertreatment system is at ambient temperature and the three-way catalyst is completely ineffective. Besides full load operation (intake air pressure is equal to ambient pressure) during engine speed up incomplete mixture preparation due to cold combustion chamber temperature has to be compensated. Therefore enrichment during engine start is required. The start enrichment is usually calibrated with a factor based on the fuel mass for stoichiometric combustion; the so called start enrichment factor.

Misfiring during engine speed up or higher start enrichment as necessary result in engine raw emissions with a high pollutant concentration. Figure 32 shows the resulting relative fuel masses for stoichiometric engine operation and engine cold start at 20 °C; both based on the required fuel mass for stoichiometric engine operation with E0 fuel. For stoichiometric engine operation the fuel mass has to be adjusted to compensate differences in air requirement and density, as mentioned above.

The fuel quantity displayed in Fig. 32 is based on a cold start configuration with high fuel pressure and stratified single injection close to ignition timing. The minimum fuel pressure for release of the first injection is 50 bar. Valve timings are chosen to deliver a small amount of residual gas. Details are listed in Table 6 and shown in Fig. 33.

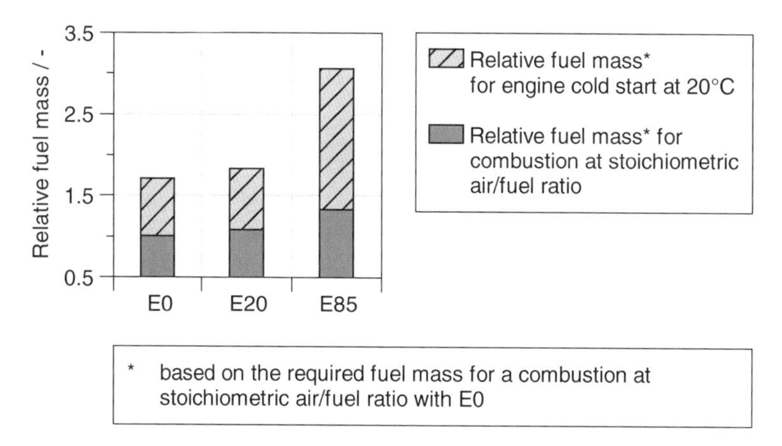

Fig. 32 Relative fuel mass for engine cold start at 20 °C

Table 6 Cold start calibration

Minimum fuel pressure (bar)	50 bar
Ignition timing (°CA BTDC)	10
Injection timing (°CA BTDC)	50
Start enrichment factor	
E0 and E20 (−)	1.7
E85 (−)	2.3

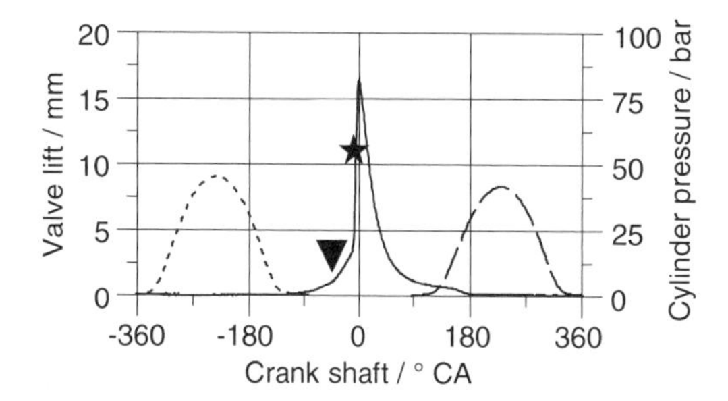

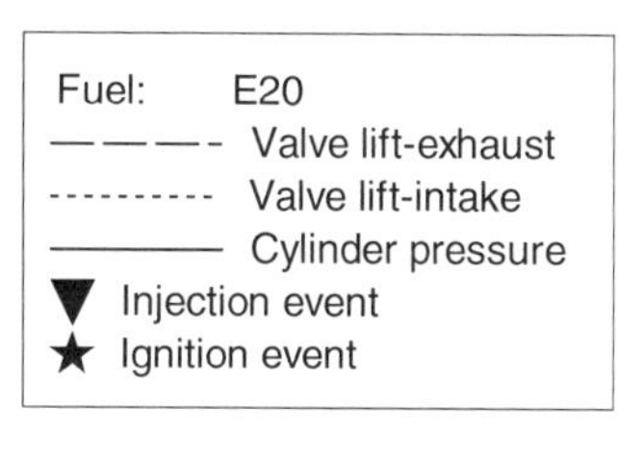

Fig. 33 Cold start calibration

5.3.2 Catalyst Heating Calibration

After a successful engine start the operation mode switches to catalyst heating to ensure pollutant reduction by the three-way catalyst as soon as possible. Therefore, a late combustion is calibrated to increase the heat released into the exhaust aftertreatment system by means of a higher exhaust gas temperature and mass flow. This is achieved via a double injection strategy and late ignition timing, see Fig. 34.

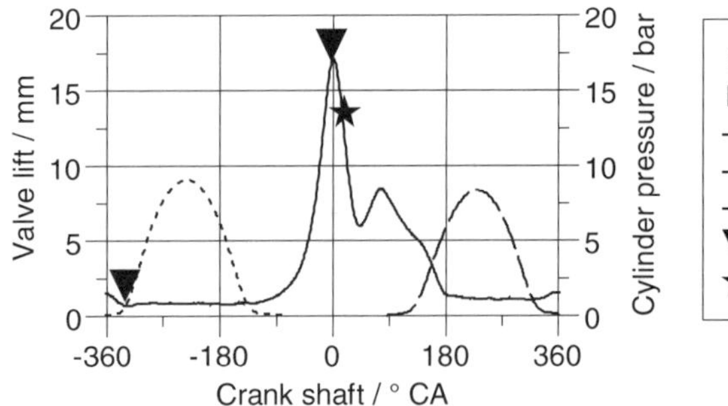

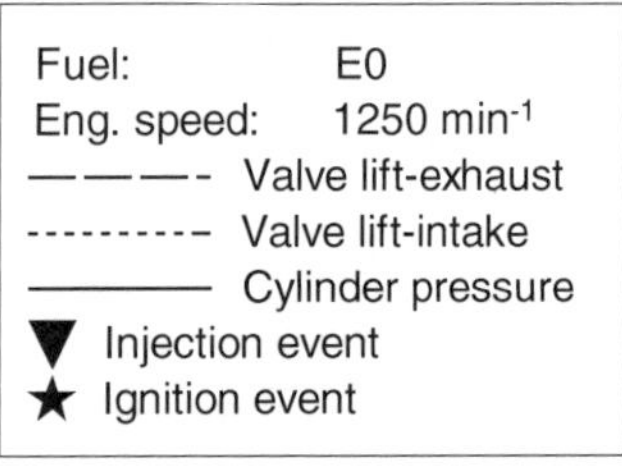

Fig. 34 Catalyst heating calibration

During this engine operation mode, engine idle speed is increased to $n = 1250\ \mathrm{min}^{-1}$ to and the relative air/fuel-ratio is set to 1.03. This reduces hydrocarbon (HC) emission and carbon monoxide (CO) emission during catalyst heating without significantly increasing NO_X emissions compared to a stoichiometric air/fuel-ratio. However, ignition timing during catalyst heating has to be calibrated fuel dependant to compensate differences in combustion stability and to deliver maximum possible exhaust gas temperature and exhaust heat flux respectively. Dependency of ignition timing on pollutant emissions and exhaust heat flux is displayed in Fig. 35.

Due to the lower exhaust heat flux with ethanol blended fuels, the ignition timing has to be retarded. The increased exhaust gas temperature and the lean combustion enable oxidation of HC emission in the exhaust system upstream three-way catalyst. The nitrous oxide (NO_x) emission increases with retarded ignition timings but due to colder combustion the NO_X emissions of the ethanol blended research fuels stay below the emissions of E0 [42]. The resulting ignition timings during the catalyst heating phase in the dynamic test cycle and the exhaust gas temperature upstream catalyst is shown in Fig. 36. The spark retard increases with higher ethanol content. The exhaust gas temperature upstream TWC is nearly identical for all research fuels. With these ignition timing curves nearly identical NO_x emissions and lower HC emissions for the ethanol fuel blends are expected.

The exhaust gas analysis in the dynamic cycle investigations approve presumptions on basis of the ignition timing variation at cold boundary conditions, see Fig. 37. The HC[1] and CO emissions of the ethanol blended research fuels are lower upstream and downstream three-way. During catalyst heating in the dynamic cycle, the NO_x emissions of E0 and E20 act quite similar and the expected effect of reduced NO_x emissions of the ethanol blended fuel occurs only with E85 fuel. Especially the higher output during engine load at 20 s is decreased.

Pollutant conversion by the TWC starts for all fuels at identical timings except the reduction of NO_x. At the end of the catalyst heating duration the reduction with

[1] Differences in the gas density of the research fuel blends are not considered in HC emission analysis.

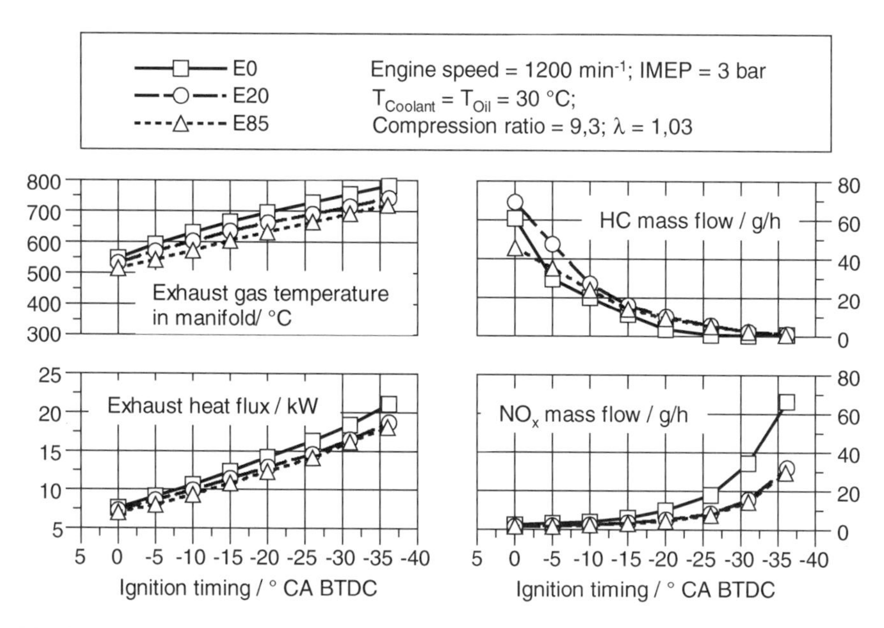

Fig. 35 Ignition timing variation under cold conditions [63]

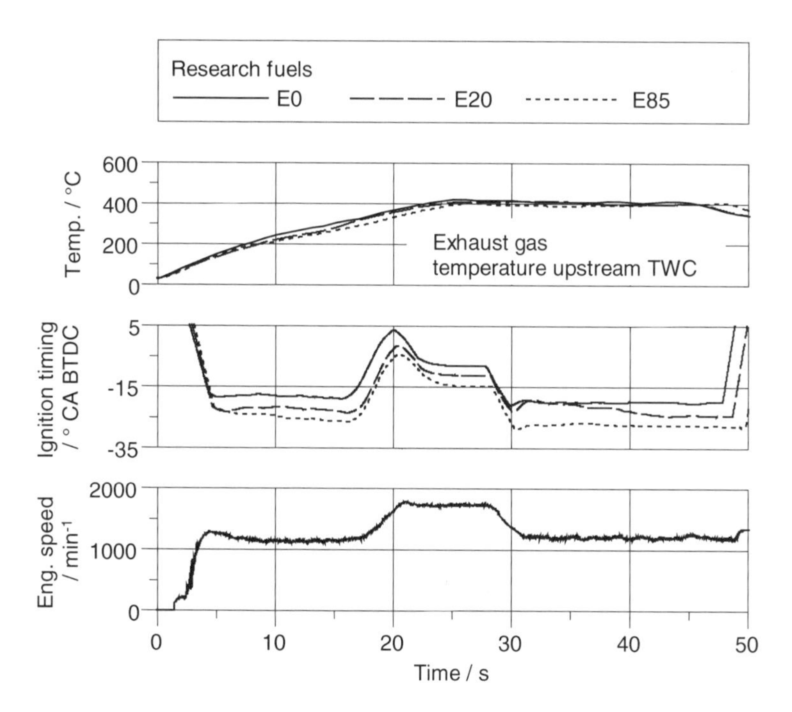

Fig. 36 Ignition timing and exhaust gas temperature upstream TWC during catalyst heating in dynamic engine cycle

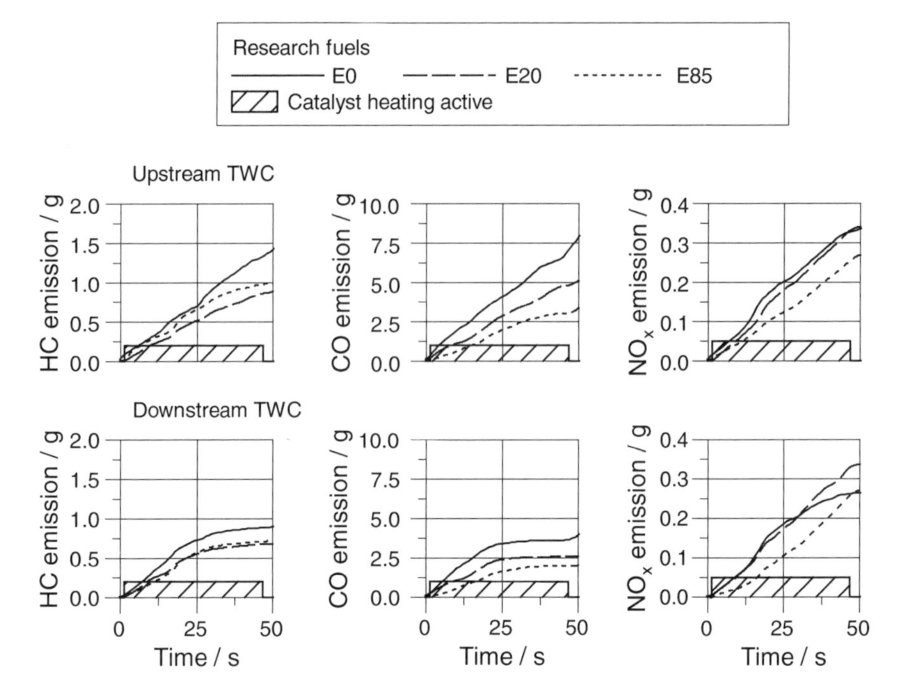

Fig. 37 Cumulated pollutant emissions during catalyst heating

E0 fuel already begins. Hence, the advantages of the ethanol blended fuels concerning colder combustion and therefore lower NO_x emissions do not deliver any benefit during catalyst heating.

5.3.3 Ignition Timing

To achieve maximum carbon dioxide (CO_2) emission reduction engine calibration has to be adapted to the fuel characteristics to the maximum extend possible. Especially the higher knock resistance due to higher octane numbers and cylinder charge cooling caused by the higher enthalpy of evaporation has to be considered here [43]. Figure 38 shows the different ignition timings of the research fuel blends in the effective engine speed range and engine load range of the dynamic test cycle.

The ignition timing is calibrated via the point of 50 % mass fraction burnt; except operation points with knock limitation. In these operation points the ignition timing was calibrated with a safety margin of 3° CA to the ignition timing resulting in knocking combustion.

Obviously the spark advance at higher engine loads increases with the ethanol content and the dependency of the ignition timing from engine load is strongly reduced for E85 fuel in the displayed range of engine speed and net indicated mean effective pressure (IMEP).

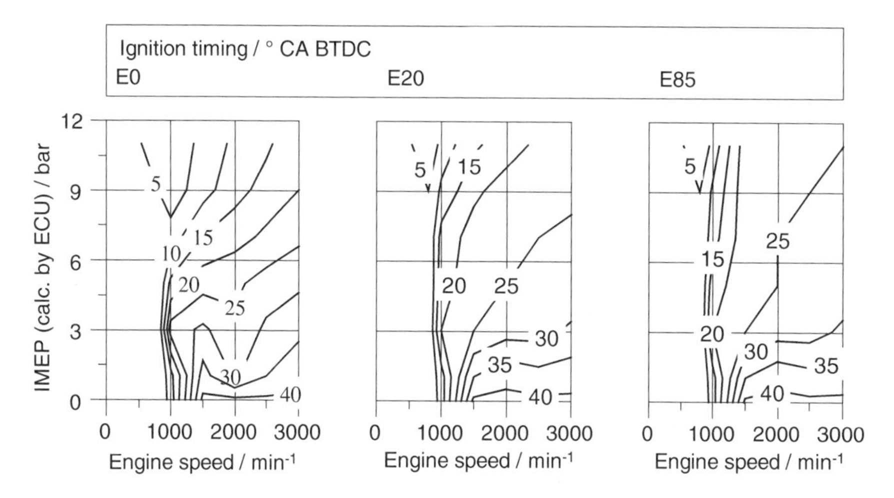

Fig. 38 Ignition timing calibration for the research fuel blends

Figure 39 shows the resulting ignition timing course in the extra-urban part of the dynamic cycle. During all accelerations and in some points of constant driving the spark advance is higher for the ethanol containing research fuels.

5.4 Results of the Dynamic Cycle Investigation

The consideration of the differences in fuel properties of the research fuels leads to individual calibrations of the research engine. For the engine cold start the differences in fuel vaporization have to be considered as well as strategies to heat-up the TWC and because of fuel consumption reasons ignition timing is optimized. These measures result in three engine operation strategies. In the following the pollutant emissions and CO_2 emission are displayed and differences are discussed.

Figure 40 displays the total pollutant raw emissions for the complete cycle. HC emissions are reduced with increased ethanol fuel content and therefore E85 fuel emits the lowest amount of the tested fuels. On the one hand, E0 fuel contains a higher amount of aromatics and olefins compared to ethanol blended fuels resulting in higher HC emissions [44, 45]. On the other hand, there is a hypothesis that ethanol fuel blends are oxygenated fuels which support HC oxidation efficiency. Besides that the lower FID response factor has a major influence on the measured HC emissions of the ethanol fuel blend. Hence, this results in nearly identical HC emissions for all research fuels.

The total NO_X emissions upstream TWC resume the same trend as during catalyst heating. The emission level with E0 and E20 fuel is nearly identical but the emissions with E85 fuel are lower. This is caused by the high enthalpy of

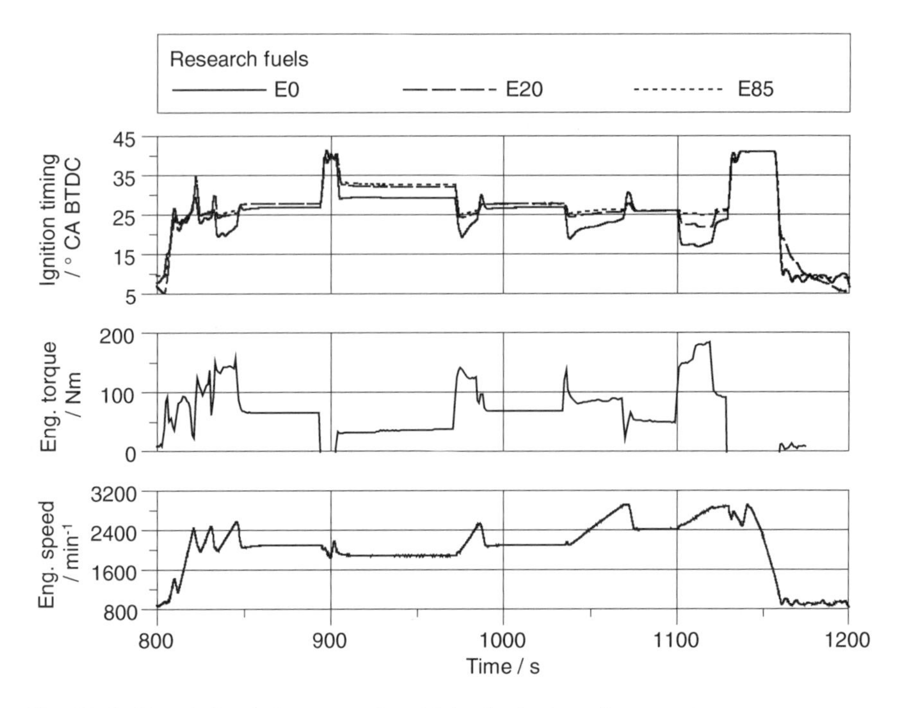

Fig. 39 Ignition timing during extra urban driving in the dynamic test cycle

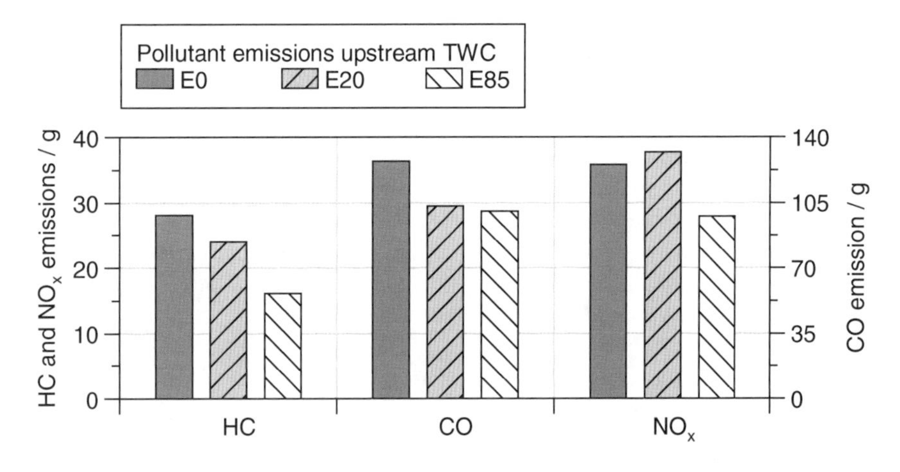

Fig. 40 Total pollutant emissions upstream TWC

vaporization and the lower adiabatic flame temperature of the fuel which results in lower in-cylinder temperatures in the burned gas [42, 46–50].

The usage of a rapid prototype engine control unit (ECU) equipped with an universal prototype software lead to pollutant emission results downstream TWC with low significance compared to exhaust emissions upstream TWC. This is due to the lack of some important functions and data sets; e.g. precise lambda control during dynamic engine behavior and compensation of transient effects. Considering an ideal exhaust gas conversion by the TWC after completion of catalyst heating (worst-case estimation: no emission slip after 100 s) exhaust gas emission as displayed in Fig. 41 would follow.

The different measured pollutants all end up in an identical range for all research fuels. The differences to the exhaust gas emissions upstream TWC are based on different catalyst light-off behaviors refer to Fig. 37.

The sum of measures in engine calibration and the different fuel properties result in the CO_2 emissions displayed in Fig. 42. The accumulated course of the CO_2 emissions shows that after catalyst heating the benefit of the ethanol blended fuels increases constantly. Here the reason is the difference in knock limitation between the research fuels; refer to Figs. 38 and 39.

The achieved total CO_2 reduction in the dynamic cycle investigation with the E20 fuel blend is 32 and 77 g for the E85 fuel blend. These values are equal to a CO_2 reduction of 1.7 and 3.8 %. However, without adaptation of the vehicles fuel tank the maximum driving range for the consumer is reduced due to the increase of the volumetric fuel consumption. Again, the reasons here are the lower stoichiometric air requirement and the different fuel density. Table 7 summarizes these results considering the theoretical driving range of the dynamic test cycle of 11 km.

The results of the dynamic test cycle investigations show a CO_2 reduction potential for fuel blends with high ethanol content. This tank-to-wheel analysis can be enlarged by the analysis of the fuel supply chain; the well-to-tank path. Hence, the complete pathway from fuel supply to combustion, well-to-wheel, has even a higher green house gas reduction potential. Figure 43 shows the green house gas

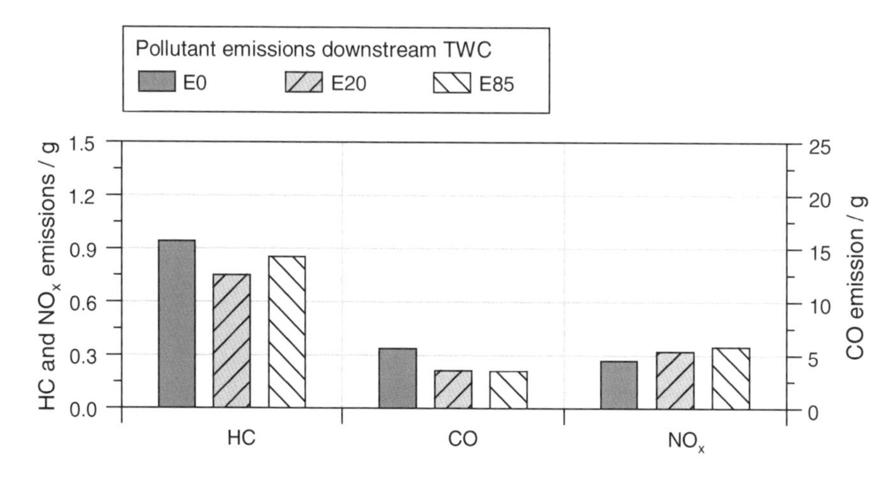

Fig. 41 Predicted pollutant emissions downstream TWC

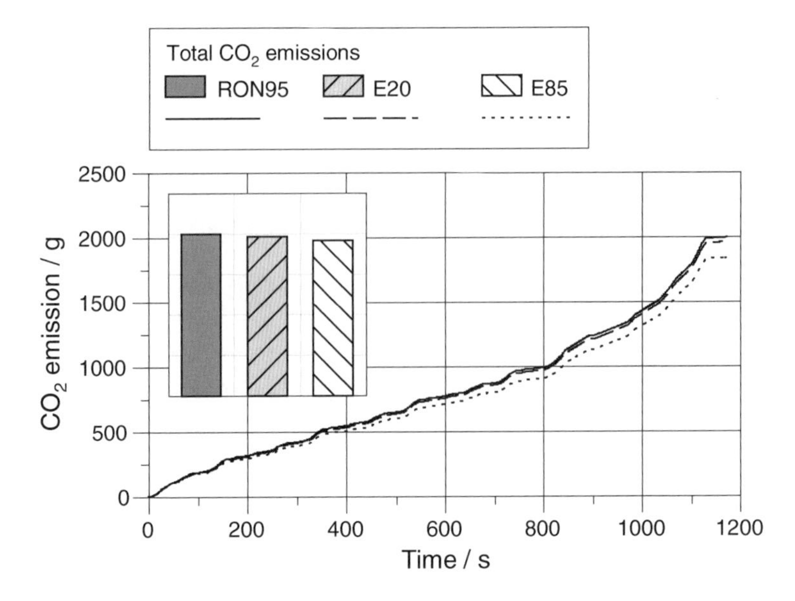

Fig. 42 Total CO_2 emissions in dynamic cycle

savings for different fuel resources and production pathways. For some resource/ pathway combinations a reduction around 80 % is possible.

5.5 Summary and Outlook

With the presented measures ethanol fuel blends allow the reduction of CO_2 emissions by usage of the fuel properties. The higher knock resistance and the high heat of vaporization allow a spark advance increase at higher part load and full load. CO2 emission reductions of 1.7 % for E20 and 3.8 % for E85 have been achieved. However, the cold start behavior and the catalyst heating especially at cold ambient conditions are challenging and considered detailed analysis and optimization.

In future investigations CO_2 emissions reduction can be enhanced. Based on the high knock resistance of the ethanol fuel blends, especially E85, the next step for CO_2 emission reduction could be an increase of the compression ratio without retarding ignition timing. This leads to an improvement of the thermal efficiency of the constant volume cycle [51].

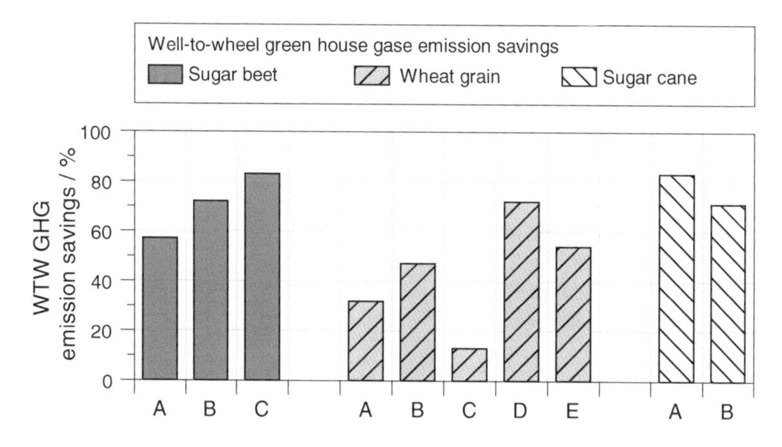

Fig. 43 Results for selected ethanol pathways [50]

Table 7 CO_2 emissions and volumetric fuel consumption

Fuel	CO_2 emissions (g/km)	Compared to E0 (%)	Fuel consumption (l/100 km)	Compared to E0 (%)
E0	181		7.8	
E20	178	−1.7	8.3	+6.4
E85	174	−3.8	10.6	+35.9

6 Impact of Biomass Derived Fuels on the Characteristics of Diesel Particulate Matter

The previous section describes the impact of ethanol fuels on the combustion and gaseous emission behavior of a gasoline direct injection engine. Over the decades, diesel engines with their high combustion efficiency, driving performance, durability and fuel economy have achieved an increased customer acceptance. Diesel engines with their relatively high particulate and NOx emissions have come under scrutiny with strict emission regulations being invoked. Also, to reduce the consumption of fossil fuels along with its emissions, bio-fuels as candidates of alternative fuels are being investigated. Particulate matter is one of the most harmful emission components from diesel exhaust. Till date particulate emission standards were mass based. Recently due to new legislative norms interest in other measures, i.e., size, number, surface and composition etc. have greatly increased. The regeneration characteristics of diesel particulate filters are mainly determined by the properties of the stored soot. Particulate formation conditions control particle structure and number concentrations. Concurrently, with the focus of research shifting towards the usage of bio-fuels as alternatives for fossil fuels, it is of vital importance to understand the effects of the these bio-fuels on particulate emissions.

Different studies have shown that fuel from biomass bears great potential to reduce combustion generated particulate matter emissions. The composition of these engine fuels and their molecular structure impacts in-cylinder particulate formation conditions which in turn controls the particulate characteristics. Soot from different fuels has different micro-structural characteristics- changing their burning behaviour in Diesel Particulate Filters (DPF).

The current research focuses on the effect of "biomass derived fuels" on "Particulate matter characteristics". The PM samples are analysed for chemical composition using solvent extraction methods, in addition, thermo-gravimetric analysis was used to determine the mass fraction of volatile PM fractions. To ensure that the findings are relevant for modern automotive business, testing programme is being carried out on a EURO6 compliant High Efficiency Combustion System (HECS) designed for modern passenger car applications [31].

6.1 State of the Art

Climate change impacts and criteria pollutant emissions have a juncture at black carbon or soot. With the global warming potential factor of black carbon at about 2000X versus CO_2 on a mass basis, about 20–25 % of the carbon footprint from an unfiltered diesel vehicle is from black carbon. Remediation of diesel soot today is primarily done to minimize the adverse health effects, but the climate forcing impact could further increase interest [52].

According to the European emission norms, targets are set for each of the regulated emissions. Table 8 shows the emission standards in the European Union for passenger cars [53].

As shown in the Table 8, new European emission legislation (Euros 5 and 6) forces the transportation sector to comply with ultra-low PM mass emissions and additionally specifies a limit for particle number emissions, which therefore derives the motivation for extensive research on particulate matter characteristics. The above reasons, in view of the emission regulations and the effect of particulate emissions on human health have necessitated the need for extensive research on exhaust particulate matter characteristics.

Table 8 EU emission standards for diesel passenger cars, g/km

Emission stage	Implementation date	CO	NOx	Particulate matter limit	Particulate number limit
Euro 4	Jan-05	0.5	0.25	0.025	
Euro 5a	Sep-09	0.5	0.18	0.005	
Euro 5b	Sep-11	0.5	0.18	0.0045	6 × 10^11 #/km
Euro 6	Sep-14	0.5	0.08	0.0045	6 × 10^11 #/km

Particulates occur during combustion with an extreme air deficiency. This is typical for combustion in diesel engines due to the heterogeneous mixture. Particulate formation is generally caused by the thermal cracking of the fuel molecules under air deficiency. This leads to the splitting of hydrogen molecules, with acetylene as an intermediate, which polymerizes to carbon rich macro-molecules and then agglomerate into the final particulates. The complex details of the kinetic process during formation of the particulates have not yet been solved [54].

Diesel particulates are composed of elemental carbon particles which agglomerate and adsorb other species to form structures of complex physical and chemical properties. Diesel particulates have a bimodal size distribution. They are a mixture of nuclei mode and accumulation mode particles, as shown in Fig. 44. Nuclei mode particles are very small—according to most authors, their diameters are between approximately 0.007 and 0.03 μm (micron).

Accumulation mode particulates are formed by agglomeration of primary carbon particles and other solid materials, accompanied by adsorption of gases and condensation of vapors. They are composed mainly of solid carbon mixed with condensed heavy hydrocarbons, but may also include sulfur compounds, metallic ash, cylinder wear metals, etc. [55].

Concurrently, with the focus of research shifting towards the usage of bio-fuels as alternatives for fossil fuels, it is of vital importance to understand the effects of the new bio-fuels blends on particulate emissions. While the benefits of oxygenated diesel blends are evident, thorough understanding of mechanisms that bring about the reductions in PM are not still understood. Many researchers have indicated that fuel oxygen content is the main factor affecting the PM emissions [56, 57]. For example, the results of Miyamoto et al. are often cited which show a decrease in Bosch number (Bosch smoke number can be taken as a gauge of the "black" smoke density. It is scaled arbitrarily from 0 to 10) that is well correlated to the fuel oxygen content, with smoke levels becoming zero at an oxygen content approximately 30 %

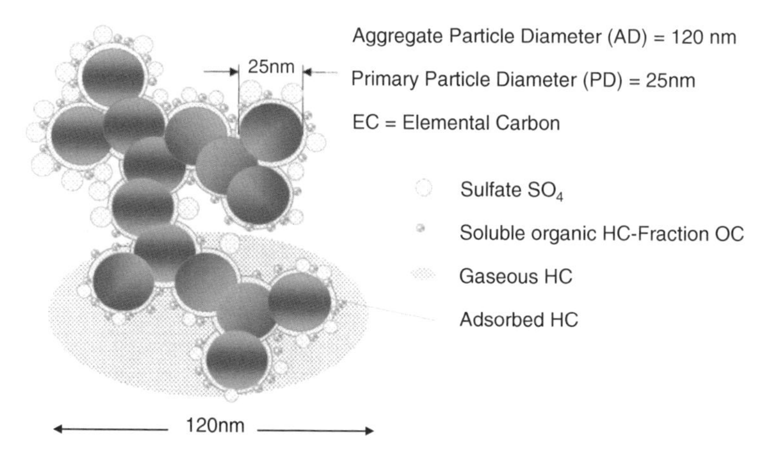

Fig. 44 An example of the schematic of diesel particulate matter [66]

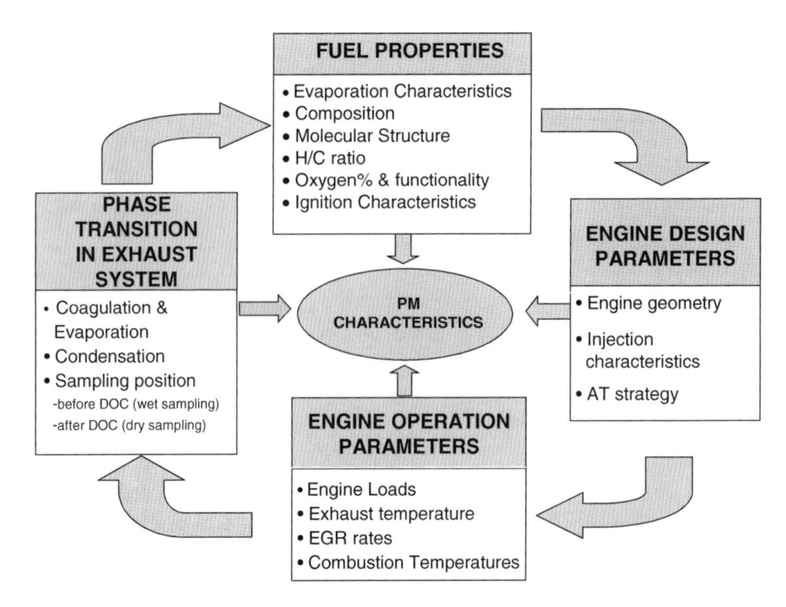

Fig. 45 Factors influencing PM characterization

by weight [58]. However, others have concluded that there are important differences depending on the chemical structure or volatility of a given oxygenate. To further investigate the mechanisms, a number of investigators have carried out numerical f chemical kinetics in the primary soot formation region. These studies provide additional insight into the nature of PM reduction with oxygenated diesel blends.

It has been seen that the engine operating conditions and boundary conditions of the investigations play a vital role in the consistency of results for a comparison of fuels and their emissions. The main influencing parameters which may affect the characteristics of the diesel particulate matter have been summarized in the Fig. 45

6.2 Experimental Set-Up and Test Procedure

The experiments are being carried out on a single cylinder diesel research engine. The main characteristics of which are listed in Table 1.

6.2.1 Engine Operation Points

The impact of biomass derived fuels on the particulate matter characteristics are being evaluated at four different part-load operating points. The key calibration parameters for the mentioned operation points are summarized in Table 2. Three of the four part-load points are located within the reference area of the New European

Driving Cycle (NEDC) while the highest load point at 2,400 rpm and 14.8 bar IMEP is outside or just inside this range depending upon the vehicle to which the engine is fitted. The strategy of closed-loop combustion control (as discussed in earlier sections) was simulated at test bench by keep the centre of combustion as constant for each fuel under tests. For each part load operation point the particulate sampling was conducted at Euro 6 indicated specific NOx emissions level. The further details about the equivalent indicated specific emission values corresponding to EU emission standards on NEDC are mentioned in publication from Müther M [31].

6.2.2 Test Fuels

The initial experimental tests were performed using the following two fuels:

- ULSD: Ultra low sulphur diesel fuel (sulphur <10 mg/kg)
- 2-MTHF/DBE: A blend of 70 % vol. 2-MTHF with 30 % vol. Di-n-butyl-ether

The composition and main properties of above mentioned fuels are summarized in Table 3.

6.2.3 Particulate Mass and Composition

Particulate samples were collected on a 47 mm Pallflex membrane filters (Teflon coated glass fiber filters T60A20) to determine mass of the total particulate matter emissions upstream and downstream Diesel Oxidation Catalyst (DOC). DOC was employed to eliminate the effect of soluble organic fractions, which is contributed mainly due to the condensed hydrocarbons from un-burnt fuel and lubricating oil. Further chemical analysis of particulate samples was conducted with the help of extraction with cyclohexane to determine soluble and non-soluble organic fractions. The gravimetric determination of the mass was carried out using an analytical microbalance (precision 1 μg), which is installed and operates in the specially designed climatic chamber/weighing room.

6.2.4 Thermo-Gravimetric Analysis

To determine the chemical/physical particulate properties, particulate samples were collected on quartz fiber filters (Quartz fiber filters are very stable at high temperatures. The filters were tempered at >700 °C, to avoid any foreign material contamination, before using them for collecting PM samples) from the undiluted exhaust at constant volumetric flow using a 190 °C heated line and 110 °C at the sampling filter. The soot load mass was determined by weighing the filter under controlled conditions. The particulate samples from Quartz fiber filters were

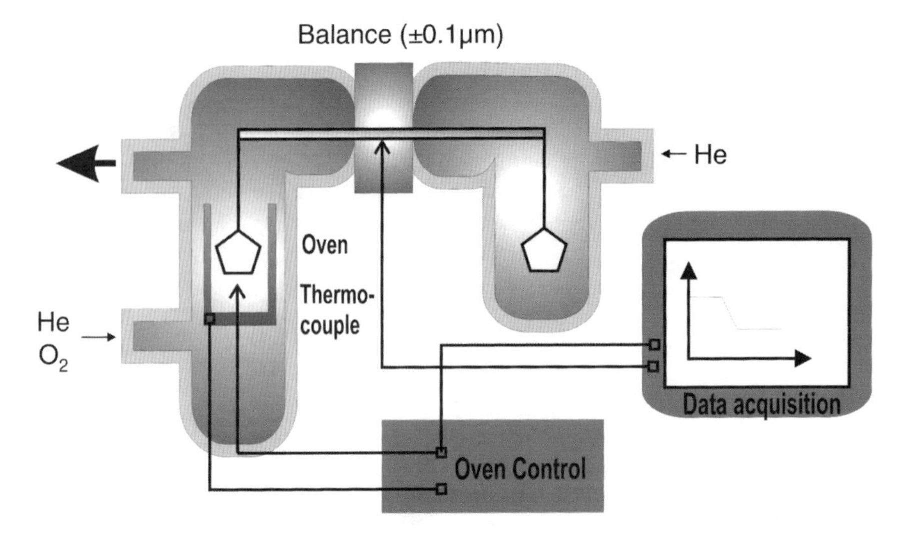

Fig. 46 Schematic diagram of the TGA equipment operation

characterized analytically using thermo-gravimetric analysis (TGA). The following procedure was used to conduct TGA (analyzer: Thermo Cahn TG 2121).

- Scan from 30° => 50 °C at 25 °C/min—dynamic (inert gas- Helium 6.0)
- Isothermal at 50 °C for 3 min
- Scan from 50 °C => 700 °C at 25 °C/min—dynamic (inert gas- Helium 6.0)
- Isothermal at 700 °C for 60 min
- Isothermal at 700 °C for 15 min (addition of 50 % Oxygen @2.0 bar)

Figure 46 shows the schematic of TGA equipment operation. By continuously determining the mass during the heating up phase, the temperature-dependent mass losses caused by evaporation were recorded until a temperature of 700 °C by passing an inert gas Helium in the absence of oxygen. In the final step 50 % Oxygen was added for 15 min. at steady state 700 °C to ensure the complete oxidation of soot. A typical PM components fractionation from TGA analysis is depicted in Fig. 47.

6.3 Test Results

The main test results of PM mass emissions, chemical speciation and volatile fractionation have been summarized in the following sections.

6.3.1 PM Emissions

Figure 48 shows PM emissions and filter smoke number from 2-MTHF/DBE fuel relative to ULSD at higher part load operation point. The particulate measurements

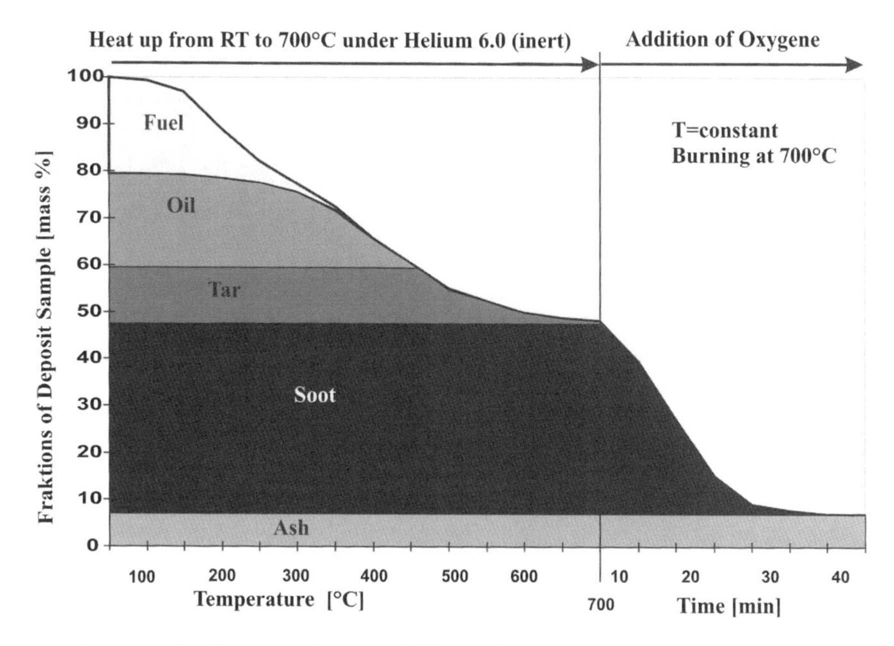

Fig. 47 Schematic diagram of the typical TGA PM fractionation

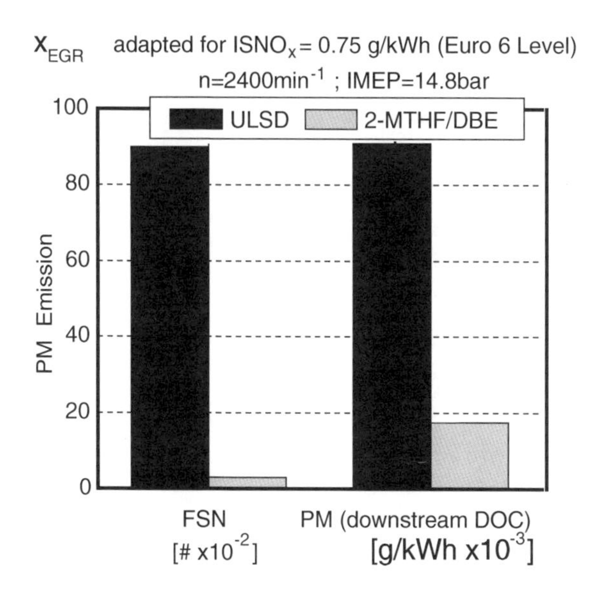

Fig. 48 PM Mass Emission and FSN (Filter Smoke Number) from the investigated fuels

were conducted under similar engine operating conditions by maintaining ISNOx emissions to a level corresponding to Euro 6 emission levels and centre of combustion was also maintained constant for each fuel under test programme. It was observed that FSN was reduced to nearly 95 % with using 2-MTHF/DBE fuel relative to ULSD fuel and a similar trend was observed with particulate mass emissions (collected over glass fiber filters), where we found more than 80 % reduction in the total PM mass downstream diesel oxidation catalyst. The reduction in the FSN and PM mass emissions with 2-MTHF/DBE fuel can be attributed to oxygen content in fuel and its better evaporation characteristics as compared to ULSD. Furthermore the CN of 2-MTHF/DBE is significantly lower than ULSD fuel (Table 3) which results in higher ignition delay. In a nutshell, the benefits in PM emissions are caused by a more favorable local air/fuel ratio and therefore an improved mixture preparation as a consequence of the oxygen content within the fuel and the longer ignition delay.

6.3.2 PM Chemical Speciation

Figure 49 shows the composition of particulate matter upstream DOC (engine out) and downstream DOC with baseline ULSD fuel and 2-MTHF/DBE. It is quite evident from the figure that the SOF content from engine out PM with operation on 2-MTHF/DBE was almost 50 % higher than baseline ULSD fuel. Although total

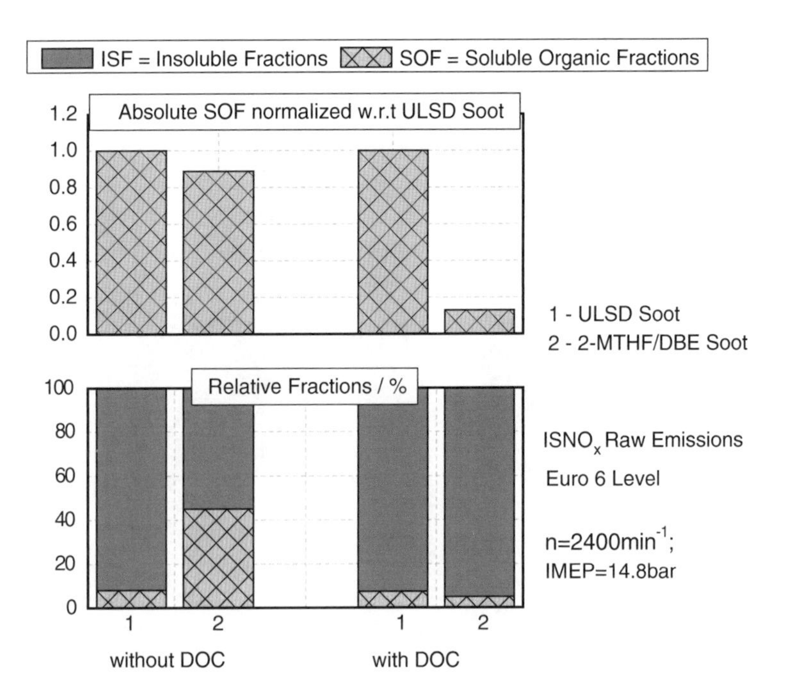

Fig. 49 Chemical speciation of PM

Table 9 Chemical speciation of PM in %

	ULSD		2-MTHF/DBE	
	SOF	ISF	SOF	ISF
Engine out (without DOC)	14.2	85.8	64.5	35.5
Downstream DOC	7.9	92.1	2.5	97.6

PM emissions from 2-MTHF/DBE are more than 80 % less than baseline diesel fuel (Fig. 49). The reason for higher relative SOF of 2-MTHF/DBE for PM sampling upstream DOC is due to the condensation of un-burnt hydrocarbons in the exhaust stream which comes from un-burnt fuel and lubrication oil. But same behaviour was not observed for the sampling downstream of DOC, where SOF from 2-MTHF/DBE was almost completely oxidized due to exothermic reactions over DOC. The results of PM chemical speciation are summarized in Table 9.

6.3.3 PM Volatile Fractionation

Based on the present TGA, total PM mass is basically divided into three main fractions i.e. volatile fraction, tar and elemental carbon. The volatile fraction can be further subcategorized as high volatile (from un-burnt fuel) and low volatiles (mainly from lube oil). During heating of the samples by following the presented temperature ramp programme, volatile material is separated into two temperature ranges. VF1 represents volatile material from fuel and VF2 represents volatile material from lube oil. Since VF1 depends on the fuel properties therefore its temperature range is dependent on the fuel evaporation characteristics. In case of Diesel VF1 lies in the range of 50 °C $< T <$ 220 °C and similarly in the case of 2-MTHF/DBE due to its lower boiling characteristics, VF1 lies $T <$ 80 °C. A very low volatile part of PM mass fraction is Tar, which lies typically in the temperature range of 500 °C $< T <$ 700 °C in TGA curve. The remaining fraction of PM which burned—off completely with supply of oxygen is elemental carbon (EC). The various components of PM are summarized in Table 10 as below:

PM mass fractionation results from TGA for exhaust gas from ULSD and 2-MTHF/DBE are presented in Figs. 50 and 51. In this case the PM samples for TGA investigations were collected downstream DOC. The solid elemental carbon fraction of the total PM is lowered with 2-MTHF/DBE fuel in comparison to ULSD particulates. The decreased EC results with using 2-MTHF/DBE are consistent with

Table 10 Classification of PM components from TGA investigation

PM fraction	VF1	VF2	Tar	EC
Typical range	50 °C $< T <$ 220 °C	220 °C $< T <$ 500 °C	500 °C $< T <$ 700 °C	@ 700 °C
Remarks	Fuel specific high volatiles	Volatile matter mainly from lube oil	Low volatile matter	Soot

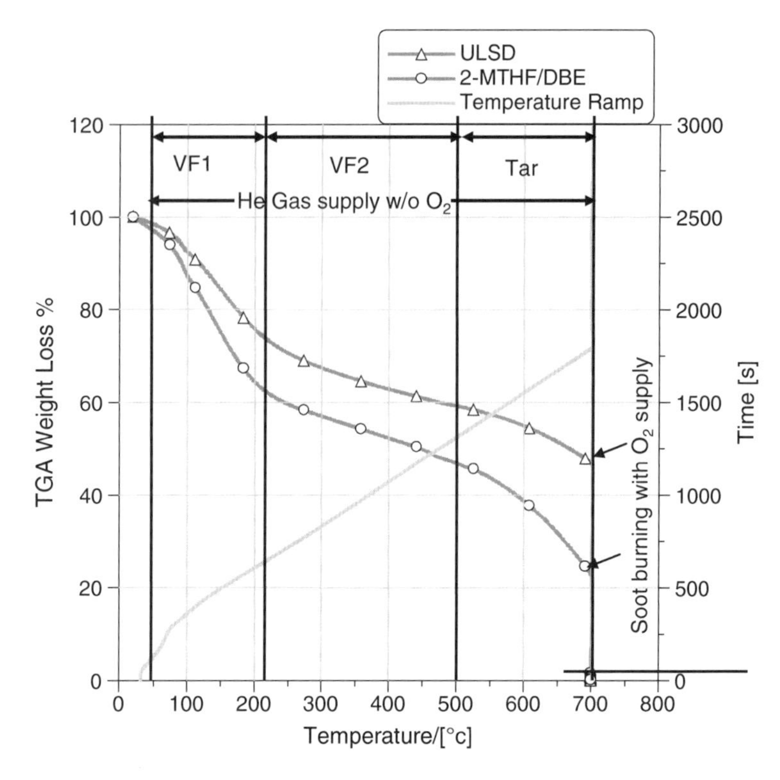

Fig. 50 TGA relative mass loss curves using PM samples on quartz filters

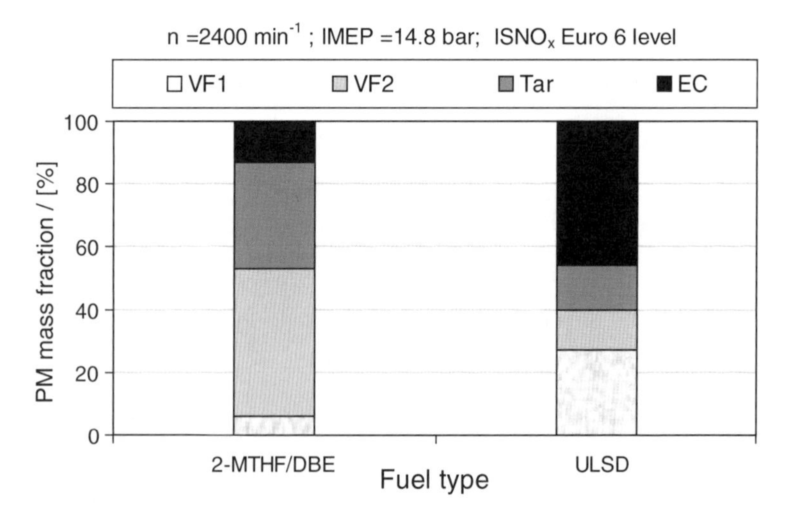

Fig. 51 PM volatile fractions from mass loss curves (relative values)

Table 11 TGA PM fractionation results in %

Fuel type	VF1	VF2	Tar	EC
ULSD	27	13	14	46
2-MTHF/DBE	6	47	34	13

earlier studies of A. William & Co-workers with oxygenated fuels [59]. The relative organic carbon fraction (VF1 + VF2) of 2-MTHF/DBE particulates is relatively higher then ULSD. The VF1 of PM from 2-MTHF/DBE appears to be negligible in comparison to ULSD (see Table 11). Due to early evaporation characteristics of 2-MTHF/DBE, this behavior can be attributed to the oxidation of fuel dependent high volatile fractions in DOC.

6.4 Major Findings and Further Research

The aim of these investigations is to evaluate the impact of biomass derived fuels on the characteristics of particulate matter emissions.

- PM mass emission results indicate almost soot free combustion in the case of 2-MTHF/DBE. Even at high load operation point the potential of PM mass reduction is more than 80 % relative to ULSD.
- PM chemical speciation results shows a tendency for higher SOFs from PM sampling upstream DOC with oxygenated fuel. The relative SOF fraction of 2-MTHF/DBE upstream DOC is significantly higher than ULSD. Although the absolute value of this SOF fraction from 2-MTHF/DBE is negligible due to very low total PM emissions with 2-MTHF/DBE
- It was decided to collect PM samples for further investigations downstream of DOC to eliminate the effect of SOF in the soot burning rate behavior and other further soot micro-structure investigations. This configuration of exhaust system is very close to real world applications as virtually all modern diesel operated vehicles are installed with oxidation catalyst.
- Thermo-gravimetric investigation suggests higher OC (organic carbon) fractions in the PM from 2-MTHF/DBE fuel however ULSD PM was detected with higher EC (elemental carbon).

Above investigations suggest that the molecular structure of fuel affects the PM formation i.e. mass emissions as well as its composition. Based on the initial investigations as described in this section, further extensive research programs have been initiated with a special focus on

- Specific surface area, micro-structure and determination of degree of graphitization of soot samples with using different generations of biomass derived fuel.
- DPF regeneration behavior with respect to active and passive regeneration characteristics.
- PM number emissions and mass based size distribution.

6.5 Summary and Outlook

The application of renewable fuels challenges the engine development process in all its facets. The different properties of new biofuels such as the boiling properties, the ignition behaviour, the viscosity, density and the polarity make adjustments of the yet existing engine system necessary. These adjustments refer to all fields of the engine development, from engine mechanics over engine control software down to the combustion itself. In this publication, a large variety of different aspects that are directly influenced by the use of modern biofuels is presented.

In the first place, the software aspect of biofuel application in modern combustion engines is evaluated. This puts special requirements on the control software in terms of variant management, code complexity and safety requirements. Hence, a flexible control system is being developed to enable innovative algorithm development. The top level software architecture is defined while the function implementation is ongoing. In order to avoid software erosion with increasing functionality over time, a continuous quality inspection framework was developed. For a comprehensive control improvement, the control software must cover all functionalities being necessary for all development steps from the test bench to vehicle integration.

Based on the boundaries set by the basic software design, a control algorithm is presented that allows the use of modern biofuels in modern, passenger cars diesel engines by adjusting only the injection control structure. Therefore, an algorithm based on Iterative-Learning-Control theory is developed that, by multiple injections, allows the realization of nearly identical combustion traces for fuels of very different properties. By that, especially the noise emissions can be reduced while the benefits of the use of oxygenated fuels (especially very low soot emissions) can be fully utilized. With algorithms as the one presented, even biofuels with extremely different properties than that of the standard fuels can be used in the yet existing well known combustion engines. Hence, the software and control structure development act as an enabler for further, biofuel research.

In the next step, one of the effects of biofuel application on the engine mechanics is evaluated. Especially in areas in which a direct contact between oil, fuel and engine structural material exists the varying fuel properties may have an important effect on the functionality and lifetime of the engine. The contact zone between piston ring and cylinder liner is one of the most critical zones in the tribological contact areas of the engine and is therefore investigated with a specifically developed model. This model allows the simulation of the piston ring movement during engine operation, taking into account the lubricity behaviour of the oil/fuel film on the cylinder liner wall.

In the last section, examples of biofuel application for both Diesel and gasoline engines are presented. For the Diesel engine, the effect of fuel properties on the soot emissions and the composition of particulate matter in the exhaust gas are analyzed. The results indicate the possibility of an almost soot free combustion for a tailor-made fuel from Biomass, called 2-MTHF/DBE. PM chemical composition and volatile fractionation investigations indicates relatively higher shares of organic carbon and significantly lower share of elemental carbon relative to diesel fuels.

Also, the consequences of ethanol-use in gasoline engines were investigated. It was shown that ethanol fuel blends allow the reduction of CO_2 emissions by use of the fuel properties. The higher knock resistance and the high heat of vaporization allow a spark advance increase at higher part load and full load. CO2 emission reductions of 1.7 % for E20 and 3.8 % for E85 have been achieved. However, the cold start behavior and the catalyst heating especially at cold ambient conditions are challenging and consider detailed analysis and optimization.

Overall, this publication outlines the vast diversity of challenges that new biofuels cause when being used in combustion engines. The outlined topics and approaches provide a holistic picture of how these new challenges can be solved. In the future, the executed and planned work in the described fields will contribute to the further penetration of biofuels in everyday life.

Acknowledgments The presented research is funded by the research cluster "Fuel production with renewable raw materials" (BrenaRo) at RWTH Aachen University (http://www.brenaro.rwth-aachen.de). This work was performed as part of the Cluster of Excellence "Tailor-Made Fuels from Biomass", which is funded by the Excellence Initiative by the German federal and state governments to promote science and research at German universities (http://www.fuelcenter.rwth-aachen.de). This article was funded by Forschungsvereinigung Verbrennungskraftmaschinen e.V. (FVV, Frankfurt) and the Bundesministerium für Wirtschaft und Technologie BMWi via Arbeitsgemeinschaft industrieller Forschungsvereinigungen e.V. (AiF), (IGF-Nr. 15402).

References

1. Yilmaz, H.: Advanced Combustion Concepts -Enabling Systems and Solutions (ACCESS) for High Efficiency Light Duty Vehicles. DOE Vehicle Technologies Program Review, Arlington (2011)
2. European Parlament and Council of the European Union: EU Directive 443/2009 (EG 2009c) (2009)
3. United States Environmental Protection Agency, Department of Transportation: Light-duty vehicle greenhouse gas emission standards and corporate average fuel economy standards—final rule (2010)
4. European Parlament and Council of the European Union: EU Directive 28/2009 (2009)
5. United States Energy Information Administration: Annual Energy Review 2009 (2009)
6. Dingel, O. (ed.): Gasfahrzeuge, III: Die Schlüsseltechnologie auf dem Weg zum emissionsfreien Antrieb?. Expert, Renninger (2008)
7. Bosch GmbH, R. (ed.): Ottomotor-Management. Vieweg, Wiesbaden (2005)
8. Flaschke, T.: Biomasse getankt? Die Anpassung der Motorsteuerung an die momentan vorliegenden Kraftstoffeigenschaften. Erfinderaktivitäten 2009. Deutsches Patent- und Markenamt, Berlin (2009)
9. Jaasma, S.: Technical status and customer experiences with the Vialle LPG direct injection system—Lpdi. In: Proceedings Gas Powered Vehicles—The Alternative to E-Mobility? Berlin (2010)
10. Ballauf, J.: Audi AG: 2.0 TFSI flexible fuel motor im Audi A4. In: C.A.R.M.E.N. Symposium (2011)

11. Herz, K., Huber, M., Fautz, O., Gessmann, T.: System and SW engineering for gas powered vehicles. In: Proceedings Gas Powered Vehicles—The Alternative to E-Mobility? Berlin (2010)
12. Verband der Deutschen Automobilindustrie—VDA: Automotive SPICE® Process Assessment Model (2010)
13. International Organization for Standardization—ISO: ISO/IEC 25000 software engineering—software product quality requirements and evaluation (SQuaRE)—guide to SQuaRE (2005)
14. Richter, K., Jersak, M.: Eine ganzheitliche Methodik für den automatisierten Echtzeit-Nachweis zur Absicherung hoch integrierter, sicherheitskritischer Software-Systeme. In: Keller, H. et al (Ed.) Proceedings Automotive Safety and Security
15. Kindel, O., Friederich, M.: Softwareentwicklung mit AUTOSAR. dpunkt, Heidelberg (2009)
16. International Organization for Standardization—ISO: ISO/DIS 26262 road vehicles—functional safety. Draft International Standard (2009)
17. Hungar, H., Reyzel, E.: SW-Entwicklung und Zertifizierung im Umfeld sicherheits-kritischer und hochverfügbarer Systeme: Bedeutung modellbasierter und formaler Ansätze für effiziente Entwicklung und Zertifizierung. In: Proceedings Software Engineering (Workshops) (2008)
18. Lederer, D.: Systematische Software-Qualität mittels einer durchgängigen Analyse- und Teststrategie. In: Proceedings "Software im Automobil" (2010)
19. Richenhagen, .J, Schloßer, A., Orth, P.: Continuous Integration for automotive model-based control software development. Proceedings AUTOREG 2011—Steuerung und Regelung von Fahrzeugen und Motoren. VDI, Wiesbaden (2011)
20. AUTOSAR: Explanation of Application Inter-faces of the Powertrain Domain. AUTOSAR Release 4.0 (2009)
21. Schnorbus, T.: Ansätze für ein zylinderdruckgeführtes Einspritzmanagement beim Dieselmotor. Dissertation RWTH Aachen (2009)
22. Lezius, U., Schultalbers, M., Drewelow, W., Lampe, B.: Abstandsbasierte Klopfregelung in zylinderdruckgeführten Steuerungen für Ottomotoren. Motortechnische Zeitschrift (MTZ), 10/2008, vol. 69 (2008)
23. Hadler, J., Rudolph, F., Dorenkamp, R., Stehr, H., Hilzendeger, J., Kranzusch, S.: Der neue 2,0-l-TDI-Motor von Volkswagen für niedrigste Abgasgrenzwerte—Teil 1. Motortechnische Zeitschrift (MTZ) 05/2008, vol. 69 (2008)
24. Seebode, J.: Dieselmotorische Einspritzratenformung unter dem Einfluss von Druckmodulation und Nadelsitzdrosselung. Dissertation Universität Hannover (2004)
25. Graglia, R., Catanese, A., Parisi, F., Barbero, S.: Die neue Dieselmotor-Steuerung von General Motors. Motortechnische Zeitschrift (MTZ) 02/2011, vol. 72 (2011)
26. Kolbeck, A., Schnorbus, T., Kremer, F., Müther, M., Cracknell, R., Rose, K.: Control strategies for different fuels in advanced combustion systems. In: Proceedings 8th International Colloquium of Fuels 2011, Stuttgart (2011)
27. Hinkelbein, J.: Verbrennungscharakteristikregelung mittels Einspritzverlaufs-modulation bei direkteinspritzenden Dieselmotoren. Dissertation RWTH Aachen (2010)
28. Hinkelbein, J., Sandikcioglu, C., Pischinger, S., Lamping, M., Körfer, T.: Control of the diesel combustion process via advanced closed loop combustion control and a flexible injection rate shaping tool. SAE 2009-24-0114 (2009)
29. Kremer, F., Schaub, J., Pischinger, S., Hinkelbein, J., Lamping, M., Körfer, T.: Verbrennungsratenregelung—ein entscheidender Schritt zur weiteren Optimierung von CO2, Emission und NVH. In: Proceedings Haus der Technik: 7. Tagung Diesel- und Benzindirekteinspritzung 2010. Berlin (2010)
30. Kremer, F., Schaub, J., Pischinger, S., Hinkelbein, J., Kolbeck, A., Steffens, C., Körfer, T., Lamping, M.: Verbrennungsratenregelung—Baustein zur weiteren Komfortsteigerung CO2-optimierter Dieselmotoren, 10. Tagung Motorische Verbrennung—Aktuelle Probleme und Lösungsansätze 2011. München, Germany (2011)
31. Müther, M.: Einflüsse alternativer Kraftstoffe auf die dieselmotorische Verbrennung. Dissertation RWTH Aachen (2009)

32. Janssen, A., Kremer, F., Baron, J., Müther, M., Pischinger, S., Klankermayer, J.: Tailor-made fuels from biomass for homogeneous low temperature diesel combustion. Energy and Fuel. doi:10.1021/ef2010139
33. Priebsch, H.H., Herbst, H.M.: Simulation des Einflusses von Kolbenringparametern. Motortechnische Zeitschrift (MTZ) **60**, 772–779 (1999)
34. Spilker, T.: Ein virtuelles Kolbenringmodell zur Bewertung des Ölbedarfs in geschmierten Kolbenverdichtern. In: Proceedings 15. Workshop Kolbenverdichter 2011. Kötter Consulting Engineers (2011)
35. Hewitt, G.F., Shires, G.L, Bott, T.R.: Process Heat Transfer. McGraw-Hill, New York (1994)
36. Reid, R.C., Prausnitz, J.M., Poling, B.E.: The Properties of Gases and Liquids, 4th Edn. McGraw-Hill, New York (1987)
37. Pint, S., Schock, H.J.: Design and development of a software module for analysis of three dimensional piston ring wear. SAE 2000-01-0920, SAE International (2000)
38. Ortjohann, T.: Simulation der Kolbenringdynamik auf Basis expliziter FEM-Software. Dissertation, RWTH Aachen (2006)
39. Burnett, P.J., Bull, B., Wetton, R.J.: Characterization of the ring pack lubricant and its environment.J. Eng. Tribol., Part J. **209**, 109–118 (1994) (Proceedings of the Institution of Mechanical Engineers)
40. Thirouard, B.: Characterization and modeling of the fundamental aspects of oil transport in the piston ring pack of internal combustion engnines. PhD thesis, Massachusetts Institute of Technology (2011)
41. Budde, M., Ehrly, M., Jakob, M., Wittler, M., Pischinger, S.: Simulation and optical analysis of oil dilution in diesel regeneration operation. JSAE 20119241, Society of Automotive Engineers of Japan (2011)
42. Yoon, S.H., Ha S.Y., Roh, H.G., Lee, C.S.: Effect of bioethanol as an alternative fuel on the emission reduction characteristics and combustion stability in a spark ignition engine. In: Proceedings of IMechE Vol 223 Part D: Journal of Automobile Enigneering, pp. 941–951 (2009)
43. Kapus, P.E., Fuerhapter, A., Fuchs, H., Fraidl, G.K. Ethanol direct injection on turbocharged si engines - potential and challenges. SAE International, 2007-01-1408 (2007)
44. Heywood, J.B.: Internal Combustion Engines Fundamentals. McGraw-Hill Book Company, New York (1988)
45. Schuetzle, D., Siegl, W.O., Jensen, T.E., Dearth, M.A., Kaiser, E.W., Gorse, R., Kreucher, W., Kulik, E.: The relationship between gasoline composition and vehicle hydrocarbon emissions: a review of current studies and future research needs. In: Symposium on Risk Assessment of Urban Air: Emissions, Exposure, Risk Identification and Risk Quantification, Stockholm (1992)
46. Celik, M.B.: Experimental determination of suitable ethanol-gasoline blend rate at high compression ratio for gasoline engine. Appl. Thermal Eng. **28**, 396–404 (2008)
47. Al-Hasan, M.: Effect of ethanol-unleaded gasoline blends on engine performance and exhaust emission. Energy Convers.Mgmt. **44**, 1547–1561
48. Ceviz, M.A., Yuksel, F.: Effects of ethanol-unleaded gasoline blends on cyclic variability and emissions in an SI engine. Appl. Thermal Eng. **25**, 917–925 (2005)
49. Nakata, K., Shintaro, S., Ota, A., Kawatake, K., Kawai, T., Tsunooka, T.: The effect of ethanol fuel on a spark ignition engine. SAE technical paper, 2006-01-3380 (2006)
50. Li, L., Liu, Z., Wang, H., Gong, C., Su, Y.: Combustion and emissions of ethanol fuel (E100) in a small SI engine. SAE technical, 2003-01-3262 (2003)
51. Pischinger, S.: Verbrennungskraftmschinen I (2009)
52. Timothy, V.J., Corning Inc.: Review of diesel emissions and control. SAE International, 2010-01-0301 (2010)
53. Delphi (2010–11) Worldwide emission standards, passenger cars and light duty vehicles. www.delphi.com/pdf/emissions/Delphi_PC.pdf
54. Pischinger, S.: Internal combustion engines II, 3rd Edn. RWTH Aachen, Aachen (2009)
55. www.dieselnet.com

56. Bhardwaj, Om P., Abraham, M.: A comparative study of performance and emission characteristics of a CRDe SUV fueled with biodiesel and diesel fuel. SAE paper no. 2008-028-0075 (2008)
57. Bhardwaj, Om P. et. al.: Field trials of bio diesel (B100) and diesel fuelled common rail direct injection EURO III compliant sports utility vehicles in Indian Conditions. SAE paper no.2008-028-0077 (2008)
58. Miyamoto, N.: High therman efficiency and low noise diesel combustion with oxygenated agents as main fuel. SAE paper 980506, SAE International (1998)
59. Williams, A. et. al.: Effect of biodiesel blends on diesel particulate filter performance. SAE Paper No. 2006-01-3280 (2006)
60. United States Environmental Protection Agency, Department of Transportation: Renewable fuel standard program (RFS2) regulatory impact analysis (2010)
61. Schleuter, W.: Zukünftige Herausforderungen an die Elektronik im Automobil. dSpace Anwenderkonferenz. Paderborn (2010)
62. AUTOSAR: Layered Software Architecture. AUTOSAR Release 4.0 (2009)
63. Kar, K., Last, T., Haywood, C., Raine, R.: Measurement of vapor pressures and enthalpies of vaporization of gasoline and ethanol blends and their effects on mixture preparation in an SI engine. SAE Int. J. Fuels Lubr. **1**, 132–144 (2008)
64. Thewes, M.: Untersuchung und Bewertung von alternativen Kraftstoffe für den Einsatz in modernen DI-Ottomotoren (2009)
65. Sehr, A.: FEV TurboDISI SG—the combination of Fun2Drive and reduced CO2-emissions (2008)
66. Norbert Metz, B.M.W, Group Traffic and Environment: Workshop Sources and Impact of Urban Air Quality. Venice (2004)

Enzymatic Degradation of Lignocellulose for Synthesis of Biofuels and Other Value-Added Products

Helene Wulfhorst, Nora Harwardt, Heiner Giese, Gernot Jäger, Erik U. Zeithammel, Efthimia Ellinidou, Martin Falkenberg, Jochen Büchs and Antje C. Spiess

Abstract Wood is a renewable source for biofuels and chemicals. An efficient pretreatment is required to destroy the highly ordered and complex structure of wood fibres and to improve their enzymatic degradability. To understand the effectiveness of pretreatment on enzymatic degradability, high-throughput analysis of cellulose kinetics using insoluble cellulosic substrate is required. The BioLector technology enables online monitoring of scattered light intensity and fluorescence signals during the continuous shaking of cellulose samples in microtiter plates. It is used to monitor the hydrolysis of three different cellulosic substrates catalysed by a commercial cellulase preparation from *Trichoderma reesei* (Celluclast). Moreover, the reduction of crystallinity and particle size is a key determining factor for an efficient hydrolysis of cellulose particles in heterogeneous system. To increase the sugar release, crystallinity and particle size were decreased by the dissolution of spruce wood in the ionic liquid EMIM Ac resulting in high conversion and reaction rates. Additionally, the enzymatic action on lignin model substrates is characterised using an activity assay and cyclic voltammetry.

Symbols and Abbreviations

A [a.u.]	Absorbance
C_{RS} [mol/cm^3]	Concentration in Randles-Sevcik equation
D [cm^2/s]	Diffusion coefficient
E [J]	Particle energy
E° [V]	Redox potential
E_{pa} [V]	Anode potential
E_{pc} [V]	Cathode potential
F [C/mol]	Faraday constant

H. Wulfhorst · N. Harwardt · E.U. Zeithammel · E. Ellinidou · A.C. Spiess (✉)
Aachener Verfahrenstechnik, Enzymprozesstechnik, RWTH Aachen University, Aachen, Germany
e-mail: antje.spiess@avt.rwth-aachen.de

H. Giese · G. Jäger · M. Falkenberg · J. Büchs
Aachener Verfahrenstechnik, Bioverfahrenstechnik, RWTH Aachen University, Aachen, Germany

M. Klaas et al. (eds.), *Fuels From Biomass: An Interdisciplinary Approach*, Notes on Numerical Fluid Mechanics and Multidisciplinary Design 129,
DOI 10.1007/978-3-662-45425-1_14

R [J/mol/K]	General gas constant
T [K]	Temperature
V [V/s]	Scan speed
V_{max} [g/L/h]	Maximum reaction rate
a [cm^2]	Electrode surface
c [g/L]	Product concentration
f [1/s]	Frequency
h [J s]	Planck's constant
i [A]	Electricity
i_p [A]	Peak current
i_{pa} [A]	Peak anodic current
i_{pc} [A]	Peak cathodic current
k_f [1/M/s]	Catalytic constant
K_m [mol/L]	Michaelis-Menten constant
l [m]	Path length
n [–]	Number of transferred electrons
q [–]	Scattering vector
r [m]	Characteristic length
α_M [–]	Mie size parameter
δ [°]	Scattering angle
ε [L/mol/cm]	Extinction coefficient
λ [m]	Wavelength
ABTS	2,2'-azino-bis(3-ethylbenzothiazoline-6-sulphonic acid)
Btu	British thermal unit
gal	Gallon
HBT	1-Hydroxybenzotriazol
lb	Pound
MTP	Microtiterplate
TEMPO	(2,2,6,6-tetra methylpiperidin-1-yl)Oxidanyl
VA	Veratryl alcohol

1 Introduction

Biomass is becoming increasingly important as a renewable raw material for biofuel production. Among the different methods for biofuel production, bioethanol and biodiesel production and the biomass-to-liquid (BTL) processes have been established. These industrial processes are the first steps on the road to sustainable mobility. They offer an alternative to fossil fuels; however, these processes compete for agricultural lands with food production.

Additionally, the BTL process has the disadvantage of very high production costs. For these reasons, there is a growing demand for the development of new processes for the production of fuels from biomass. The lignocellulosic residues and waste materials offer a suitable source of raw materials without competing with food production. Examples of lignocellulosic residues and waste materials are crop and wood residues.

Lignocellulose is a natural composite material contained in plant cell walls. Lignocellulose consists of 40–50 % cellulose, 15–25 % hemicellulose and 15–30 % lignin, depending on the plant. Cellulose is a polysaccharide composed of many monomeric glucose units combining to form 4 nm cellulose fibrils, which aggregate into larger microfibrils of about 500 nm. These microfibrils are surrounded by the polysaccharide hemicellulose and the phenolpropanoid polymer lignin. The molecular structure of lignin is defined as an amorphous, polyphenolic material which consists of the three phenylpropanoid monomers, coniferyl alcohol, p-coumaryl alcohol and sinapyl alcohol. Lignin has a complex structure consisting of many different types of bonds, such as the most frequently occurring ether β-O4 bond and the carbon-carbon bonds β-5 and β-β. Lignin can be enzymatically degraded by oxidative enzymes. The removal of lignin is a crucial step in enzymatically degrading cellulose and hemicellulose and the conversion of lignocellulosic biomass into biofuels. In addition, lignin can be converted in other value-added products such as vanillin.

In this study, the enzymatic depolymerization of the pretreated lignocellulose is investigated. Appropriate methods for determining the activity of the enzymes involved are developed and the lignin- and cellulose degrading enzymes are characterized.

2 State of the Art: Pre-treatment and Hydrolysis of Lignocellulose

The general biochemical process of biofuels production from lignocellulose can be classified in three main steps: pre-treatment of the raw material, saccharification and fermentation or chemical transformation of the obtained sugar. Developments in the field of lignocellulose pre-treatment can be attributed to the paper manufacturing industry, since wood has been the sole source of raw material for paper production. The disintegration of wood is possible in different ways: Digestion can be mechanical, thermo-mechanical, chemical or chemo-thermo-mechanical. The mechanical pre-treatment is performed first to comminute the wood. A combination of the mechanical method with the other physical methods, such as steam explosion and hydrothermolysis, enables reduction of the particle size and almost complete disintegration of the wood fiber. However, the wood substrate still contains many impurities after this step. Moreover, the high temperatures in thermo-mechanical and hydrothermal treatment usually lead to an undesirable degradation of carbohydrates.

To overcome these disadvantages, physical and chemical pre-treatment are combined. This is usually an acid (sulfite process) or base hydrolysis (sulfate or Kraft pulping process) of the fiber, which allows the production of regenerated cellulose and lignin from lignocellulose and thereby promotes the subsequent enzymatic hydrolysis. Another effective method to improve the enzymatic degradability is the ammonium fiber expansion (AFEX). It is based on the application of high ammonium concentrations under the conditions of steam explosion (20 bar and 100–120 °C) [1]. The disadvantages of this method include expensive equipment and high costs for the use and recycling of toxic chemicals.

The cleavage of hemicellulose and lignin bonds makes it possible to fractionate the wood and concentrate the lignin. Another method for the digestion of lignocellulose is the Organosolv [2, 3] process. It combines an organic or aqueous/organic solvent mixture with an acid catalyst (HCl or H_2SO_4). Acetone, ethylene glycol, or tetrahydrofurfuryl alcohol can be used as a solvent. A further development of the Organosolv process is the Organosolv-Ethanol process. With this process, also known as the Alcell® digestion method from the Canadian pulp and paper industry, industrial hardwood is fractionated. This further development enhances the enzymatic breakdown of carbohydrates and leads to a significantly higher yield of high quality lignin [4–6]. Organocatalytic procedures represent another alternative to achieve a fractionation of wood for the enzymatic hydrolysis [7].

An alternative to the described acid/base hydrolysis and the use of organic solvents are ionic liquids. Ionic liquids are liquid salts consisting of asymmetric ions with a delocalized charge. This composition results in the characteristic low melting temperature of ionic liquids, which offers decisive advantages in the industrial application of this method. Most ionic liquids consist of organic or inorganic cations and organic anions, such as hexafluorophosphate, halogenides or organic acid anions, formate, acetate and lactate. The specific selection of the ions can be used to specifically tune properties of ionic liquids such as melting temperature, hydrophobicity, acid–base character and polarity. Due to their high polarity, some ionic liquids are able to dissolve cellulose. During the dissolution the H-bonds within the microfibrils are broken. The mechanism of the dissolution is not yet fully understood, but it is certain that an effective decrystallization of cellulose and a partial separation of lignin can be achieved. The first studies on the application of ionic liquids for the dissolution of cellulose were published and patented by Swlatovski et al. [8, 9]. Meanwhile, a number of ionic liquids have also proven to be suitable for timber disintegration [10]. They disintegrate the wood and allow for an effective separation of lignin from wood [11].

2.1 *Enzymatic Production of Sugars from Regenerated Cellulose*

Pretreatment with ionic liquids increases the availability of cellulose for subsequent enzymatic hydrolysis. A key objective of the development is to obtain enzyme activity despite low water content and solvent impurities [12]. The influence of ionic liquids on the activity of the enzymes is not yet sufficiently understood. Yet, different ionic liquids affect enzymatic activity. Systematic studies on commercially available Celluclast (cellulase from *Trichoderma reesei*) have shown that cellulase is inhibited by ionic liquids, but there is no irreversible deactivation of the enzyme. In the presence of ten percent (v/v) ionic liquid there is a 70 % decrease in Celluclast activity [13]. Only very few bacterial cellulases can tolerate concentrations of about 30 % of the various ionic liquids [14]. The ions of ionic liquids have a negative effect on enzyme activity [13, 14].

In order to increase the enzyme activity and stability even at high concentrations of ionic liquid, directed evolution, and methods of rational design are applied. By metagenomic studies and SeSaM technology [15], the decisive motifs were localized in the N-terminal part of the protein and in the cellulose binding domain of the cellulases [14]. In conclusion, however, it should be noted that currently neither the enzymes from *T. reesei* nor bacterial cellulases can be applied for cellulose hydrolysis in the presence of high concentrations of ionic liquids.

Therefore, hydrolysis of regenerated cellulose is of special interest in the development of new production processes for sugar from wood. In this case, the efficiency of the enzymatic reaction mainly depends on the composition and structural properties of the treated cellulose [16–18].

2.2 Mechanism of Cellulose Depolymerization

The hydrolysis of cellulose is based on a synergistic effect of various cellulases. For hydrolysis of cellulose to glucose the following three enzymes are involved: endoglucanases, exoglucanases and β-glucosidases. The endoglucanases split the cellulose molecule predominantly in the amorphous regions. The chain length of the cellulose molecule is reduced, and oligosaccharides and small amounts of cellobiose and glucose are formed as end-products. Exoglucanases or cellobiohydrolases (CBH) cleave off cellobiose from the ends of the cellulose chains, a dimer of β-1,4-glucosidic linked glucose units. Depending on which end of the cellulose chain is cleaved by a cellobiohydrolase, two types of cellobiohydrolases can be differentiated. Cellobiohydrolase I (CBH I) attacks the reducing end of the chain and cellobiohydrolase II (CBH II) attacks the non-reducing end. β-glucosidases complete the degradation of cellulose to glucose by hydrolyzing cellobiose to glucose [19].

The mechanism of the cellulases is based on an acid-base mechanism. Here, the hydrolytic cleavage of the β-glycosidic bonds of the cellulose chains by the cellulases occurs by two different cleavage mechanisms: Retaining or inverting. In both cases, water is directly involved in the reaction, which is required for the nucleophilic attack at the anomeric center of the molecule and the cleavage of the bond. The mixture of ionic liquid and water is also used as solvent for the substrates and products and thus influences the mass transfer and reaction rate.

Most cellulases that are able to hydrolyze undissolved cellulose contain two protein domains: a cellulose binding domain (CBM), and a catalytic domain. The clamp-like binding domain of cellulase enzymes allows adsorption to the cellulose surface. After the adsorption, the cleavage of the β-glycosidic bonds by the catalytic domain takes place based on the retaining and inverting mechanism [20].

2.3 Enzymatic Depolymerization of Lignin

The "blue" enzyme laccase (EC 1.10.3.2) is a known copper-containing oxidoreductase, which is involved in the natural degradation of lignin. Laccases have been traditionally used in the pulp and paper industry for delignification, to combine monomers to dimers and oligomers, and to oxidize a variety of alcohols to their corresponding aldehydes or ketones. Laccases contain four copper ions in their active site, where the reduction of molecular oxygen takes place. Table 1 shows potential products of lignin and their current market-value [21].

3 Establishment of the BioLector Technique to Monitor Cellulose Hydrolysis

Different pretreatment methods affect the properties of the cellulose particles and thus the enzymatic activity. When searching for suitable enzymes for optimal hydrolysis, it is important to characterize the enzyme kinetics. Often, investigations are made by using modified soluble substrates [22], or by analyzing the samples for *the* glucose or reducing sugars content [23]. These methods do not take into account relevant effects for the future process such as enzyme adsorption and limitation through the available substrate surface.

The adsorption of the enzyme on the available surface area of the cellulose particles determines the hydrolysis rate. Therefore, the reaction kinetic should be characterized based on the solid cellulose properties, such as particle size. The enzymatic reaction with native cellulose substrate using offline analysis also requires high cost and effort. To select potentially suitable enzymes for industrial applications, new online techniques for high-throughput screening for the reaction systems with high solid content are needed. For this reason, an online small scale monitoring system was used to develop a new screening method for the comparative determination of enzyme activity, which characterizes the reaction kinetics in turbid suspension.

Table 1 Potential products derived from the conversion of lignin and their current market values [21]

Products derived from lignin	Market value
Process heat and power	$6/106 Btu
Syngas	Variable
Syngas products	
Methanol/dimethylether (minuscule)	$0.80/gal
Ethanol/various alcohols	$1–$3.5/gal
Fischer-Tropsch liquids	$1.5–$2/gal
Various liquid fuels	$1.3–$2/gal
Hydrocarbons	
BTX and higher alkylates	$2/gal
Cyclohexane	$2.20/gal
Styrene	$0.70/lb
Phenols	
Phenol	$0.55–$0.65/lb
Phenol derivatives	$0.70–$2.00/lb
Cresols, resorcinol	>$1.5/lb
Oxidized products	
Vanillin	$5.90/lb
DMSO	<$1/lb
Aromatic acids	$0.40–$0.50/lb
Aliphatic acids	$0.45–$0.65/lb
Quinones	>$1/lb
Cyclohexanol/al	>$0.75/lb
Macromolecules	
Carbon fiber	$3–$5/lb
Polymer compounds	$1–$2/lb
Fillers, diluent polymer	<$1/lb

Provided by the Pacific Northwest National Laboratory for the U. S. Department of Energy

3.1 Nephelometric Methods and Particle Size Measurement

The enzyme activity in turbid suspensions can be detected by measuring substrate consumption, which leads to a change in substrate color or clarification of turbid suspensions. In this case, the initial reaction rate of the decrease in turbidity can be detected spectrophotometrically and correlated to the enzyme concentration. This has been verified by analysis of the enzymatic reaction of the resulting soluble products (amino acids or reducing sugars) [24, 25]. Furthermore, some authors used the measurement of particle size distributions in order to track the hydrolysis catalyzed by enzymes [26, 27]. The change in particle morphology during the enzymatic reaction can be detected by microscopy. However, it is difficult to quantify the morphology, taking into account reproducible sampling and statistical analysis of the particles [28]. Nummi et al. [29] detected the cellulase activity as a decrease of the absorbing cellulose particles. Alternatively, the substrate hydrolysis, i.e. degradation of the particles, can be measured using a nephelometer. In the nephelometric methods, the scattered light of the particles surface is measured at an angle of 90° at 600–1,000 nm.

The following physical principles explain the mechanisms of the scattered light measurement in the degradation of cellulose particles in turbid suspension. The scattering of light is defined as non-directional change in direction of light on a surface [30]. The incoming light is scattered by the microscopic structure of the sample. This variation can be detected as a function of the angle φ between the irradiating beam and the scattered light. In general, the structures which can be dissolved by the scattered light are of the same order as the inverse of the scattering vector q (Eq. 1):

$$q = \frac{4\pi \sin\left(\frac{\varphi}{2}\right)}{\lambda} \tag{1}$$

The smaller the wavelength λ (m) or the larger the energy of the particles, the smaller is the structure, which can be examined with the light scattering experiment. Therefore, laser light scattering can be used for the investigation of colloidal systems. The energy E of a particle is:

$$E = hf \tag{2}$$

Here, h (J s) is Planck's constant and f (1/s) is the frequency resulting from the ratio of the speed of light (m/s) and the wavelength. The scattered light intensity is measured as a function of the scattering vector q. It is proportional to the total scattered light diameter which depends on the particle shape and size. In addition, the interaction between the particles affects the measurement of scattered light.

The turbid cellulose suspensions are coarsely dispersed systems. The dispersed phase consists of solid particles. The incoming light is scattered at the surface of all particles suspended in the medium. The scattering amplitude is proportional to the particle volume, and therefore, the large particles scatter more light than small particles [31]. The light scattering due to different particle sizes is related to the particle radius and the shape of the particle, the wavelength and the refractive index ratio between the solvent and the dissolved particles.

If the particles are much smaller than the wavelength, it is assumed that each small volume element, including the particle, is illuminated with light of the same wavelength. The light scattering of such small particles is described by the Rayleigh theory. It represents a limited case to the theory of scattering developed by Mie [32]. For larger particles, the difference in refractive index between the particles and the solvent are large. If the particles are much larger than the wavelength of the irradiating light refraction and reflection phenomena occur. According to the Mie theory, the scattered light is split into two orthogonally polarized components which are a function of the refractive index m, the Mie size parameter α and the scattering angle φ. Both scattered light components are considered in the Mie scattering function $I(\varphi)$. Mie scattering is only weakly dependent on the wavelength and is independent of it (the incident light appears white) when the dimensions of the particle gets larger than λ [32]. The size parameter α_M (–) (Eq. 3) defined by Mie represents the ratio of the characteristic length r (m) of the scattering particle to the stimulating wavelength.

$$\alpha_M = \frac{2\pi r}{\lambda} \tag{3}$$

For very large values of α_M, the forward scattering is enhanced and a lateral intensity maximum occurs. Since the dimensions of the cellulose particles are much larger than the wavelength of the irradiating light, this case applies to the measurement of scattered light in cellulosic suspensions. In conclusion, the change of the particle size, shape, and state of aggregation can be monitored by light scattering or absorption photometry. These techniques are sensitive and non-invasive.

3.2 Scattered Light Measurement Using BioLector

The BioLector technique allows for the measurement of the light scattering and fluorescence in microtiter plates under shaken conditions [33]. The samples to be analyzed are pipetted into a microtiter plate, which is sealed with a self-adhesive film to prevent evaporation. The microtiter plate is then mounted onto an orbital shaker and the wells can be observed from the bottom by a fluorescence spectrometer via optical fibers. This is done with a defined wavelength and a fixed inclination angle of the fiber. The optical fiber can be moved to a well position measured by an x-y-positioning device. A computer processes the measured signals, and controls the entire process. In Fig. 1 the schematics of a BioLector are shown.

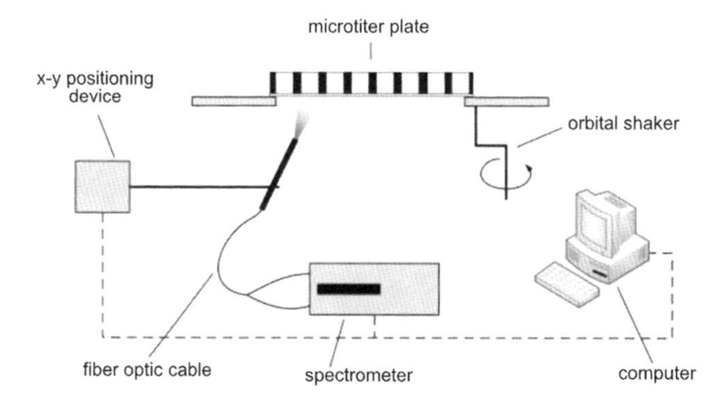

Fig. 1 Schematic representation of the BioLector [36]

In contrast to other methods the shaking of the microtiter plate does not have to be stopped during the measurement. Taking samples during the reaction is not necessary and sources of error are avoided. During cellulose hydrolysis, an interruption of shaking could lead to sedimentation of particles and thus limit the mass transfer in the wells. Aside from the simple implementation, a high degree of homogenization by continuous shaking and a larger linear range compared to a conventional photometer are some of the benefits of light-scattering measurement with the BioLector. This system has been successfully used for screening in clear media. In turbid media, screening has not been adequately studied using this system. Therefore, the objective of the following experiments was to determine the extent to which the BioLector technique is suitable for screening of cellulase activity in turbid suspensions.

3.2.1 Online Measurement of Cellulose Hydrolysis by BioLector

The hydrolysis of cellulose by cellulases can be monitored online with the help of BioLector technique (Fig. 2). As substrate, a suspension of three different types of cellulose, α-cellulose, Sigma Cell 101 or Avicel in 0.1 M sodium acetate buffer (pH 4.8) with a solids content of 1–10 g/l was used. Due to the light scattering on the surface of the suspended particles at the beginning of the reaction a high signal in scattered light was measured in BioLector. By adding the enzyme to the substrate and subsequent measurements of the scattered light, intensity curves can be recorded.

The cellulose hydrolysis was performed in a 96-well microtiter plate (MTP) in BioLector® at the following standard conditions: 96-well MTP, filling volume 200 μl, shaking frequency 900 rpm, shaking diameter d_0 = 3 mm, substrate α-cellulose 2.5 g/L (Sigma-Aldrich, Deisenhofen, Germany) in 0.1 M sodium acetate buffer, pH 4.8, reaction at 37 °C for 50 h, with an enzyme concentration of 0.25 g/L (Celluclast, Novozymes, Denmark).

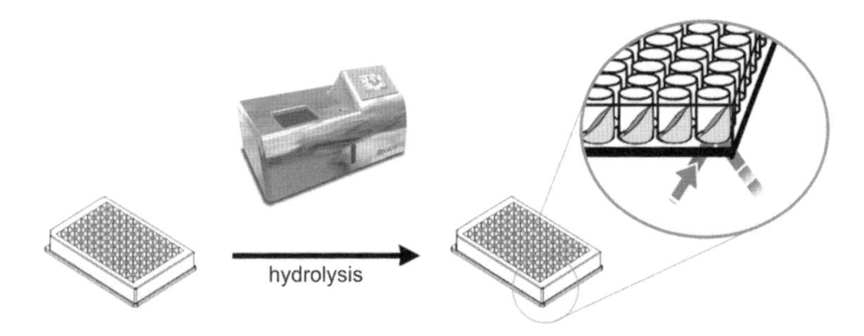

Fig. 2 Principle of online monitoring of cellulose hydrolysis in BioLector® (m2p-labs GmbH, Aachen, Germany) [37]

3.2.2 Scattered Light Signal and the Concentration of Cellulose Particles

Enzymatic cellulose hydrolysis is based on a synergistic effect of different cellulases, which split the cellulose polymers from the end (exoglucanases) or in the middle of the polymer chain (endoglucanases). These enzymes specifically hydrolyze the glycosidic bonds and break down the cellulose polymers to oligomers and monomers. Figure 3 shows that cellulose particles in a cellulose suspension

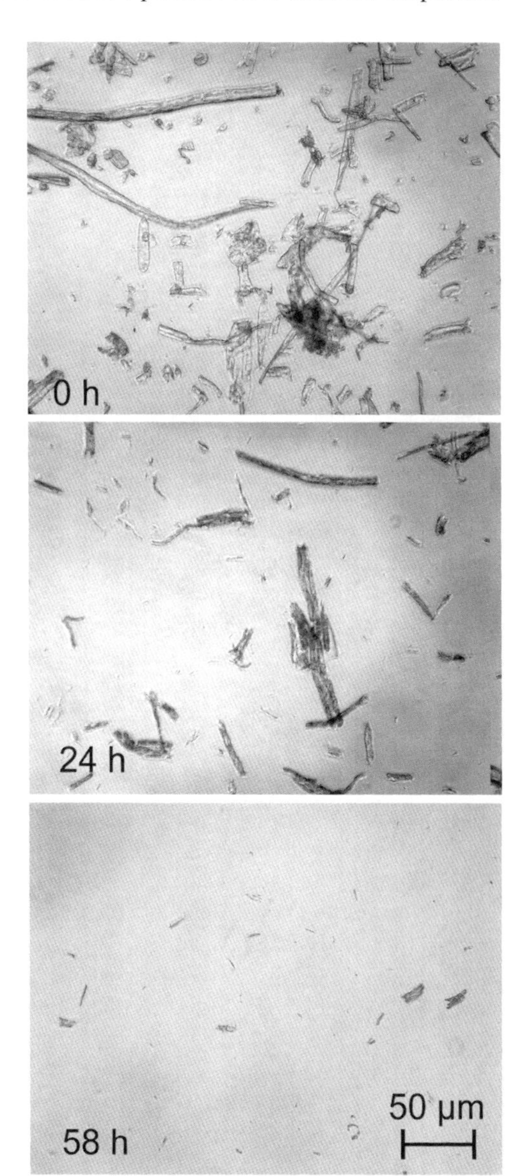

Fig. 3 Hydrolysis of α-cellulose. Decrease in particle size as a result of enzymatic activity. Optical micrograph of the sample at time 0, 24 and 58 h Nikon Eclipse E600 microscope (40x) [37]

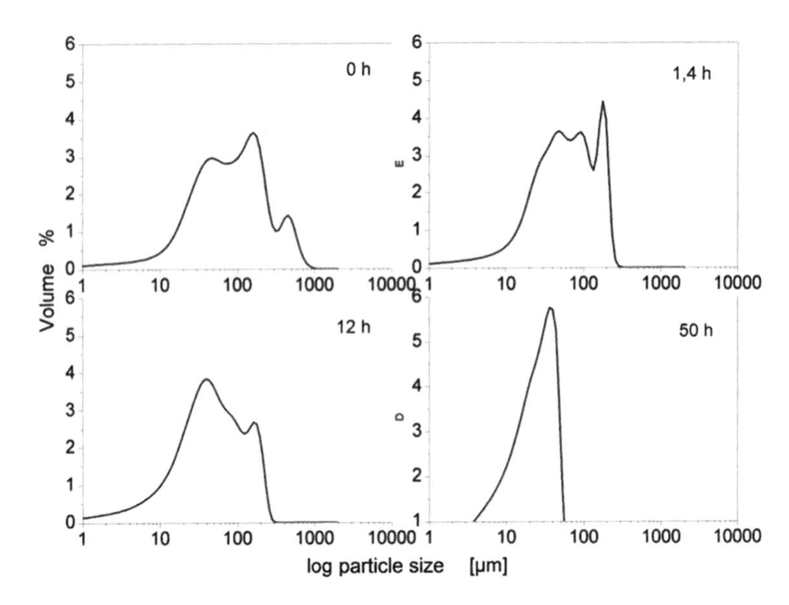

Fig. 4 Hydrolysis of α-cellulose. Particle size distribution of BioLector samples after 0, 12 and 50 h, measured with LS 13 320 (Beckman-Coulter). BioLector conditions: 2.5 g/l α-cellulose, 0.1 M acetate buffer, pH 4.8, enzyme concentration 0.25 g/l (Celluclast, Novozymes, Denmark), filling volume 200 μl, 37 °C, 50 h, 900 rpm, d_0 = 3 mm, 96-well MTP

differ significantly in shape and size. The micrographs of the samples at time points 0, 24 and 58 h show that the addition of enzyme to the reaction mixture caused a decrease in particle size and number because of the enzymatic activity.

The analysis of the particle size during the hydrolysis reaction was performed offline using a conventional light scattering detector LS 13 320 (Beckman Coulter) and with the online measurement in BioLector. In Fig. 4, the corresponding particle size distributions are presented as a function of reaction time. In the first hour, the number of larger particles decreased sharply. At the beginning of the reaction, the cellulose sample showed a broad (1–2,000 μm) particle size distribution. After 12 h, the mean diameter of the suspended particles of cellulose shifted to lower values. After approximately 50 h the larger particles were hydrolyzed and a narrow particle size distribution with an average particle diameter of about 40 μm was present.

In Fig. 5, the scattered light intensity and the mean particle size are plotted against the reaction time. A continuous decrease in the mean particle size during the hydrolysis reaction can be detected. This decrease correlates with the shape of the scattered light intensity. Analysis of the particle size distribution shows the influence of the particle size of the scattered light signal and confirms the assumption that the curve shape of scattered light is based on the change of the particle size during hydrolysis.

The scattered light intensity is proportional to the total diameter of scattered light. The diameter of scattered light depends on the particle shape and size (see

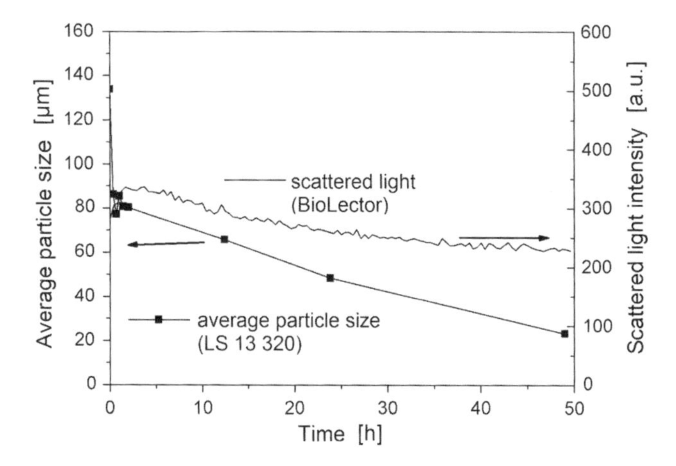

Fig. 5 Hydrolysis of α-cellulose in BioLector. LS 13 320 (Beckman-Coulter). BioLector loading conditions: 2.5 g/l of α-cellulose, 0.1 M acetate buffer, pH 4.8, enzyme concentration 0.25 g/l (Celluclast, Novozymes, Denmark), filling volume 200 µl, 37 °C, 50 h, 900 rpm, d_0 = 3 mm, 96-well MTP

Sect. 3.1). The scattered light intensity is measured as a function of the scattering vector. To define the measuring range of the BioLector and to investigate the influence of particle concentration on the BioLector signal, the light scattering was measured at various concentrations of cellulose.

Figure 6 shows the measured scattered light intensity as a function of cellulose concentration. The resulting calibration curve is linear up to 10 g/L. The cellulosic suspension in this concentration range contains a high amount of solid particles. Due to their high density, the solid particles in the unshaken condition settle relatively quickly. Therefore, a turbidometric measurement is not feasible in a conventional photometer. Unlike traditional techniques for measuring enzyme activity,

Fig. 6 Dependence of the scattered light intensity from the α-cellulose concentration [37]

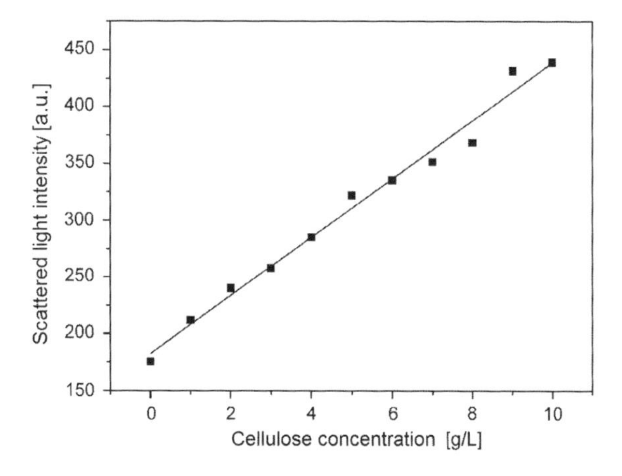

the BioLector technology provides an online measurement of the scattered light intensity signals during continuous shaking of the samples in microtiter plates. During the assay, this prevents the sedimentation of particles in the microtiter plate and ensures optimal mass transport.

3.2.3 Investigation of the Comparability of BioLector with the Conventional Reaction System Thermomixer

To demonstrate the comparability of BioLector technique with the conventional reaction system for the analysis of cellulose hydrolysis, the cellulose hydrolysis was performed in both a Thermomixer and BioLector under identical conditions. To demonstrate the comparability of the concentration of dissolved sugar by DNS assay, the glucose and cellobiose concentration was also measured using HPLC. The DNS assay measures reducing sugars of a sample by means of a color reaction. In this reaction, 3.5-dinitrosalicylic acid is reduced from the free aldehyde group of the reducing end of a sugar molecule to the corresponding 3-amino-5-nitrosalicylic acid under alkaline conditions (Fig. 7).

Due to the 3-amino-5-nitrosalicylic acid, a color change from orange-yellow to red occurs in the sample solution which can be quantified using a wavelength of 540 nm in a photometer. In a certain concentration range, a linear relationship exists (from 0.5 to 5 g/L of glucose) between the measured absorbance and the number of reducing ends [23].

The results of the DNS assays are shown in Fig. 8. The diagram represents the sugar concentration of the dissolved sugar over time in BioLector and in the Eppendorf tubes in a thermomixer. It is clearly visible that the sugar concentrations of the compared hydrolysis reactions are equal in the first 5 h. Afterwards, it can be seen that the two curves vary only slightly at the end of the reaction time. The BioLector and the Eppendorf tubes in a thermomixer offer comparable reaction conditions. Glucose and cellobiose concentration measurements by HPLC can also be used. Figure 9 shows the curves plotted over the reaction time. Here, *both glucose and cellobiose*, the concentration profile of the reaction in BioLector is very close to the concentration profile of the reaction in the Eppendorf tubes. In conclusion, the curves are comparable in BioLector and thermomixer.

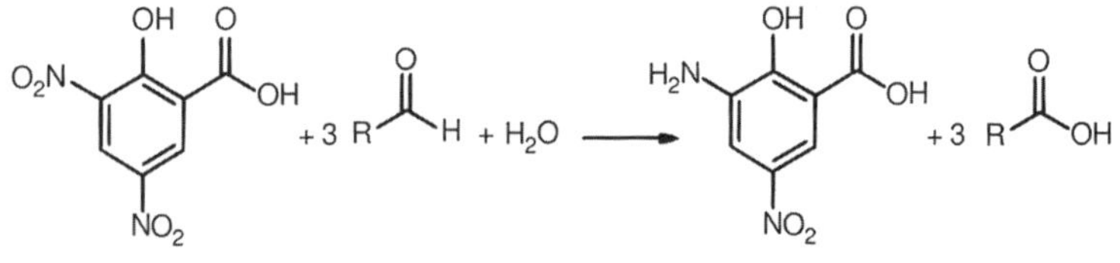

Fig. 7 Reaction of 3,5-dinitrosalicylic acid to aldehyde

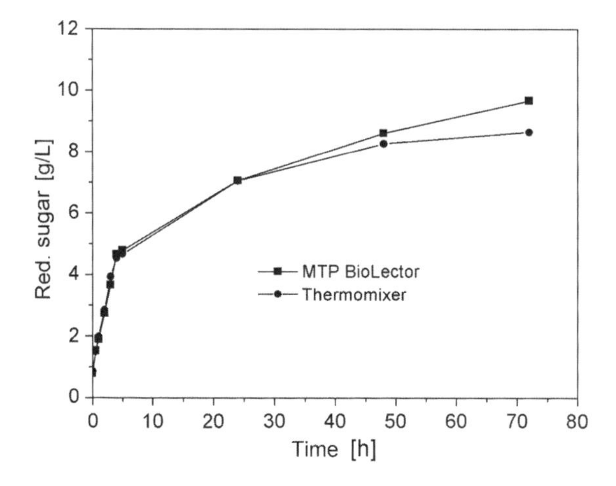

Fig. 8 DNS measurement of the hydrolysis reaction in the BioLector (MTP), Themomixer (Eppendorf tubes) Substrate: 10 g/L Sigmacell 101, enzyme: 0.25 g/L Celluclast desalted, pH 4.8, NaAC buffer, 37 °C, 900 rpm, d_0 = 3 mm

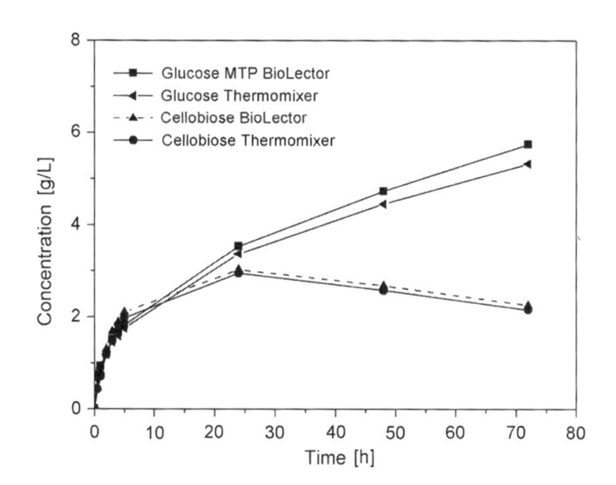

Fig. 9 HPLC measurement of the hydrolysis reaction in the reaction systems BioLector and Themomixer (Eppendorf tubes) Substrate: 10 g/L Sigmacell 101, enzyme: 0.25 g/L Celluclast desalted, pH 4.8, NaAC buffer, 37 °C, 900 rpm, instrument: HPLC

3.2.4 High-Throughput Analysis of Cellulose Hydrolysis

After comparing the BioLector with the conventional reaction system, the Bio-Lector technique is used to analyze cellulase kinetics. To characterize the enzyme reaction, the decrease in light scattering during cellulose hydrolysis in the microtiter plate was monitored. The initial reaction rate (g/L/h) can be determined from the

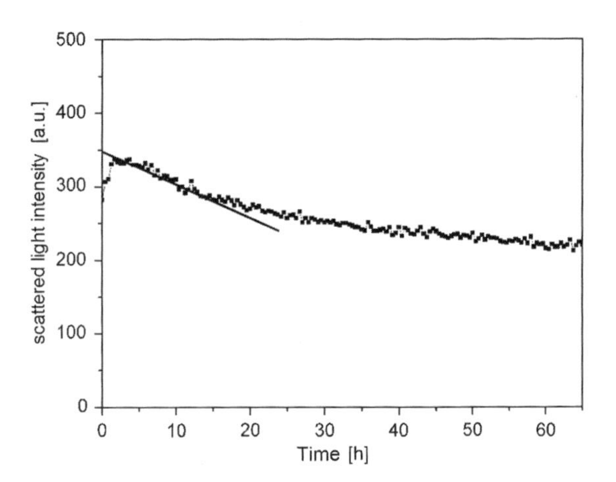

Fig. 10 Determination of the initial reaction rate of α-cellulose hydrolysis [37]

decrease in the scattered light intensity in the linear range of the curve. This relationship is illustrated for the α-cellulose hydrolysis in Fig. 10.

The influence of the pH on the enzymatic activity was assayed using the BioLector. The online monitoring of cellulose hydrolysis took place under identical conditions, and only the pH of the aqueous buffer phase was varied. The enzyme activity can be determined from the linear region of the curve. The detection of the change in enzyme activity in dependence of the pH allows a determination of the pH optimum. In Fig. 11a three hydrolysis curves are exemplified for the pH values 2.8, 5.08 and 6.35. At pH values 2.8 and 6.35 the scattered light signal falls only slightly in comparison to the approach using pH 5.08. The stronger decrease in the scattered light signal at pH 5.08 indicates a better cellulose hydrolysis and the optimum pH range for cellulase activity.

When searching for an appropriate enzyme at the process development stage choosing the optimal pH is very important. It is desirable that the hydrolysis reaction proceeds under optimal conditions. Figure 11b shows the dependence on pH of the cellulase. Based on this figure it is clear that the enzyme has an optimal activity at a pH of 4.7. This information corresponds to the value specified by the manufacturer of the cellulase preparation. Consequently, with the BioLector the pH optimum of hydrolytic enzymes was determined in a single experiment. This allows for a quantitative evaluation of the suitability of enzymes for their use in industrial processes.

To determine the kinetic parameters of the enzymes used, the reaction rates as a function of substrate concentration were measured with a defined pH and enzyme concentration. These reaction rates determined from the light scattering curves are plotted against the substrate concentration. Figure 12 shows this relationship for the three celluloses: Sigmacell, α-cellulose and Avicel PH101. All celluloses show a significant increase in enzymatic activity between 0 and 5 g/L cellulose. From 10 g/L on all three substrates drift into saturation, which shows that at higher concentrations, no increase in enzyme activity can be expected by further addition of substrate.

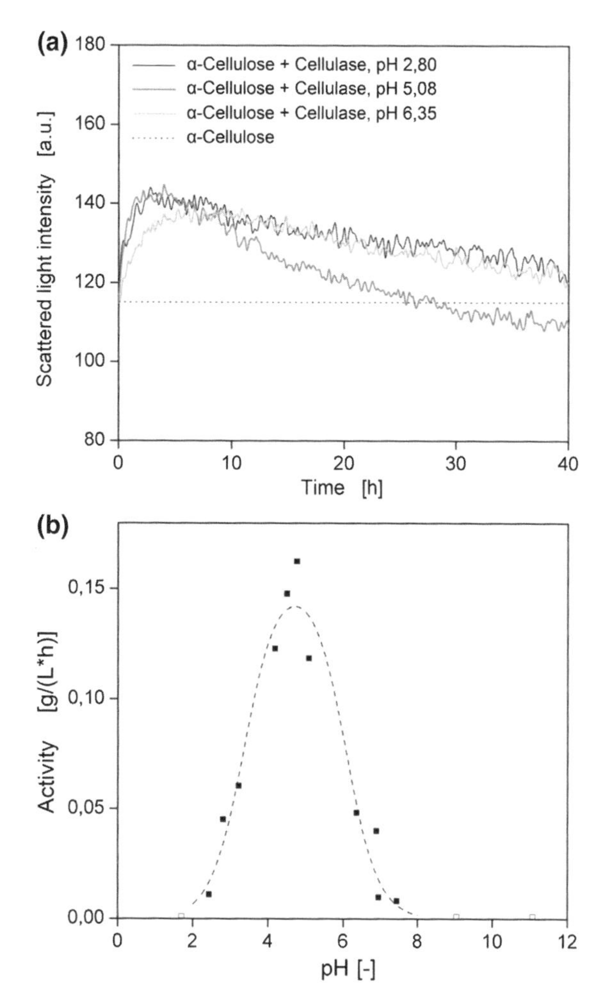

Fig. 11 Determination of the pH optimum of cellulose hydrolysis. BioLector conditions: 2.5 g/L α-cellulose, 0.1 M acetate buffer, enzyme concentration 0.25 g/L (Celluclast, Novozymes, Denmark), filling volume 200 μL, 37 °C, 50 h, 900 rpm, d_0 = 3 mm, 96-well MTP. **a** Scattered light intensity versus time at varying pH. **b** Enzyme activity as a function of pH [36]

The kinetic parameters K_M and V_{max} are determined for the three cellulosic substrates and are used to characterize the hydrolysis using the Michaelis-Menten model. In Table 2 the crystallinity index CrI and the average particle size d_p are listed in addition to the kinetic parameters. The results show that the amorphous substrate with the smallest particle size and the largest available surface area is best degraded. In contrast, the largest particles were measured with Avicel PH101, resulting in the lowest reaction rate.

The high throughput hydrolysis confirmed the influence of the crystallinity of the substrate and demonstrates that the crystallinity and the particle size of the substrate must be reduced by pre-treatment for efficient hydrolysis. For this purpose, ionic liquids provide a suitable means.

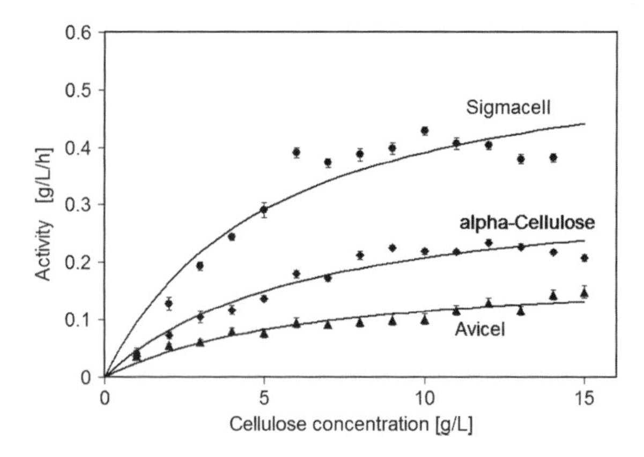

Fig. 12 Determination of the kinetic parameters of the reaction rate of cellulose hydrolysis as a function of concentration. BioLector conditions: varying concentration of Sigma cell, alpha-cellulose, or Avicel PH101, 0.1 M acetate buffer, pH 4.8, enzyme concentration 0.25 g/L (Celluclast, Novozymes, Denmark), filling volume 200 μL, 37 °C, 50 h, 900 rpm, d_0 = 3 mm, 96-well MTP [36]

Table 2 Physical properties of the cellulose substrates and Michaelis-Menten parameters determined from the light scattering curves in BioLector[a]

Cellulose	Crl (%) [38]	d_p (μm) [38]	K_m (g/L)	V_{max} (g/L/h)
α-cellulose	64	68.77	6.30 ± 1.41	0.338 ± 0.032
Avicel PH101	82	43.82	6.53 ± 1.70	0.190 ± 0.021
Sigmacell 101	Amorph	15.86	5.22 ± 1.59	0.596 ± 0.071

Celluclast 1.5 L with various cellulosic substrates, reaction conditions: BioLector, 0.1 M sodium acetate buffer (pH 4.8), 0.25 g/L Celluclast (Novozymes), T = 37 °C, VL = 200 μl , n = 900 rpm, d_0 = 3 mm, 96-well MTP

[a] Error is given as SD; *Crl* crystallinity of cellulose, d_P geometric mean particle size, K_M predominant Michaelis constant, V_{max} maximum cellulase activity

4 Improved Degradation of Cellulose by Pretreatment with Ionic Liquids

Lignocellulose has a highly crystalline structure. An effective decrystallization of lignocellulose as mentioned above can be achieved by the treatment with ionic liquids. Therefore, lignocellulose is partially dissolved in the ionic liquid EMIM Ac. Two fractions are obtained under the selected reaction conditions: a dissolved cellulose-rich fraction and an undissolved cellulose-poor wood residue. Both wood fractions are hydrolyzed using the cellulase preparation Celluclast®. The hydrolysis kinetics are analyzed by sugar analysis, DNS assay and HPLC. The first results confirm a positive effect of pretreatment of lignocellulose with ionic liquids on the

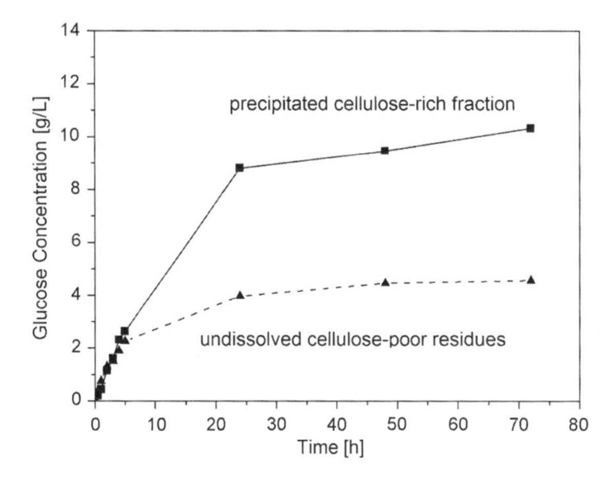

Fig. 13 Enzymatic hydrolysis of wood after treatment with ionic liquids. Reaction conditions: 10 g/L spruce pretreated with EMIM Ac, 2.5 g/L Celluclast (Novozyme, Novo Nordisk, Denmark) desalted, total volume 1.5 ml, 1,000 rpm, 45 °C

hydrolysis kinetics, since both wood fractions can be effectively hydrolyzed to sugar. Figure 13 illustrates this with reference to the rising concentration of the monosaccharide glucose.

5 Lignin Degradation

After the separation of lignin from lignocellulose, lignin is a potential source for many value-added products (Table 1). For further processing, lignin has to be depolymerized. In nature, laccase enzymes facilitate the degradation of lignin. The aim is therefore to develop a better understanding of lignin degradation kinetics.

Due to steric hindrances between the laccase and the lignin polymer, the presence of a mediator is required for an oxidative degradation. Figure 14 illustrates the laccase—mediator system. Laccase (e.g. *Trametes versicolor*) accepts electrons from molecular oxygen, reduces it to water and transfers the ions to a mediator [34]. Mediators such as ABTS (2,2′-azino-bis(3-ethylbenzothiazoline-6-sulphonic acid)), 5-hydroxyimino-barbituric acid (violuric acid) can be used. Model substrates are commonly used to investigate the degradation of specific linkages of lignin's complex structure. They provide an insight into the complete degradation process.

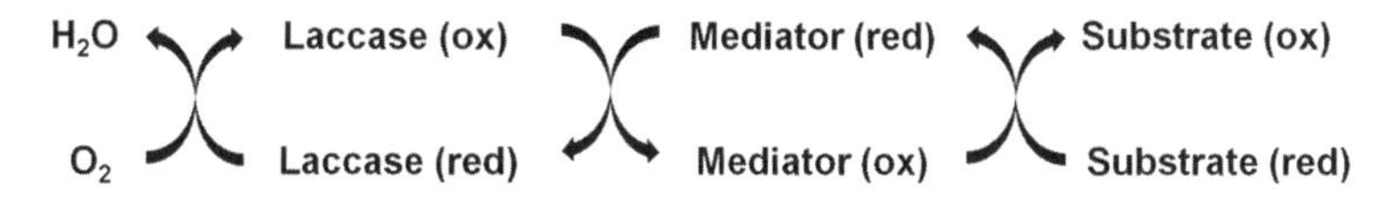

Fig. 14 Catalytic cycle of laccase-mediator system

5.1 Cyclic Voltammetry for the Analysis of Lignin Degradation

Cyclic voltammetry (BAS 100 Cyclic voltammetry device, Fig. 15) is a versatile method to investigate the electrochemical reaction between mediator and substrate. More detailed investigations can also reveal the electrochemical reaction rate between the mediator and substrate. It can also be used to determine the redox potential of laccases. The redox potential is a measure of the ability to transfer electrons from the laccase to the lignin (model) substrate. The ABTS mediator is oxidized by one of the electrodes of the device, which simulates the complete laccase–mediator–system consisting of the laccase of *Trametes versicolor* and the mediator ABTS, and then reacts with the substrate.

The device consists of three electrodes: working, counter and reference electrode. The potential is measured between the reference and the working electrode, and the current is measured between the working and the counter electrode.

The working electrode potential is increased linearly over time by the device. Once a set potential is reached, the potential of the working electrode is inverted and linearly decreases over time (Fig. 16). Scan rates are the change in potential over change in time, typically expressed in mV/s. As the cathodic potential increases with time, current typically also increases and can be measured by the device. This is shown in Fig. 17. The concentration of the analyte is depleted close to the electrode surface causing the current to decrease slightly with increasing potential. The diffusion of the analyte from the bulk solution may not be as fast. Higher scan rates tend to result in larger currents but increased resistance.

Once the peak currents were experimentally determined, the formal potential of the solution was calculated. The formal potential of the reaction is the average of

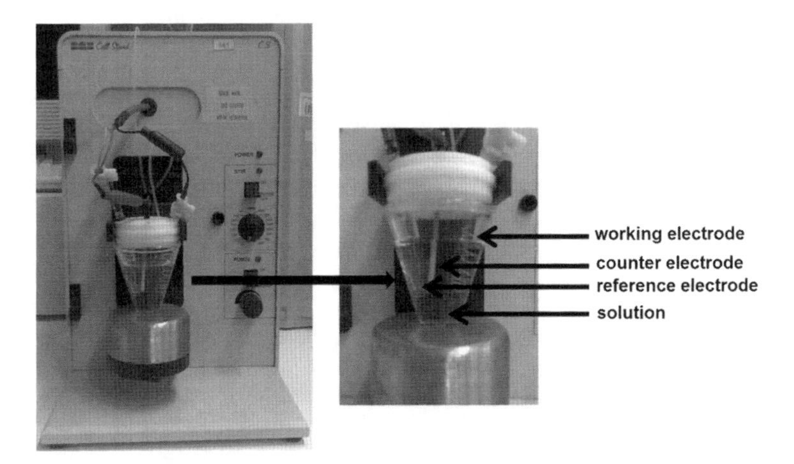

Fig. 15 *Left* BAS C-3 cell 100 cyclic voltammetry device. *Right* Counter, working and reference electrodes

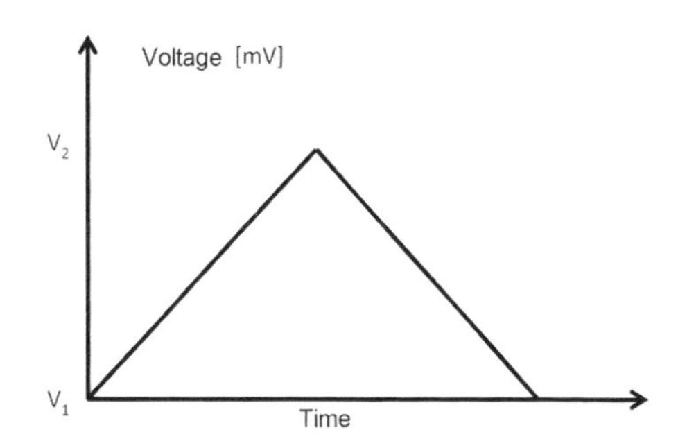

Fig. 16 Increasing and decreasing electric potential over time in the cyclic voltammetry

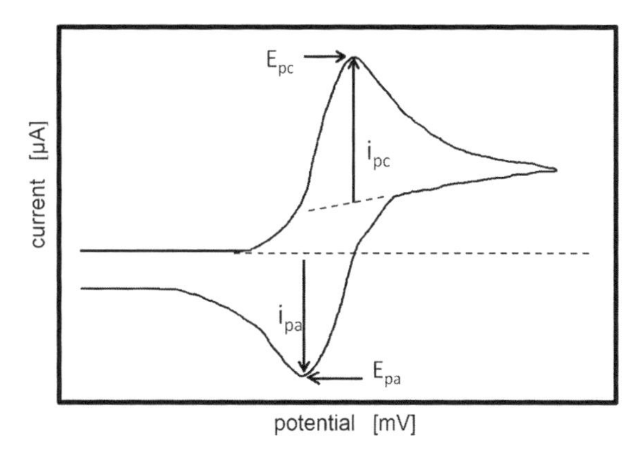

Fig. 17 Current over potential in the cyclic voltammetry

the anodic and cathodic peak potentials shown in Eq. 4, the formal potential (E° (mV)) is the average of the anodic and cathodic peak potentials.

The redox potential describes the tendency of the analyzed component to transfer electrons. This is particularly useful for identifying good mediators. The reversibility of the reaction, the ratio of the extreme values of voltage of cathode and anode may be determined with Eq. 5. Here, i_{pc} and i_{pa} correspond to the peak current at the cathode and anode, respectively. If the quotient is equal to 1, it is an ideal reversible reaction.

$$E^{\circ} = \frac{E_{pc} + E_{pa}}{2} \tag{4}$$

$$\frac{i_{pc}}{i_{pa}} = 1 \tag{5}$$

To check the suitability of ABTS as a mediator and to validate the experimental arrangement, cyclic voltammetry experiments were performed. For this purpose, buffer (0.1 M Na_2HPO_4 and 0.05 M citric acid), laccase (2 U), ABTS (1 mM) and veratryl alcohol (1 mM, as lignin model compound) have been measured individually. Figure 18 shows that the ABTS solution has a high redox potential in

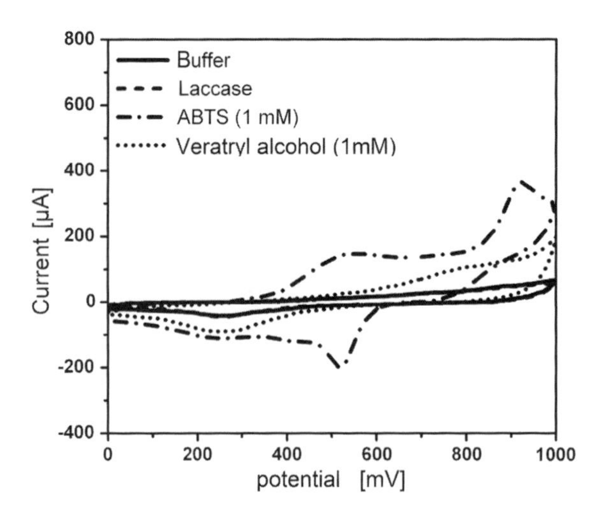

Fig. 18 Cyclovoltammograms of various solutions: buffer (pH = 4.5, 0.05 M citric acid, 0.1 M Na_2HPO_4), laccase (2 U), ABTS solution (1 mM), veratryl alcohol (1 mM). Scan rate: 20 mV/s Potential: 0–1,000 mV

comparison to the other solutions and hence has a high affinity for electron transfer. There are no peak currents for the buffer and the laccase compared to ABTS. The peak currents were determined for ABTS and can clearly be seen compared to veratryl alcohol.

The peak potentials of other mediators such as (2,2,6,6-tetramethylpiperidin-1-yl) oxyl, (2,2,6,6-tetramethylpiperidin-1-yl) oxidanyl (TEMPO), apocynin, 1-hydroxy benzotriazol (HBT) and 5-hydroxyiminobarbituric acid (violuric acid) were also determined. In addition to the peak potentials, the reversibility of the mediators in solution was measured as shown in Table 3. From the experimental data, it would also be possible to determine the diffusion coefficient using the Randles-Sevcik equation (Eq. 6). Here, i_p is the measured peak current, n is the number of the transferred electrons, a is the electrode surface area (cm^2), F is the Faraday constant (C/mol), D is the diffusion coefficient (cm^2/s), C_{RS} is the concentration of the mediator in solution (mol/cm^3), V is the scanning velocity (V/s), T is temperature (K) and R is the universal gas constant (J/mol/K).

Table 3 Redox potentials E° and reversibility of various mediators and their concentrations (c) and scan rates (mV)

Mediator	E° (mV)	Reversibility
ABTS	492	0.94
TEMPO	504	0.92
Acetovanillone	520	1.89
HBT	616	6.54
Violuric acid	702	1.12

Equation 6 shows that the concentration of the component to be analyzed is proportional to the height of the peak current. It also shows that increasing scan rates correspond to increasing current.

$$i_p = 0.4463 \cdot n \cdot F \cdot a \cdot C_{RS} \left(\frac{n \cdot F \cdot V \cdot D}{R \cdot T} \right)^{0.5} \quad (6)$$

The peak current was measured at various scan rates and veratryl alcohol substrate concentrations (of ABTS and veratryl alcohol). By measuring the current at different scan rate and substrate concentrations, the effect of the scan rate and substrate concentration on the increased current can be taken into account. Therefore, the extra increase in peak current can be attributed to the reaction between the ABTS mediator and the lignin model compound veratryl alcohol. A scheme for the reaction between ABTS and veratryl alcohol as well as the flow of current to the electrode is shown in Fig. 19.

In order to investigate the influence of the mediator concentration on the reaction, cyclic voltammetry experiments were performed with varying concentrations of ABTS. In buffer (0.1 M Na_2HPO_4 and 0.05 M citric acid) experiments with 1 mM ABTS and with 1 mM ABTS and combined with 1 mM veratryl alcohol (as lignin model component) were performed.

Figure 20 shows the corresponding results. The increased peak current at about 900 mV indicated the presence of $ABTS^+$ (oxidized ABTS mediator). Larger veratryl alcohol-to-ABTS concentration ratios correspond to higher peak currents. These higher peak currents correspond to higher reaction rates, since higher concentrations of $ABTS^+$ are present.

Cathodic peaks could not be observed (Fig. 20) due to comproportionation of ABTS. Since $ABTS^{2+}$ reacts with veratryl alcohol, it is reduced to $ABTS^+$. The solutions with veratryl alcohol have larger peak currents compared to pure ABTS solutions. The higher peak currents indicate higher concentrations of reduced ABTS which reacts with veratryl alcohol.

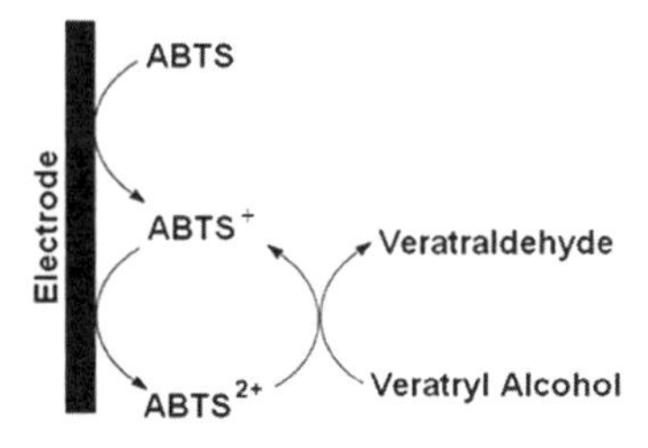

Fig. 19 Schematic representation of the reaction at the working electrode

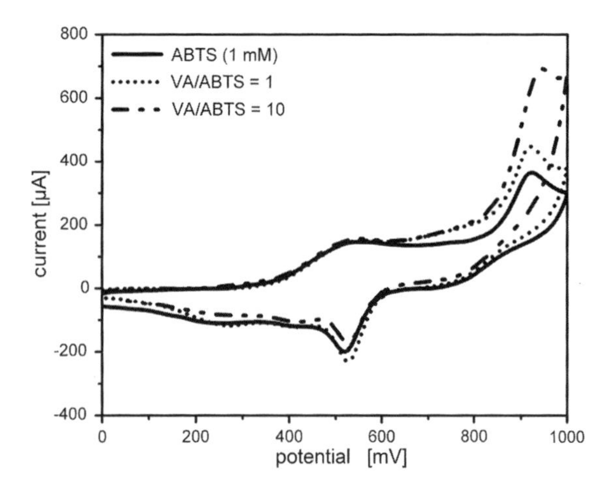

Fig. 20 (1 mM), ABTS (1 mM) with veratryl alcohol (1 mM)—(VA/ABTS = 1), and ABTS (1 mM) with veratryl alcohol (10 mM)—(VA/ABTS = 10). Scan speed: 20 mV/s Potential: 0–1,000 mV

5.2 Determination of Reaction Kinetics via Simulation

The classical method for determining the reaction rate between the mediator and substrate is based on the work of Nicholson and Shain (1964) [35]. Based on mathematical equations, they correlated experimental data at varied scan rates to reaction rates for different reaction scenarios. On the basis of the presented experimental data (Fig. 20) and compared with Nicholson [35], the catalytic constant k_f for the reaction between ABTS mediator and the lignin model component veratryl alcohol was 70 $M^{-1}s^{-1}$.

In this work, a system of equations based on physical relationships (diffusion and reaction) was established to determine the reaction rates (Eqs. 7 and 8). Experimental parameters were used as input parameters for the simulation, based on the equations for diffusion of the mediator and the substrate. In this case, the variables for the simulation and the fixed parameters include the mediator and substrate concentration C, the mediator and substrate diffusion coefficient D, the number of electrons $n = 2$, the electrode surface a, the Faraday constant $F = 96{,}485$ C/mol, the voltage potential range i, the concentration gradient between the electrode and the bulk liquid of the component to be analyzed dc/dx, and the scanning speed V (boundary condition). The peak current and its change over time were proportional to the concentration of the component analyzed (Eq. 9). From the simulated data redox potentials, reversibility and diffusion coefficients were determined for the component to be analyzed and compared to experimental data.

$$\frac{dC_{ABTS}}{dt} = D_{ABTS} \cdot \frac{d^2 C_{ABTS}}{dx^2} + k_f \cdot C_{ABTS+} \tag{7}$$

$$\frac{dC_{substrate}}{dt} = D_{substrate} \cdot \frac{d^2 C_{substrate}}{dx^2} - k_f \cdot C_{ABTS+} \tag{8}$$

The concentration of the ABTS mediator in the sample was converted into current using Eq. 9 and could be included in the simulation. Equation 9 also shows that the concentration gradient ABTS at the electrode surface is proportional to the current flows at the electrode.

$$i = n \cdot F \cdot a \cdot D \cdot \frac{dC_{ABTS}}{dx} \tag{9}$$

To determine the concentrations over time and the distance from the electrode of the component to be analyzed, the differential equations (Eqs. 7 and 8) were modeled with gPROMS (Process Systems Engineering version 3.3.0). First, the relationship between potential and flow was modeled using experimental data from cyclic voltammetry. Experimental parameters were used as inputs such as the scan rate, the potential range and the initial concentration of the mediator. The redox potential, the reversibility and the diffusion coefficients were evaluated from the experimental data. Using parameter estimation for $k_{f,}$ the model of the simulated data was adapted based on the experimental data. As seen in Fig. 21 the simulations of the cyclic voltammetry graphs of ABTS (1 mM) correspond to the experimental data. The peaks of the simulated electrical currents depend on the concentration of the analyte and scan rate (Eq. 6). The difference between the forward and reverse peak currents is due to the irreversibility of the reaction.

Cyclic voltammetric simulations of 1 mM ABTS solution combined with 1 mM veratryl alcohol also correspond well with experimental data (Fig. 22). The reaction rate ($k_f = 0.014\ s^{-1}$) was obtained by parameter estimation of the simulation compared to the experimental data.

Since a kinetic model for the understanding of the interaction between laccase, mediator and lignin has been set up, the optimization of the reaction can be

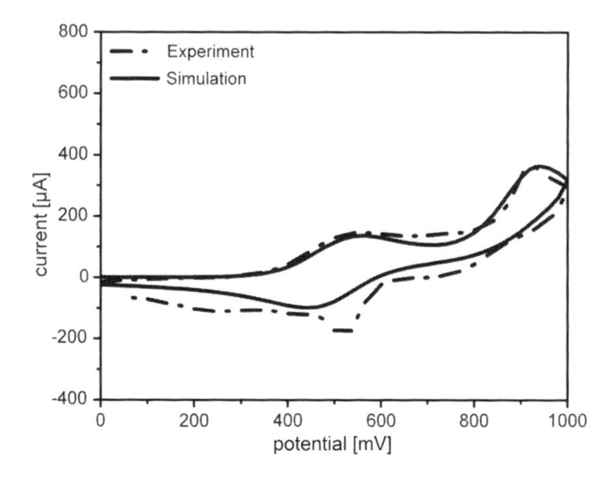

Fig. 21 Simulation of ABTS (1 mM) compared to experimental data. Scan rate: 20 mV/s Potential: 0–1,000 mV

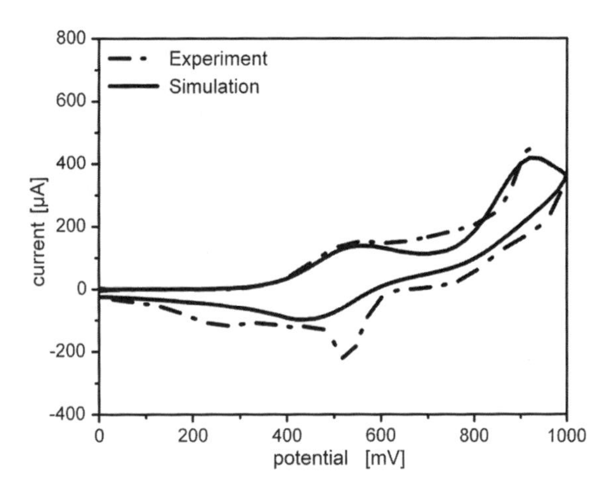

Fig. 22 Simulation of ABTS (1 mM) and veratryl alcohol (1 mM). Scan speed: 20 mV/s Potential: 0–1,000 mV

investigated. Due to the solubility of lignin in ionic liquids, the effect of ionic liquids on the stability of laccase and mediator function should be analyzed. This would support the degradation of lignin for the production of biofuels.

5.2.1 Determination of Reaction Rates

A standard activity assay was developed to measure the reaction between laccase and the ABTS mediator. The reaction can be monitored by a spectrophotometer due to the color change as the reaction progresses and the ABTS becomes oxidized. The color of the solution changes from a clear to a green color and can be measured at 420 nm, as shown in Fig. 23. The assays were performed with the photometer Biotek Synergy 4—Microtiter Plate Reader. Laccase and the ABTS mediator were

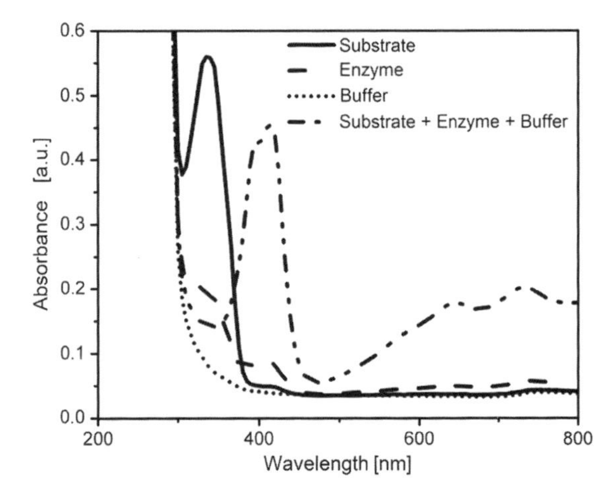

Fig. 23 Wavelength range of laccase and ABTS in buffer solution. Buffer (pH 4.5, 0.05 M citric acid, 0.1 M Na_2HPO_4), laccase: 3 mg/L ABTS concentration: 5 mM, filling volume: 200 µl, 96-well MTP, absorbance: 200–800 nm, 25 °C

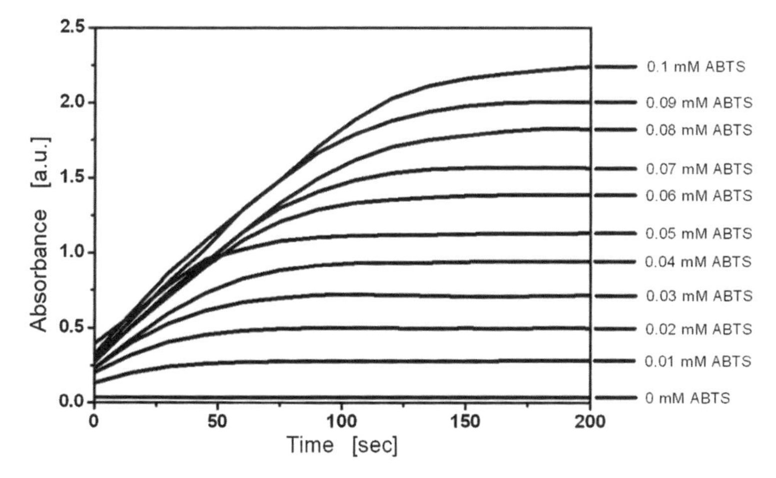

Fig. 24 ABTS activity assay at varying concentrations. Buffer (0.05 M citric acid, 0.1 M Na_2HPO_4, pH = 4.5), laccase concentration: 3 mg/L ABTS concentration: 0–1 mM. Filling volume: 200 µl, 96-well MTP, absorbance: 420 nm, 25 °C

obtained from Sigma Aldrich, the citric acid and Na_2HPO_4 buffer components were obtained from Carl Roth.

Using the Lambert-Beer's law (Eq. 10), the absorbance can be correlated with the ABTS concentration. A denotes the absorbance (a.u.), ε the extinction coefficient (3.6 L mol^{-1} cm^{-1}), c is the concentration of product (mmol/L), and l is the path length (cm).

$$A = \varepsilon \cdot c \cdot l \tag{10}$$

By varying the ABTS concentration, the reaction rate between the laccase and ABTS mediator was determined. With higher concentrations of ABTS, the reaction process occurs more rapidly as shown by a larger change in absorbance (Fig. 24).

With the change in the absorbance for increasing concentrations of ABTS the initial reaction rates at each concentration were determined. The reaction followed the Michaelis-Menten kinetics model (see Fig. 25). The V_{max} values indicate the maximum reaction rate and K_m values indicate the affinity between enzyme and substrate. This value also corresponds to the substrate concentration at half the V_{max} value. The maximum reaction rate was 0.93 uM/s and the K_m value was 0.035 mM.

6 Summary

The degradation of cellulose with cellulase was studied with conventional sugar analysis and particle sizing. During cellulose hydrolysis, the particle size and number decrease and dissolved sugars are released. The analysis of dissolved sugar

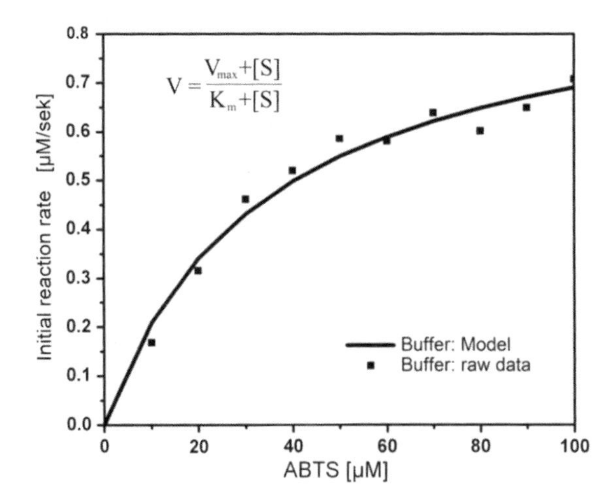

Fig. 25 Initial reaction rate as a function of ABTS concentration. Biotek Synergy 4—Microtiter Plate reader. Buffer (0.05 M citric acid, 0.1 M Na_2HPO_4, pH = 4.5), laccase concentration: 3 mg/L (Sigma-Aldrich), ABTS concentration: 0–1 mM, filling volume: 200 µl, 96-well MTP, absorbance: 420 nm, 25 °C

is not sufficient to characterize the reaction in the suspension. Therefore, in this study the BioLector technology was applied for online monitoring of cellulose hydrolysis and biochemical characterization of cellulase activity in turbid suspensions. This method is based on an online monitoring technique based on scattered light signals and allows a semi-quantitative determination of cellulase activity using native cellulose substrate.

Three different cellulosic substrates were compared with respect to their enzymatic degradability. High throughput hydrolysis shows the better degradability of the most amorphous cellulose substrate with the smallest particle size. Therefore, a pretreatment method was used in which the particle size and crystallinity were reduced. By pretreatment in the ionic liquid EMIM Ac, the hydrolysis rate and the conversion of the reaction could be significantly increased. This method for determining cellulase activity in turbid suspensions can be transferred to other particulate substrates and offers a great potential for various industrial applications.

The possibility of enzymatic degradation of lignin model compounds was investigated briefly. Veratryl alcohol, a lignin model compound, was used as the substrate and the mediator ABTS was used. Cyclic voltammetry was a suitable method to measure the reaction rates between the mediator and substrate in buffer solution. In addition to ABTS, other mediators have been considered. An activity assay was developed to determine the rate of reaction between the laccase and mediator.

7 Outlook

Further studies are necessary for a better mechanistic understanding of the observed phenomena. The established high-throughput method for the screening of cellulase is used to compare the effects of different ionic liquids on the following hydrolysis reaction. The data obtained can help to establish a kinetic model of cellulose hydrolysis in suspension, taking into account particle properties in the future. This would allow choosing a suitable treatment method in the process development step. For lignin degradation, specially modified enzymes should be used which promise better stability in ionic liquids for the pretreatment of lignin. Different ionic liquids and natural mediators should also be considered. By using other lignin model components the complexity of the lignin bonds within the lignin polymer would be taken into account. Based on experimental data, a kinetic model could be developed correlating the reaction kinetics in varying concentrations of ionic liquids.

Acknowledgments This work was performed as part of the Nordrhein-Westfalen (NRW) Research School "Brennstoffgewinnung aus nachwachsenden Rohstoffen (BrenaRo)" and the Cluster of Excellence "Tailor-Made Fuels from Biomass", which is funded by the Excellence Initiative by the German federal and state governments to promote science and research at German universities.

References

1. Mosier, N., Wyman, C.E., Dale, B.E., Elander, R., Lee, Y.Y., Holtzapple, M.: Features of promising technologies for pretreatment of lignocellulosic biomass. Bioresour. Technol. **96**(6), 673–686 (2005)
2. Holtzapple, M.T., Humphrey, A.E.: The effect of organosolv pretreatment on the enzymatic hydrolysis of poplar. Biotechnol. Bioeng. **26**, 670–676 (1984)
3. Pan, X.J., Arato, C., Gilkes, N., Gregg, D., et al.: Biorefining of softwoods using ethanol organosolv pulping: preliminary evaluation of process streams for manufacture of fuel-grade ethanol and co-products. Biotechnol. Bioeng. **90**, 473–481 (2005)
4. Stockburger, P.: An overview of near-commercial and commercial solvent-based pulping processes. Tappi J. **76**(6), 71–74 (1993)
5. Pye, E.K., Lora, J.H.: The Alcell process-a proven alternative to kraft pulping. Tappi J. **74**, 113–118 (1991)
6. Pan, X., Xie, D., Kang, K.Y., Yoon, S.L., Saddler, J.N.: Effect of organosolv ethanol pretreatment variables on physical characteristics of hybrid poplar substrates. In: Applied Biochemistry and Biotechnology, ABAB Symposium, Part 3, pp. 367–377 (2007)
7. Domínguez de María, P., vom Stein, T., Grande, P., Leitner, W., Sibilla, F.: Integrated process for the selective fractionation and separation of lignocellulose in its main components. EP 11154705.5. Filed, (2011)
8. Swatloski, R.P., Spear, S.K., Holbrey, J.D., Rogers, R.D.: Dissolution of cellose with ionic liquids. J. Am. Chem. Soc. **124**, 4974–4975 (2002)
9. Swatloski, R.P., Rogers, R.D., Holbrey, J.D.: Dissolution and processing of cellulose using ionic liquids. World Patent WO/03/029329, 2003

10. Pinkert, A., Marsh, K.N., Pang, S., Staiger, M.P.: Ionic liquids and their interaction with cellulose. Chem. Rev. **109**, 6712–6728 (2009)
11. Viell, J., Marquardt, W.: Disintegration and dissolution kinetics of wood chips in ionic liquids. Holzforschung **65**, 519–525 (2011)
12. Kragl, U., Eckstein, M., Kaftzik, N.: Enzyme catalysis in ionic liquids. Curr. Opin. Biotechnol. **13**(6), 565–571 (2002)
13. Engel, P., Mladenov, R., Wulfhorst, H., Jäger, G., Spiess, A.C.: Point by point analysis: how ionic liquid affects the enzymatic hydrolysis of native and modified cellulose. Green Chem. **12** (11), 1959–1966 (2010)
14. Pottkämper, J., Barthen, P., Ilmberger, N., Schwaneberg, U., Schenk, A., Schulte, M., Ignatiev, N., Streit, W.R.: Applying metagenomics for the identification of bacterial cellulases that are stable in ionic liquids. Green Chem. **11**, 957–965 (2009)
15. Wong, T.S., Roccatano, D., Loakes, D., Tee, K.L., Schenk, A., Hauer, B., Schwaneberg, U.: Transversion-enriched sequence saturation mutagenesis (SeSaM-Tv+): a random mutagenesis method with consecutive nucleotide exchanges that complements the bias of error-prone PCR. Biotechnol. J. **3**, 74–82 (2008)
16. Hall, M., Bansal, P., Lee, J.H., Realff, M.J., Bommarius, A.S.: Cellulose crystallinity—a key predictor of the enzymatic hydrolysis rate. FEBS J. **277**(6), 1571–1582 (2010)
17. Hall, M., Bansal, P., Lee, J.H., Realff, M.J., Bommarius, A.S.: Biological pretreatment of cellulose: enhancing enzymatic hydrolysis rate using cellulose-binding domains from cellulases. Bioresour. Technol. **102**, 2910–2915 (2011)
18. Bommarius, A.S., Katona, A., Cheben, S.E., Patel, A.S., Ragauskas, A.J., Knudson, K., Pu, Y.: Cellulase kinetics as a function of cellulose pretreatment. Metab. Eng **10**, 370–381 (2008)
19. Lynd, L.R., Weimer, P.J., van Zyl, W.H., Pretorius, I.S.: Microbial cellulose utilization: fundamentals and biotechnology. Microbiol. Mol. Biol. Rev. **66**, 506–577 (2002)
20. Mosier, N.S., Hall, P., Ladisch, C.M., Ladisch, M.R.: Reaction kinetics, molecular action, and mechanisms of cellulolytic proteins. Adv. Biochem. Eng. Biotechnol. **65**, 23–40 (1999)
21. Holladay, J.E., White, J.F., et al.: Results of screening for potential candidates from biorefinery lignin. Top Value-Added Chem. Biomass **2**, 87 (2007)
22. Ghose, T.K.: Measurement of cellulase activities. Pure Appl. Chem. **59**(2), 257 (1987)
23. Miller, G.L.: Use of dinitrosalicylic acid reagent for determination of reducing sugar. Anal. Chem. **31**(3), 426–428 (1959)
24. Chen, C.S., Penner, M.H.: Turbidity-based assay for polygalacturonic acid depolymerase activity. J Agric. Food Chem. **55**, 5907–5911 (2007)
25. Wang, P., Broda, P.: Stable defined substrate for turbidimetric assay of endoxylanases. Appl. Environ. Microbiol. **58**, 3433–3436 (1992)
26. Fan, L.T., Lee, Y.H., Beardmore, D.H.: Mechanism of the enzymatic hydrolysis of cellulose-effects of major structural features of cellulose on enzymatic-hydrolysis. Biotechnol. Bioeng. **22**, 177–199 (1980)
27. Walker, L.P., Wilson, D.B., Irwin, D.C.: Measuring fragmentation of cellulose by *Thermomonospora fus*ca cellulase. Enzyme Microb. Technol. **12**, 378–386 (1990)
28. Bowen, P.: Particle size distribution measurement from millimetres to nanometers and from rods to platelets. J. Dispers. Sci. Technol. **23**(5) (2002)
29. Nummi, M., Fox, P.C., Nikupaavola, M.L., Enari, T.M.: Nephelometric and turbidometric assays of cellulase activity. Anal. Biochem. **116**, 133–136 (1981)
30. Vogel, H., Gerthsen, C.: Physik. Springer GmbH, Berlin (1976)
31. Urban, C.: Development of fiber optic based dynamic light scattering for a characterization of turbid suspensions. Dissertation, Swiss Federal Institute of Technology, Zurich (1999)
32. Hecht, E.: Optik. Oldenburg Wissenschaftsverlag GmbH **4**, 159 (2005)
33. Samorski, M., Müller-Newen, G., Büchs, J.: Quasi-continuous combined scattered light and fluorescence measurements: a novel measurement technique for shaken microtiter plates. Biotechnol. Bioeng. **92**(1), 61–68 (2005)
34. Bourbonnais, R., Leech, D., et al.: Electrochemical analysis of the interactions of laccase mediators with lignin model compounds. Biochem. Biophys. Acta **1379**(3), 381–390 (1998)

35. Nicholson, R.S., Shain, I.: Theory of stationary electrode polarography—single scan and cyclic methods applied to reversible irreversible and kinetic systems. Anal. Chem. **36**(4), 706 (1964)
36. Jäger, G., Wulfhorst, H., Zeithammel, E.U., Ellinidou, E., Spiess, A.C., Büchs, J.: Screening of cellulases for biofuel production: online monitoring of the enzymatic hydrolysis of insoluble cellulose using high-throughput scattered light detection. Biotechnol. J. **6**, 74–85 (2011)
37. Wulfhorst, H., Jäger, G., Ellinidou, E., Büchs, J., Spiess, A.C.: Charakterisierung von Cellulasepräparationen mittels Streulicht. Chem. Ing. Tech. **82**(1–2), 117–120 (2010)
38. Jäger, G., Wu, Z., Garschhammer, K., Engel, P., et al.: Practical screening of purified cellobiohydrolases and endoglucanases with alpha-cellulose and specification of hydrodynamics. Biotechnol. Biofuels **3**, 1–12 (2010)

Printed by Printforce, the Netherlands